Mathematical Models in Economics

An Introduction

First Edition

Christopher Laincz

Drexel University

cognella®

SAN DIEGO

Bassim Hamadeh, CEO and Publisher
John Remington, Managing Executive Editor
Gem Rabanera, Senior Project Editor
Susana Christie, Senior Developmental Editor
Abbey Hastings, Production Editor
Emely Villavicencio, Senior Graphic Designer
JoHannah McDonald, Licensing Coordinator
Natalie Piccotti, Director of Marketing
Kassie Graves, Senior Vice President, Editorial
Jamie Giganti, Director of Academic Publishing

About the Cover Art

The Mathematical Bridge at Queen's College Cambridge was built in 1749 and designed by William Etheridge. It is remarkable for the number of myths surrounding it (e.g., that is based on designs by Leonardo da Vinci or Isaac Newton or that it comes from China) and for its design. The design ensures dominant forces on the timbers tangential to the arch are compression forces rather than placing force on the timbers' sides which would cause bending or breaking.

For more information please visit: https://www.queens.cam.ac.uk/visiting-the-college/history/college-facts/mathematical-bridge.

Mathematical Models in Economics: An Introduction

Contents

Preface

Did someone say math? Is it safe to come out yet?

This textbook grew out of a need to provide mathematical tools to economics students transitioning from the principles level to the intermediate level.[1] In particular, my colleagues and I observed many students struggling with how to interpret and utilize mathematical models. Furthermore, too often students came into the classes with insufficient calculus preparation, had forgotten what they had learned, were not required to take multivariate calculus or linear algebra, or, even if they had the math classes, were not comfortable using that same math in the context of an economic model.

The book you are looking at began as a 60–page handout for my intermediate macroeconomics course and was also distributed by some of my colleagues teaching intermediate microeconomics or macroeconomics. The handout expanded to about 100 pages when we added a master's–level economics program and wanted our incoming students to review the core math tools prior to starting classes. Some years later, a colleague and I were asked to revise our principles courses. From several undergraduate focus groups and our faculty colleagues, it was clear that there was a severe gap between the principles level and the intermediate level, and, for a number of reasons many of our students were not

---

[1]A ginormous thank you to Lexi Fu, Tanja Kirmse, Nomalia Manna, Zixuan Pei, Ioan Rusu, Rediet Woldesenbet, Han Zhang, and Yuyun Zhong for editing and proofreading assistance and to my father, Paul Laincz, for the entertaining illustrations.

being given an adequate grounding in the modeling and math concepts needed to thrive. Asking students to review the mathematical appendices included in most intermediate economics textbooks was similarly inadequate. The faculty were also frustrated, feeling that too much time was spent covering basic calculus and optimization rather than economic ideas and intuition. While we had a traditional Mathematical Economics course, that course was intended for students who were past the intermediate courses and wanted advanced tools, perhaps to prepare for graduate programs. Thus, that course was not an appropriate solution for the issue at hand without drastically changing the contents to serve a different type of student.

We proposed a new course to fill the gap that would act as a complement to the intermediate–level courses. The primary target audience was our second–year students majoring in economics, finance, and other related fields. The course was designed to be taken either prior to or concurrently with intermediate microeconomics. We also considered delivering the material in a "lab" or recitation section that would accompany the intermediate economics courses, but found the stand–alone course more expedient. There were, and remain, a number of options for how to deliver this material.

However we chose to deliver the material, the clear need we saw was introducing math modeling skills to help students bridge two gaps: 1) between the principles and intermediate–level economics; and 2) between math courses and economics courses. This manuscript was deliberately written to be unlike a math textbook or even most of the mathematical economics books that are out there. Learning to use the math tools, understand a model, and how to interpret equations in an economics context were the primary goals. The textbook goes through the critical tools and their interpretation largely by presenting and discussing examples. It does not take a theorem-proof approach. It avoids getting into technical discussions of the math; it does not cover economic concepts in depth, nor does it explore all their implications. That is, this textbook is not a math textbook and it is not an economics textbook. This textbook and the associated course are meant to connect the two by helping students see how economics applies math to build models and understand the world around us. The concepts are largely taught by example with the idea that the best way to learn is by practicing. Hence, lots of practice problems are provided. One can think of the approach as similiar to an intensive foreign language course where we may skip over the nuances of some grammar points and rely heavily on conversation practice.

The textbook covers basic calculus optimization, both single variable and multivariate, optimization, corner solutions, an introduction to comparative statics, with some linear algebra and integration at the end. The aim is present the math models as they appear in most mainstream intermediate economics textbooks. While the main subjects here are obviously math tools, they are math tools used in business and economic analysis, modeling, and research. The textbook covers applications of unconstrained and constrained optimization to microeconomics, macroeconomics, and, to a lesser extent, econometrics and makes connections between the tools shown and their uses. Completing this course and utilizing this textbook students prepares students to handle the math used in intermediate undergraduate or introductory master's–level economics courses.

What is expected before taking this course? It is expected that students have had classes that covered set theory, exponents, natural logs, basic calculus, algebra for solving for multiple unknowns from multiple equations, the quadratic formula, and the point-slope formula. Although this course does include calculus, the treatment here is quite rapid and is meant more to help refresh a student's memory and then move quickly to applications, rather than explain it in detail. Furthermore, it is fully expected that all students have had at least a Principles of Microeconomics course and are familiar with key concepts such as utility, opportunity cost, marginal cost, profit maximization, etc. It is also recommended that students have familiarity with Principles of Macroeconomics concepts, though not necessary.

What is covered? This textbook covers basic modeling concepts, calculus, the exponential and natural log function, unconstrained and constrained optimization in single and multivariate calculus, including a limited treatment of second-order sufficient conditions (up to two variables only), and an introduction to comparative statics. That includes the Lagrangian method for solving a constrained optimization problem. The last few chapters cover integration, basic linear algebra, and Cramer's Rule. From economics, as examples of applications, examples of the following models are included: labor-leisure trade off (Robinson Crusoe model), utility maximization with two goods, two-period consumer model from macroeconomics, profit maximization, cost minimization, capacity constraints, and adjustment costs.

What is not covered? The treatment of single–variable calculus is light and skips taking limits, continuity, and differentiability. Those are essential topics that the student should be familar with prior to reading this manuscript. The textbook stops short of going into second-order conditions for constrained multivariate problems where linear algebra makes the analysis manageable. The textbook also does not present real analysis, which is very useful for advanced microeconomics, in particular, and the it leaves out difference and differential equations that would be useful for advanced macroeconomics. However, most of these topics are rarely included at the intermediate level in economics courses, and the approach found in traditional mathematical economics textbooks for those topics may be more valuable for those more advanced concepts.

Feedback is always welcome, including correction of typos, which undoubtedly exist. Hopefully, they have been minimized and do not detract from optimally enjoying the material. **Good luck and have fun!**

Introduction: Tips for the Student on How to Use this Textbook

1. This book is best read with a notebook and pen ready to attempt the practice problems and examples sprinkled throughout the chapters. Even though some of them may appear self–explanatory, like a foreign language, practice and repetition are helpful. Nearly all of the techniques are best learned or understood if you invest a few minutes here and there trying to utilize them as you go.

2. When introduced, key terms are in bold and definitions are provided in a side box for your reference, as in this example for **economics**.

> **Definition**
>
> **Economics**: The social science that studies decision-making with unlimited wants but constrained by limited resources and how those decisions across many individuals and institutions impact society.

3. Study the graphs and figures when they appear. The visual representation of the math can be very helpful in building intuition even if the math representation remains challenging to decipher.

4. Some side boxes contain additional helpful activities. The example here asks you to use the internet to obtain a list of the Greek alphabet, which is commonly used in the mathematical models of many science fields, including economics. These are recommended to help you engage with different aspects of the economics material as you go and highlight interesting topics.

> **Making the Connection**
>
> **Greek alphabet:** Obtain a list of the Greek alphabet, both lower and upper case letters. Can you name them all? If not, practice. Keep this list and make notes in your microeconomics and macroeconomics classes as to what the letters represent. For example, Π is commonly the symbol for profits in microeconomics.

5. Finally, a third type of side box, the Quick Check, invites you to try some basic exercises using the material being discussed. In all cases the answers can be found at the end of the chapter, so you can check your progress. If you are getting these wrong, reread corresponding sections and work through the existing examples.

> **Quick Check**
>
> Find dy/dx from $y = 8x^8$.
>
> Find dQ/dL from $Q(L) = AL^\rho$.
>
> Answers are at the end of the chapter.

1 Models, Parameters, and Variables

1.1 Why Math Models?

Math is the language of logic. Economics as a field relies heavily on math to answer questions and understand the world around us. Economists build "models" of some interesting aspect of the real world using math to represent how agents (consumers, firms, governments, etc.) interact with each other. We use these models to analyze and predict the consequences of those interactions or policies that influence agents' behavior.

The term "model," as used here, is akin to the fact that a toy car is a model of a real car. It looks like the real thing, can do some of the things a real car does, but is smaller and has limits. The model may have wheels that actually roll like a real car. Some model cars may have a motor of some kind such that the toy can move like a real car. How does the model help us? For example, suppose you are just learning to drive. Parallel parking is one of the trickier techniques to learn when driving. A student driver can gain some understanding of the technique by using the model/toy car and mimicking the motions with a bird's eye view. An **economic model** is meant to do the same for some specific aspect of human decision-making.

The models we build are in many ways similar to the models physicists build, with one crucial difference being that our models are populated by decision-makers. Economic models are meant to capture key features of the real world and, often and importantly, the choices these agents make and the constraints they face. In addition, by writing down our theories in the form of models, we make explicit all the assumptions that go into the model, which may affect the results. That provides a great advantage in our approach because anyone can then examine the theory and understand which assumptions drive the results. This

> **Definition**
>
> **Economic model**: The International Monetary Fund defines an economic model as a simplified representation of reality, designed to yield hypotheses about economic behavior that can be tested.

approach stands in contrast to other fields that use the written language to convey their ideas. The written language is full of vague terms and nuance, can be easily misinterpreted, and suffers from the fact that words change their meaning over time. Math is crisp, precise, has well-defined rules and is largely immutable. Moreover, by using math we can see very clearly if the internal logic is consistent and correct. The heavy reliance on math does lead to some limitations. Math does not come easily to everyone. Math places a rigid set of rules on analyses and thus makes conveying the concepts to a general audience challenging.

The standard principles-level textbook definition of economics usually reads something like *the science of understanding how society reconciles unlimited wants with limited resources.* That phrase sounds like a common set of calculus problems collectively called "constrained optimization." In constrained optimization one tries to maximize or minimize something (e.g. , profits, utility, resources) while staying within some kind of limitation or requirement such as technological capability, production capacity, time limit, or budget. As such, constrained optimization techniques have a special place in economic analysis. We frequently, as in almost always, see our agents as facing a constrained optimization problem that they "solve." The solution shows us their actions, i.e. their predicted behavior and we examine what affects that optimal (best) solution and how different types of agents (e.g. , firms and consumers) interact when they are all solving their own distinct optimization problems. Thus, much of this textbook is devoted to building the tools that deal with models using constrained optimization techniques and how to interpret them.

International Monetary Fund (IMF). The IMF is a great source for learning about economics in general and what is happening around the world economically. Here is a QR code link to the IMF home page:

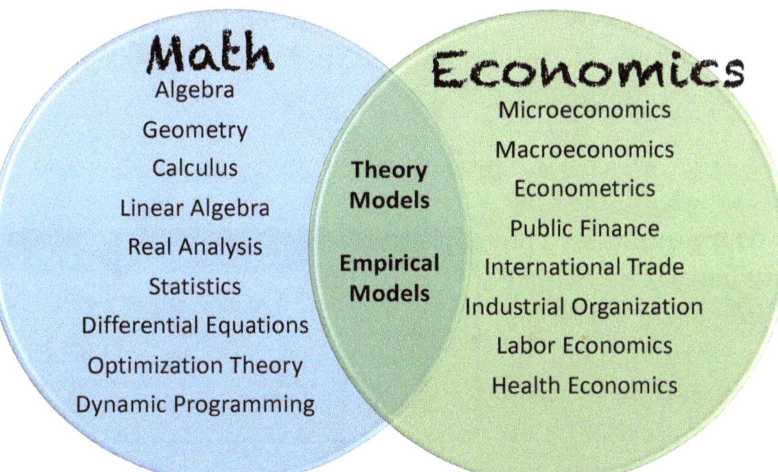

1.2   What Is in a Model?

Mathematical models are built to answer questions. In our mathematical models, we include **variables**, symbols that are usually alphabetic or from the Greek alphabet, that represent particular quantities relevant to the question we are asking. To help illustrate, we will build a commonly used model in both micro-economics and macroeconomics, the *Robinson Crusoe Model of Labor Supply*.[2] This model addresses the question "How should a person optimally

> **Definition**
>
> **Economic variable**: A quantity in an economic model that can take many different values. Variables are typically represented with a letter, sometimes a Greek letter, to indicate that many possible values are possible.

---

[2]Robinson Crusoe is a reference to the novel by Daniel Defoe pubished in 1719. The book centers around the main character being stranded on an island and figuring out how to survive. In essence, he

allocate their limited time between working to obtain consumption goods versus enjoying leisure time?" In effect, the model provides an understanding of what factors affect the supply of labor at the individual decision-maker level.

1.2.1 Constraints

Suppose that we want to understand how a person, let's call her Margo, divides her time between work and "not work," which we will call leisure. We need variables to represent the amount of time spent in each activity. When we assign a variable, we commonly use the phrase, "Let X be blah blah blah." In this case, let H be the hours Margo works. Also, let N be the hours not worked or we can call N the hours of leisure. H and N are variables in the model we are building.

Morever, they are both **choice variables**, meaning that our agent actively chooses the values of these variables in the problem. H and N can also be referred to as **endogenous variables**. The value of an endogenous variable is determined within the model. Choice variables are endogenous variables *but not all* endogenous variables are choice variables. That means their quantities are determined within the context of the model. What we mean by that is through solving for an optimal solution or an equilibrium of some kind, we will figure out through the logic of math what those quantities are.

For example, suppose Margo runs off to the fish market, and we have a model of supply and demand at the fish market. The quantity of fish Margo buys is both a choice variable and an endogenous variable. However, the price of fish, from Margo's point of view, is not a choice she makes. However, if our model represents all consumers' demand at this market and the decisions of the vendors at the fish market, the price of fish will be determined through supply and demand, as you learned in microeconomics. Thus, the price of fish is an endogenous variable, but no one consumer gets to choose that price.

That differs from an **exogenous variable**, which is determined by forces not captured in the model itself. Continuing our example of Margo at the fish market, the quantity of fish caught and supplied depends on the weather the fisherman faced that morning at the shore. The weather is exogenous. The temperature could be many dif-

> **Definition**
>
> **Constraint**: A mathematical representation of a limited resource such that choices of variables must not violate the constraint. Examples inlcude a consumer's budget constraint, a worker's time constraint, or a firm's production capacity.

> **Definition**
>
> **Choice variable**: An economic variable that an agent in the model determines by making a decision. Examples include, how many fish a consumer buys, how much time a worker spends looking for a new job, how much effort a worker puts into their current job, and a firm's choice how many workers to hire.

> **Definition**
>
> **Endogenous variable**: a variable that is determined within the model. Choice variables are endogenous variables *but not all* endogenous variables are choice variables.

---

forms the simplest of economies and is faced with basic trade-offs. The 'Robinson Crusoe' model has been a workhorse in economics textbooks for decades.

ferent values, but none of our agents in the model choose the temperature, and none of their actions that morning have an impact on the weather.

Unfortunately for Margo, time is limited. Let T be the total amount of time available to Margo. That could be 24 hours in a day, or 7 days in a week, or 365 days in a year. But for now let's just think of T as the total time **endowment**. T is not a choice variable, and it is not endogenous. It is an exogenous variable from the point of view of the agent.

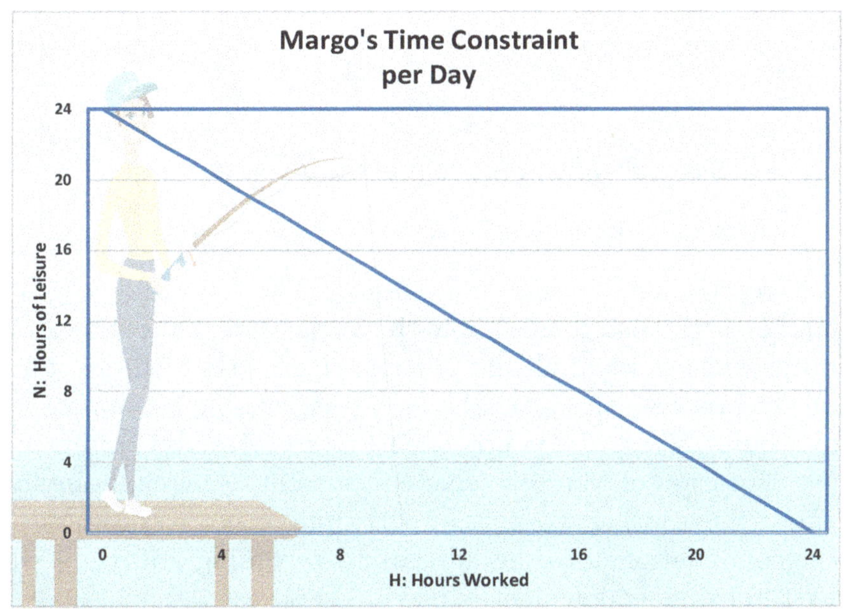

Figure 1.1: Margo's time constraint

T is an exogenous variable. It has a value, and we can explore how changing it affects choices, but its value is given by nature and accepted by the agents who have no control over it. Nothing Margo does can change the number of hours in a day. That leads us to the following equation:

$$T = H + N.$$

That equation reads the total time endowment must equal the hours worked plus the hours not

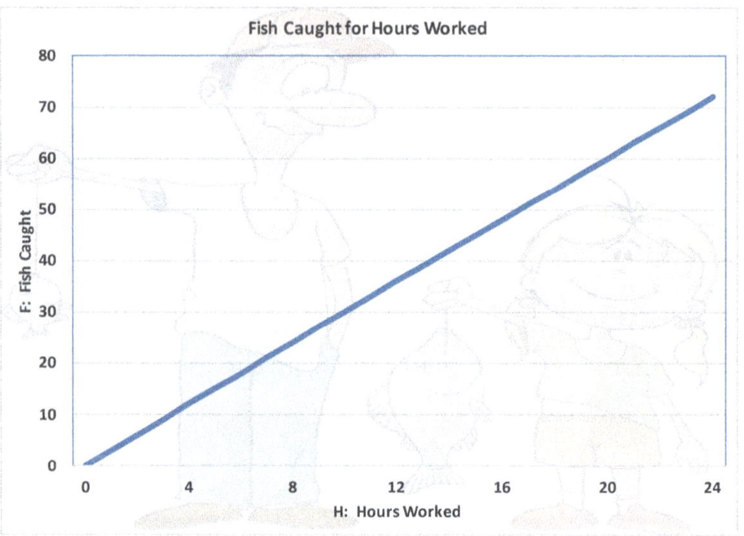

Figure 1.2: Margo's production function

worked. In short, it represents the time constraint. The left-hand side shows how much is available, and the right-hand side shows how that time can be used. As a constraint, the equation provides a set of all the *possible* choices that Margo can make. To see the set of options clearly we can graph the equation with hours worked, H, on the x-axis and hours of leisure, N, on the y-axis. See Figure 1.1. Every point on the line represents a possible allocation of Margo's time endowment. The endpoints of the line represent the extremes: She's either working all the time or not working at all. Points in between represent combinations of time allocated to working and time allocated to leisure such that her total time endowment is used.

Suppose that Margo's **production function** is

$$F = 3H$$

which means for every unit of time, say an hour, Margo spends fishing she catches three fish. Now we can make a graph of the production function with hours fishing, H, on the x-axis and fish caught on the y-axis. See Figure 1.2.

Futhermore, we can combine the time constraint and the production function into one equation that shows the trade–off between fish and leisure time. Rearrange the time contraint to be: $H = T - N$ and substitute that into the production function to get

> **Definition**
>
> **Endowment:** Refers to something given by nature to an agent at the start of the problem. The endowment is something that comes from outside of the model. The endowment is pre-determined: The agent starts the problem with the endowment and we are not concerned with how the agent got the endowment, though we can ask how changes in the amount of the endowment affect choices. In a consumer choice model, the consumer may be endowed with a level of wealth to spend on goods.

> **Definition**
>
> **Production Function:** A mathematical representation that shows how the quantity of inputs used (e.g. , raw materials, capital, labor, time, energy) translate into the amount of the output (e.g., cookies, fish, widgets, bubble tea, car washes) the agent (firm) generates.

$$F = 3(T - N).$$

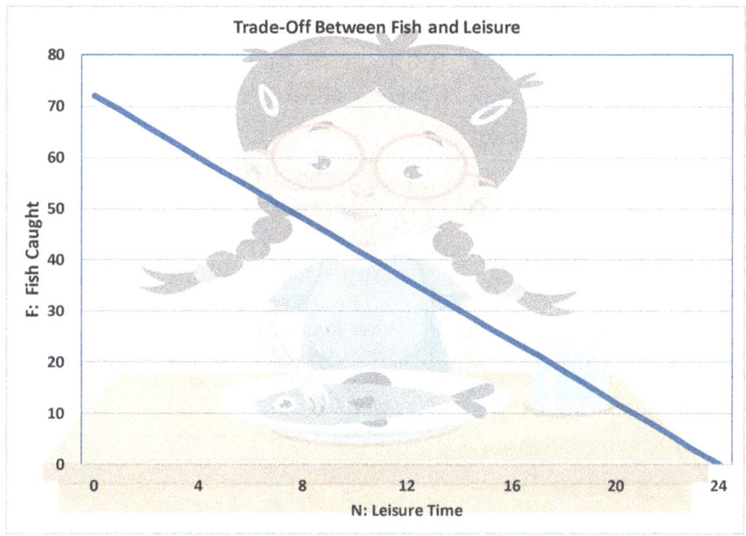

Figure 1.3: Margo's budget constraint

The graph of that constraint reveals the trade–off between fish to consume and leisure time, N. The graph in Figure 1.3 shows the trade–off visually. Look carefully at the equation for a moment. Suppose Margo wants one more hour of time to play with her cat instead of fishing. That's more leisure time. If N increases by one unit that means she will catch three less fish. That change reveals the *marginal product* of Margo's time; we will have more to say on the subject of marginal changes in the next chapter.

1.2.2 Objective Functions

The time constraint equation and the budget constraint equation say nothing about what choice would be best or **optimal**. In economics, we usually assume that individuals always do their best to achieve their goals and desires given the information and resources they have. We assume

> **Definition**
>
> **Optimal solution:** A solution to a problem that attains the "best" value feasible (maximum or minimum) subject to the constraint(s) and parameter(s) of the agent's choice problem.

that agents choose to *optimize* by using their resources to achieve something they want. That does not mean they do not make mistakes. Agents may also be misinformed or lacking information to make the best choice. However, we assume they will always do their best given what they know.[3]

But what do they want? Firms' success is determined by profits: revenue minus costs. In the case of individual consumers or workers, we typically measure their well-being by utility. From Perloff's microeconomics textbook, "Economists summarize a consumer's preferences using a *utility* function, which assigns a numerical value to each possible bundle of goods, reflecting the consumer's relative ranking of these

---

[3]Behavioral economics explores settings where different cognitive biases can alter those choices and how they would differ from the standard optimizing behavior we normally assume.

bundles. "[4] The higher the utility, the better off the consumer. is In our example, Margo values leisure time and fish. The more of each, the better off she is, so the utility value assigned will rise with more fish and rise with more leisure time. Margo's goal, her objective, will be to maximize this utility value by making choices.

So, in the case of the firm, it wants to *maximize* profits. In some analyses the firm knows what it needs to produce, so it seeks to *minimize* the costs of producing. The consumer wants to *maximize* her utility. Profits and utility are the objectives. The mathematical expressions of these concepts are called **objective functions**. These are the values the agents want to optimize by making choices to achieve a minimum or maximum depending on the context. The objective function shows what choice variables directly affect the value the agents want to optimize.

For example, suppose a fishing rod firm's revenue is given by the price, P, of fishing rods times quantity sold, Q, minus the costs of producing those fishing rods. Let the cost of producing those fishing rods be represented by the following function: $C(Q) = \alpha Q^2$ where $\alpha > 0$. The meaning of the parameter α will be addressed later. Then the profits, represented by π, will be given by revenue minus costs; in other words, the firm's *objective function* is

$$\pi(Q) = PQ - \alpha Q^2.$$

That is the expression the firm rod firm wants to maximize by choosing the quantity, Q, of fishing rods to sell.

In the case of Margo, she wants to maximize her utility, her happiness. She gets utility from eating fish, F, and from leisure time, N. So, let's represent Margo's utility as $U(F, N)$, which reads *utility is a function of the amount of fish consumed and the amount of leisure time*. Let us further imagine that the explicit functional form of Margo's utilty function is

$$U(F, N) = F^{1/2}N^{1/2}.$$

That is Margo's *objective function*, which she is trying to maximize. In fact, she is trying to maximize her utility function *subject to* her constraint $F = 3(T - N)$. Formally we would write out Margo's decision problem as

$$\max_{N,H} U(F, N) = F^{1/2}N^{1/2}$$

$$\text{subject to}$$

$$T = H + N$$
$$F = 3H.$$

---

[4]Perloff, J. , <u>Microeconomics</u> 8th edition, Global Edition, Pearson Publishing, New York, 2018, p.98.

The first line shows the objective function and indicates whether the goal is to max(imize) or min(imize) the objective function. This line is followed by the phrase "subject to," which tells us that constraints or other conditions apply. In

> **Definition**
>
> **Constrained optimization**: Mathematical methodologies and tools for finding the minima and/or maxima of a function subject to a limitation, which may itself be a function.

this problem, both the time constraint and the production function follow. Notice that underneath the *max* operator is a list of Margo's choice variables: H and N. She has control over these and will select their values. On the other hand, T is not listed there because, as noted before, it is exogenous. Margo has no control over the number of hours in a day. F is an endogenous variable in that Margo has control over its value, but she controls it by choosing H and N. We could rewrite the problem such that she chooses F directly with her choice of H in the background. We will look at those options in more detail later, after building the optimal choice framework a bit more in the subsequent chapters.

Before moving on, please see the definition of **equilibrium**. We include this point here because it is important to note that optimal and equilbrium are distinct concepts in math and economics. Equilibrium values are the resulting values that solve the problem, but they may not be optimal from the point of view of one or

> **Definition**
>
> **Equilibrium:** A solution to a problem that involves equalizing multiple requirements or equations. For example, in a supply and demand analysis, the resulting price is said to be an equilibrium price. That does not mean it is *optimal*.

more agents or a "social planner" who is responsible for everyone's welfare.

For example, suppose that in Margo's town the petrochemical plant determines that they should optimally produce 4,444 kilotons of chemical agent double XX. That decision is optimal from the point of value of the petrochemical plant. The price each kiloton sells for is $100, which is an equilibrium price determined by supply and demand. The petrochemical plant would choose a higher price, but since they compete in a market it is not their choice. Moreover, the production of chemical agent double XX pollutes the lakes and harms the fish that Margo likes to catch and eat. The petrochemical plant does not take that effect into consideration since it has no bearings on their profits. From a *social* point of view the amount of petrochemicals being produced is not optimal. A social planner, given power over the petrochemical plant's choices, who also considers the impact on the fish and the welfare of the town, would choose a different, lower amount of petrochemical production to achieve a socially optimal level that accounts for the trade-off.

This example is one of the concept economists call *externalities*, which are beyond the scope of this textbook. However, externalities frequently mean that equilibrium results and optimal results differ. That means there is scope for economic policy to improve welfare. The distinction between equilibrium and optimal outcomes is an important subject of study and economic policy. As you move forward be sure you look for the difference and do not confuse these two concepts.

1.3 Levels of Generality

We will reserve the steps to solve Margo's problem for a later chapter, but re-examine the problem for a moment. Where did the 3 come from in the production function? Why are there 1/2 exponents in the utility function? Why does the utility function have that math expression of two things multiplied together with exponents on each? Are the exponents equal for some reason? Look back at the example of the firm's cost function. Why does it have a Freek letter, α, instead of a regular letter? What is α?

There are two parts to answering these questions. First, as we build models to answer questions, sometimes we write down very specific models with exact numbers, and other times we are more general and use variables to represent those values. So, there is a level of *generality* that we need to consider when building a model. In an important sense, the more general the model, the better. The more general it is, the more situations the math logic will hold true (i.e. the theory is more widely applicable). However, as always, there is a trade-off. More general models may be harder to work with or too abstract to produce meaningful insights.

For example, suppose Margo comes up with a theory of world peace based on how many fish the world's cats get to eat. In her theory for achieving world peace she specifies the following mathematical relationship between cats and fish:

$$C = 567.89F^{1/3} - 73.4F^{1/2} + 99F^{-3/4} + 42$$

where C is the number of cats and F is fish. That is, Margo's theory applies if the relationship between fish and cats is described by that equation exactly. That function is very specific. It describes an exact relationship such that for every number of fish, there is exactly one value of the number of cats that fits the equation. If the real world does not fit that equation exactly, then Margo's theory does not help.

Margo spends more time on her theory and *generalizes* that equation to the following: $C = aF^\alpha - bF^\beta + wF^\gamma + z$, where a, b, w, and z are all greater than zero and α, β, and γ must all lie between -1 and 1. Now that relationship covers many different possibilities, including the original one, but it also rules out some others. That is, this new version is more general than the first version and thus holds true in many more situations. As such, Margo's theory is more widely applicable, and there is greater hope for world peace. At the same time, suppose we wanted to graph that equation and explore it more carefully. To graph it we would have to pick values, exact numbers, for each of the variables and parameters. Moreover, as we change some values, the graphs could look vastly different and generate major differences in interpretation, with only small changes in the numbers picked. So, while generality makes the model more applicable, it also makes it more challenging to visualize and analyze the relationship.

Now suppose Margo fully generalizes that relationship to $C = f(F)$, which just says cats are a function of fish. There are no specific functional forms as in the polynomial we had before and no parameters or other variables to consider or create additional limits on the range of values they can take. So, now Margo's theory of world peace would hold if cats were a function, *any* function, of fish. Her theoretical model is much more widely applicable now because it encompasses an enormous range of possibilities. At the same time, how would you visualize or analyze that relationship? You can't say that an increase in fish leads to more cats because the function could be decreasing. Any graph of a function would work because everything is permissible, so we can tell almost nothing about the relationship between the variables. Moreover, suppose we take Margo's theory to the data. We observe the number of cats and fish in 1944 but also that the majority of the world was not at peace. That is evidence against Margo's theory, sadly. We will return to the concept of specificity versus generality after covering basic calculus, because calculus helps us tremendously in creating models with a great degree of generality yet describe key features in the relationship between variables so as not to be overly general.

Going back to the questions raised above about why the choice of the number 3 and α, the second part of the answer is that we distinguish between variables and **parameters** often by designating parameters with Greek letters. That is not a hard-and-fast rule but does hold pretty

generally in economics, as presented in textbooks and most research. What is a "parameter?" A parameter is a specific value in our models that is not determined within the model in any way. Parameters are exogenous and fixed. For example, in Margo's production function, $F = 3H$, the 3 is a parameter with a specific value. It is thought of as exogenous. You can think of it in this context as resulting from the fishing technology that Margo is using or Margo's skill in fishing, or both. However, in the context of our model, Margo cannot change this value. The parameter is not a choice, nor does any other feature of the model determine or affect its value.

We could make this production function more general and instead write it as $F = \gamma H$, where the Greek letter gamma (γ) takes the place of the 3. We might also wish to specify a range of values that γ can take. Why limit the range? Some values will make economic sense and others will not. Suppose $\gamma = 0$. That implies $F = \gamma H = 0H = 0$, or Margo catches zero fish no matter how long she spends fishing. While anyone who has ever gone fishing has felt that way at some point, it makes the model totally uninteresting because Margo choosing to spend any time working provides no benefits. So, she would never make that choice, and the model provides no insights. Moreover, if $\gamma < 0$, it becomes impossible or absurd to interpret that as a *production* function. With a negative value, Margo loses fish for every hour she spends fishing. How would you interpret having *negative* 9 fish if she worked for 3 hours? Perhaps you could use your imagination and make up some story, but recall the purpose of our model is to answer the question how individuals should optimally divide their time between work and leisure. Thus, zero or negative values are unrealistic and do not contribute to our understanding. With that in mind, we put restrictions on the range of parameter values and variables themselves to rule out cases that make no economic sense ($\gamma < 0$) or represent situations that yield results from which we learn nothing ($\gamma = 0$).

Therefore, in this part of the model, we would assume that $\gamma > 0$. That would allow us to ask additional questions of our model. Suppose that γ represents Margo's fishing technology, in other words, it represents the quality of her fishing rod. Now if she acquires a better fishing rod with a higher value of γ, we can compare how she chooses work and leisure with better technology as compared to the old technology. The real world analogy is what if workers have access to a better technology, will they work more or less? That is a very relevant economics question.

As another example, look again at the profit maximization problem in Section 1.2.2. In that example, α is a general parameter that governs the costs of producing. We assumed, without explanation, that $\alpha > 0$. Why do you think that assumption makes sense? Similarly, the Q in the cost function is raised to the power of 2. The 2 is very specific. We could have generalized that a bit more by writing the cost function as αQ^ρ, where $\rho > 1$. Now ρ could take on the value of 2 as originally specified, but it can also take on many other values. Note also the range here was selected such that ρ is greater than one. That is not arbitrary. Can you think of why that makes economic sense? You should see this concept in microeconomics.

1.4   What Makes a Good Model?

You might be asking at this point, what makes for a *good* model? A good model helps us answer the question it was designed to address. Refer back to the definition of an economic model at the start of this chapter. A good model should have testable implications. That is, and not all models have this feature, we would like to test the underlying theory in the model with real world observations (data). If the model is too general, we will have a difficult time determining clear predictions to test. If the model is extremely specific, we may not be able to find the situation that fits the model to test, or we may just reject the model as false every time.

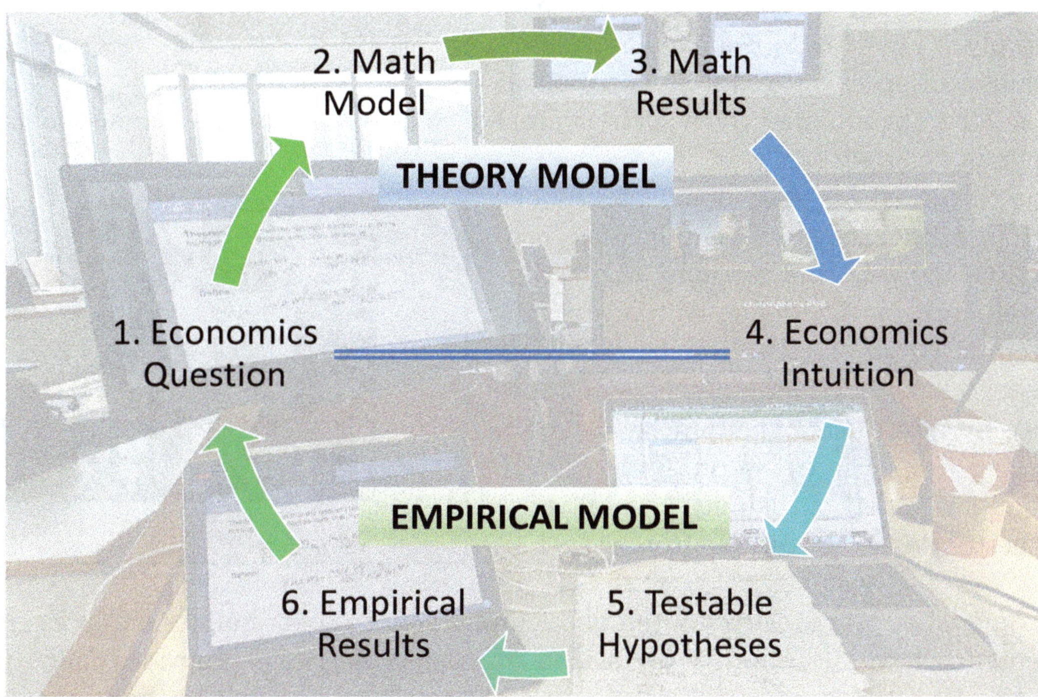

Figure 1.4: Economics cycle of research life

We build theoretical models, like those you will encounter in this textbook and your microeconomics and macroeconomics courses, to answer specific questions. In Figure 1.4, the starting point is always the question, step 1. What do we want to learn? In our example with Margo, the question might be "How does improved technology affect a worker's labor supply?" That is potentially an interesting question because better technology might make the worker more productive and able to earn more, therefore increasing labor supply. On the other hand, better technology might mean the worker earns a higher income and but works less so as to enjoy more leisure time. Which is it? We then build a mathematical model, step 2, to represent the choices the worker faces by giving her an objective function and constraints. The model should contain those forces we believe are important to the relevant question. For example, we have the productivity parameter, γ, in the model meant to represent the level of technology. The inclusion of the technology variable allows us to explore how Margo's choices respond to changes in technology.

Once we have our model set up, we use the logic of math (calculus, algebra, etc.) in step 3 to work through to results that show us an answer in math terms. We then translate that math in step 4 into economic intuition, which we can relay in words. Perhaps our answer to Margo's labor supply is something like, "If Margo is poor, an improvment in technology will increase her labor supply, but if Margo is relatively wealthy, an improvement in technology will mean she decreases her labor supply." That would is a classic trade-off between a substitution effect and an income effect that you may have encountered (or will encounter) in your economics courses.

Odds are, the principles of economics textbooks you learned from focused on discussing the economic intuition and why but without the underlying rigorous mathematical logic, bascially jumping from step 1 to step 4. Typical intermediate-level courses develop modern economics as a proper social science by introducing the underlying math logic and going deeper into the questions raised. This textbook is focused on helping you fill those gaps with the tools used in getting from step 1 to step 4.

But that model of Margo is so limited and simple, how can you believe it represents complex human behavior? Another dimension of models is the assumptions made to get the model started. There is a long-standing discussion about how realistic the assumptions should be relative to the ability to make accurate predictions. For example, Milton Friedman won a Nobel Prize in economics for a model in which he assumed consumers live forever.[5] Clearly, the assumption is empirically not valid, but the predictions generated by the model helped us understand consumer behavior much, much better than before, and his work from the 1960s in this area still heavily influences how macroeconomics is understood and taught. While it may be desirable to have realistic assumptions, because these will be less likely to be invalidated by data, a model that is too realistic can quickly become impossible to solve. Simplifications can make a model manageable such that we can use the rules/logic of math to understand the implications of those assumptions. Moreover, even if we make a hugely simplifying assumption such as "people live forever," we can go back and check to see how much this assumption drives the results we get. Models are meant to answer specific questions; models are not meant to do everything. Think back to the toy car analogy from the start of this chapter. If you are worried that the small toy car with a remote control electric motor won't give you a good prediction for a real car because it is too small, not gasoline powered, does not have a real driver inside, and so forth, you could try adding all these elements. That could prove difficult without access to an automobile factory, and eventually your "model" will become so complex as to defeat the entire purpose of building a model which is to simplify and focus on the key points relevant to your specific question.

Furthermore, to check how good a model really is in helping us understand human decision-making and its consequences, the next step is to take those predictions to the real world using data. In Margo's labor supply problem, we form a testable hypothesis based on the economic results and intuition from the theoretical model. We suggested the model results in an increase in labor supply with better technology for those who have low incomes but a decrease in labor for those with high incomes. Suppose we gather data on a set of workers, their income levels, other characteristics (e.g., age, education, etc.), and we have a change in technology the workers experienced at their workplace. For example, suppose we have data on Alaskan fisherman

Economics questions: Think back to what you learned in your prior economics classes. What results did you learn? What questions did those results answer? Try to write down at least three questions and results from you remember from previous classes. What do you think are the important features that go into those theoretical models? Who are the agents? What constraints do they face? What features of the real world are important for how people (or firms) make those decisions? Suppose you were skeptical that a result you learned is really true. What data might you gather to see if it holds in the real world?

over a time period with the introduction of advanced sonar technology on their fishing boats. In steps 5 and 6 in the diagram, we use the math logic found in statistics and econometrics to see if the data show a pattern consistent with the theoretical results. Those steps are beyond this textbook, but if you take econometrics, that is exactly what you will learn how to do. Once we have our empirical results, maybe they fit with the theoretical predictions or maybe they do not. Often they may fit, but not in all cases, which leads to the question "Why not?" That new question forms a new investigation, and we start all over again. Or maybe we find the theory does fit well for Alaskan fisherman, but that does not mean we know if it works well for baristas in Seattle or palm oil plantation workers in Sumatra or marketing professors in Barcelona. That is another question and therefore more research is needed. Maybe we get data on baristas

---

[5]Officially it is referred to as "The Nobel Memorial Prize in Economic Sciences."

in Seattle and the emprical results are completely opposite to the predictions of the model. Why? How can we build a model to capture the differences in human decision making behavior? Are there policies that might improve their welfare? What are they? That is what economic research and economic policy work is about.

One final note on generality and the choices you will see in your textbooks and problems. When textbooks and instructors build problems for you to work on to help you understand both the methods and the insights, it is often easier for everyone to write down very specific models. That makes the math work you, as a student, have to do easier, because it often makes finding an exact value for an answer easier. But do not miss the forest for the trees when doing such problems in microeconomics or macroeconomics or any other course for that matter. The problems are built to help you understand the intuition and logic. No one is committed to the idea of Margo's fishing production function being $F = 3H$ as an exact representation of the real world. That function can be easily generalized to capture broader ideas. But the specific values and type of function are helpful when first learning to work with these models because the key economic forces are illustrated by these types of examples very well. That is, the closing advice from this first chapter is: Throughout your coursework, focus on the ideas the equations represent and do not get caught up in the details by wondering if these are exactly like anything in the real world.

1.5 Exercises

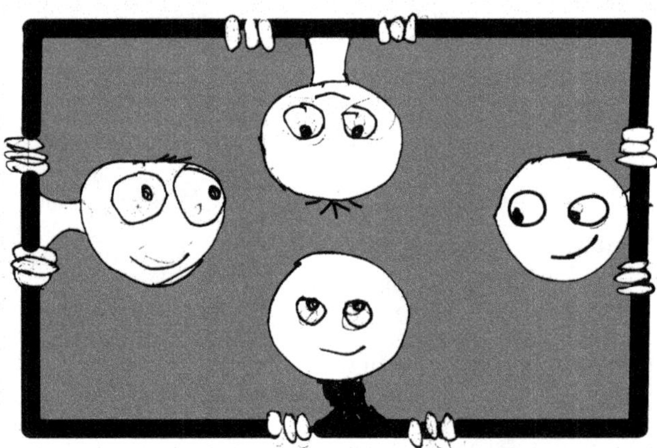

1.5.1 Quick Check Answers

a) If $\rho > 1$, the graph would be something like the following, with the curve going up at a steeper and steeper rate.

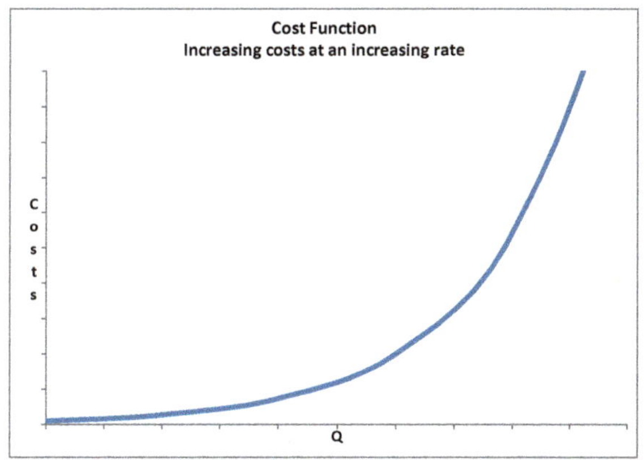

b) If $0 < \rho < 1$, the graph would be something like the following, with the curve going up but getting flatter and flatter.

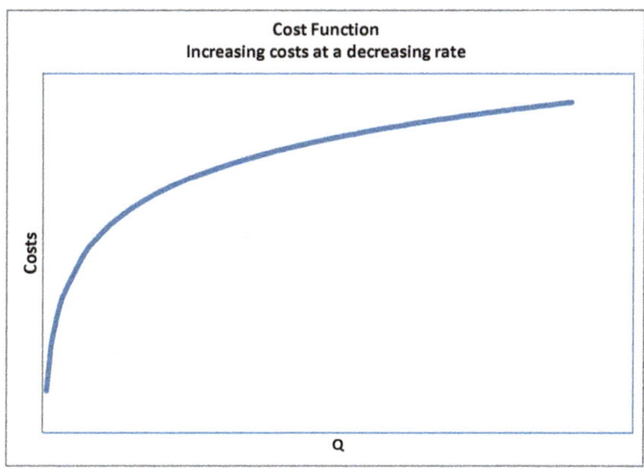

c) In (a), we would say that the production costs are increasing at an increasing rate. In (b), the costs are increasing but at a decreasing rate. Those shapes are referred to as *convex* and *concave*, respectively. We will have a lot to say on this subject in Chapter 5.

1.5.2 Practice Problems

1. A critical skill in working with economic models is recognizing which variables are *choices* made by the agents in the model and which variables are not. To help build that intuition, consider the following scenario: Margo is out of food. She is going to go to a store to get food to eat. In the following list, which of these are choices that Margo can make, and which variables are *exogenous* (determined outside of Margo's control)? Which ones are endowments?

    (a) Which store to go to

    (b) The distance to the nearest store

    (c) The total amount of time in the day

(d) How to get to the store: walk, bicycle, car, bus, or skateboard

(e) How many fish to buy

(f) The amount of cash she has

(g) Her income from last month

(h) The price of the fish

(i) The amount of time she spends shopping

(j) The sunny weather today

(k) Getting cookies or not

(l) The brands of cookies available at the store

(m) Bringing her cat to the supermarket or not

(n) The tax rate on buying fish

(o) Using cash, credit card, a mobile app, or a debit card for payment

(p) The brands of credit cards or mobile apps accepted by the store

Instructions for problems 2–7: Consider the models in the following problems and answer all of the following questions. What is the objective function? What are the constraints (if any)? If the constraint(s) is not written as an equation, write out the proper equation for it. What are the choice variables? What are the endogenous variables? What are the exogenous variables? What are the parameters? Finally, write down the optimization problem in the same format as Margo's problem (p.19) as follows:

$$\max_{N,H} U(F, N) \;=\; F^{1/2} N^{1/2}$$
$$\text{subject to}$$
$$T \;=\; H + N$$
$$F \;=\; 3H.$$

2. Margo has given up fishing and taken up growing vegetables instead. Her utility function is now $U(V, N) = V^{1/3} N^{2/3}$, where V is the vegetables she eats and N is leisure time with her cat. H is hours spent working in her garden growing vegetables. Let the number of hours in the day again be represented by T. Her vegetable production function is $V = \nu H$. Answer the general questions above. Also, can you think of any reasonable restrictions on the range of ν?

3. Yumika works for a health insurance company and likes to spend her time visiting Seoul, Korea. Her utility function is given by $U(K, H) = AK^{1/2} - BH^{1/2}$, where K is the number of days spent in Seoul and H is the days worked each year at her job. Furthermore, traveling to Korea costs money. She can afford to spend 1 day in Seoul for every 30 days worked at her company, $K = (1/30)H$. Answer the general questions above.

4. Hiro purchases only two goods: apples (a) and kumquats (k). His utility function is given by $U = 5a + 3k$. He has a budget of \$40 while the price of apples is \$3 per pound and the price of kumquats is \$6 per pound. Answer the general questions above.

5. Dylan consumes pizza (Z) and burritos (B). His utility function is given by $U(B, Z) = B^\alpha Z^\beta$ and his budget is $W = P_B B + P_Z Z$, where W is his wealth level and P_B and P_Z, are the prices of burritos and pizza, respectively. Answer the general questions above. Also, can you think of any reasonable restrictions on the range of P_B and P_Z? What about α and β?

6. Antoni has to decide how much to consume in the present and how much to save for his future. His utility function is given by the following: $U(c_1, c_2) = \ln(c_1) + \beta \ln(c_2)$, where c_1 is consumption in the present (period 1) and c_2 is consumption in the future (period 2). His budget constraint is given by the following:

$$c_1 + \frac{c_2}{1+r} = a + y_1 + \frac{y_2}{1+r}$$

where r is the real interest rate, a is Antoni's initial endowment of wealth, y_1 is his first period income, and y_2 is his second period income. Answer the general questions above. Also, what do you think the interpretation of β is? What restrictions might make sense here? Why?

7. Capitalist Cannibals Corporation (CCC) has acquired a monopoly on cat food. Because they have a monopoly, the price they can sell at is determined by the quantity they sell, as given by the following price function: $P = A - BQ$. Their cost of producing cat food is given by $C(Q) = cQ^\gamma$. Answer the general questions above. Also, consider A, B, c, and γ. What reasonable restrictions might apply to their range?

2   The Derivative and the Marginal Interpretation

The derivative of a function has an important interpretation in economics. It is the marginal change in the value of the function with respect to a marginal change in some variable. By now in your economics courses you have heard phrases such as marginal utility, marginal benefit, marginal cost, marginal product of labor, and so forth. These concepts are written down formally and logically in calculus through the derivative.

This chapter starts with interpreting an example calculus problem using the term "marginal" everywhere it applies. Using graphs, hopefully you will see the marginal approach in this simple example. The chapter then reviews the rules of taking derivatives, with a special emphasis on the chain rule, which can be tricky but appears widely in commonly used functions in models of economic analysis.

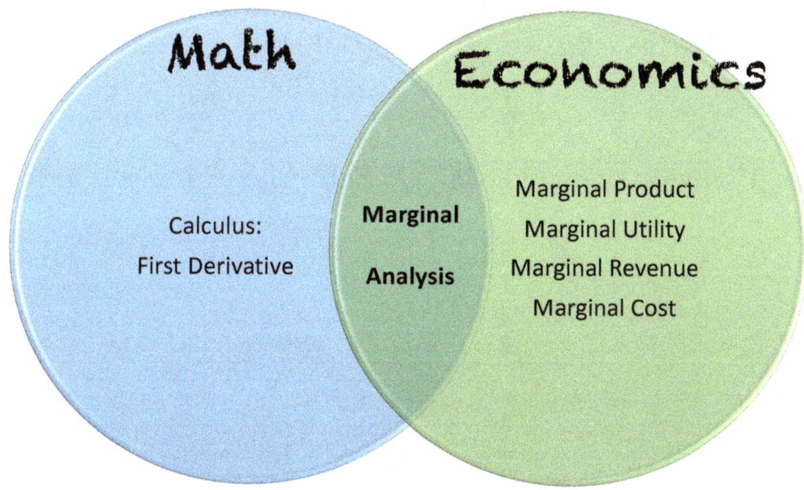

2.1 The Derivative Is the Margin

Before explaining (or reviewing for some of you) the derivative and the calculus rules of derivation, let's consider a simple example from microeconomics. Suppose Luffy has a monopoly on selling burgers on the island. The following equation is his profit maximization problem:

$$\max_{Q} \Pi = (10 - Q)Q - 2Q$$

where Π represents profits. The price of burgers, P, is determined by $P = 10 - Q$ and thus represents a downward sloping demand curve. (Try graphing $P = 10 - Q$ to convince yourself it is a demand curve.) Q is the quantity of burgers that the mo-

> **Definition**
>
> **Monopolist:** A firm that has no competitors in selling a particular product.

nopolist will choose to maximize his profits, and the cost per unit of Q produced is 2. In this problem, the monopolist has one choice, Q. Q can be chosen and is therefore an endogenous variable. Note that Q is written below *max* in the expression, indicating it is the variable chosen to maximize profits. The entire expression in the equation is the objection function, as we learned in the previous chapter.

Since $P = 10 - Q$, then $(10 - Q)Q$ is really price times quantity, or the total revenue of the firm. $2Q$ is the total cost to the firm. The difference between total revenues and total costs is profits. To find the level of Q that maximizes profits, we take the derivative with respect to Q. After rewriting slightly by multiplying the Q through the expression for price we have

$$\max_{Q} \Pi = 10Q - Q^2 - 2Q.$$

Now, taking the derivative we get

$$\frac{d\Pi}{dQ} = 10 - 2Q - 2.$$

The derivative must be equal to zero to achieve maximum profits for reasons we will explore. Setting that derivative equal to zero, we call this expression the **first-order condition**:

$$10 - 2Q - 2 = 0.$$

Now we have one equation with one unknown, Q. Solve the expression for Q to get

$$Q = \frac{10 - 2}{2} = 4.$$

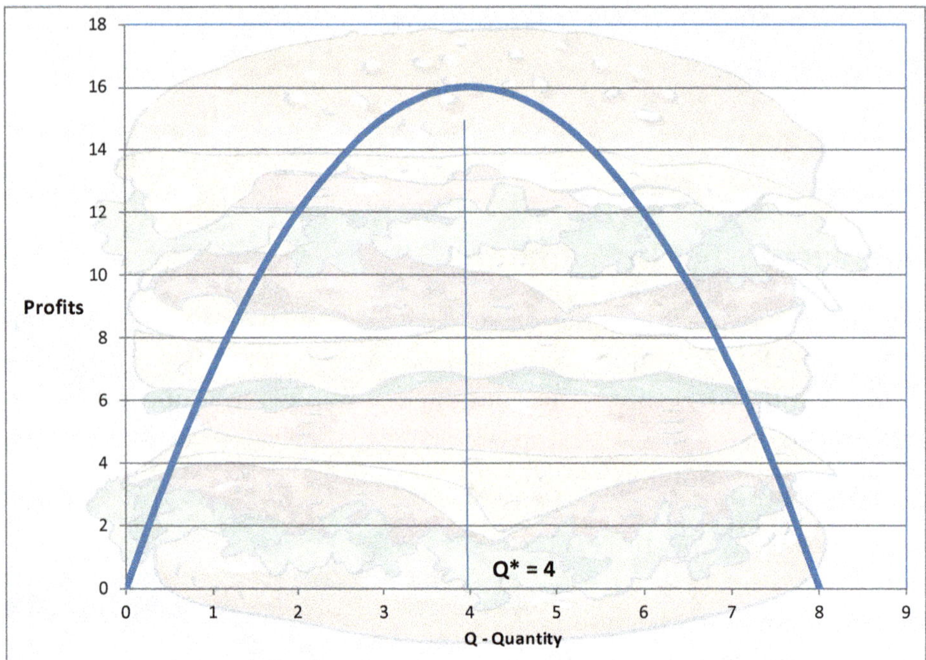

Figure 2.1: Luffy's monopoly profit function.

So the quantity of burgers to sell that maximizes profits is $Q^* = 4$. The asterisk ($*$) indicates the optimal choice of Q. Indeed, if we graph the profit function with Q on the x-axis and Π on the y-axis, we can see that 4 is the maximizing choice. See Figure 2.1.

Now let's go back to the first-order condition.

$$\underbrace{10 - 2Q}_{\text{Marginal Benefit}} \quad \overbrace{-2}^{\text{Marginal Cost}} \quad = 0$$

There are really two parts to this derivative. The first part comes from the revenue portion of the profits: $(10 - Q)Q$. That derivative is $10 - 2Q$ and it represents the *marginal revenue*, the additional amount of revenue gained by in-

> ### Definition
>
> **First-Order Condition (FOC):** The necessary condition for a point to be an extreme point (a minimum or maximum) of a function is that the point must make the first derivative equal to zero. $\frac{dy}{dx} = 0$.

creasing quantity by an infinitessimally small amount. The second part of the derivative is the -2, which comes from the cost portion of profits, $-2Q$. That represents the *marginal costs*, the additional costs incurred by increasing quantity by an infinitesimally small amount. All first-order conditions will have this structure: a **marginal benefit** of some kind and a **marginal cost** of some kind. The profit function is maximized where the marginal benefit exactly equals the marginal cost (a concept you should have encountered in principles of economics).

$$\underbrace{10 - 2Q}_{\text{Marginal Benefit}} = \underbrace{2.}_{\text{Marginal Cost}}$$

We can also see this graphically by separately plotting the marginal benefit component and the marginal cost component.

The downward sloping line is $10 - 2Q$ and represents the extra revenues the firm gets from producing that level of Q. The horizontal line at 2 is the marginal cost. At any level of Q, the additional cost is always 2, so it is a straight, flat line. Notice that they intersect precisely at $Q = 4$, where the marginal benefit equals the marginal cost.

> **Definition**
>
> **Marginal Benefit:** The additional benefit (utility, profits, revenue, etc) an agent obtains from choosing a marginal amount more of some good or action. For example, choosing to produce a little more would mean more revenue for the firm and how much more revenue is the marginal benefit.

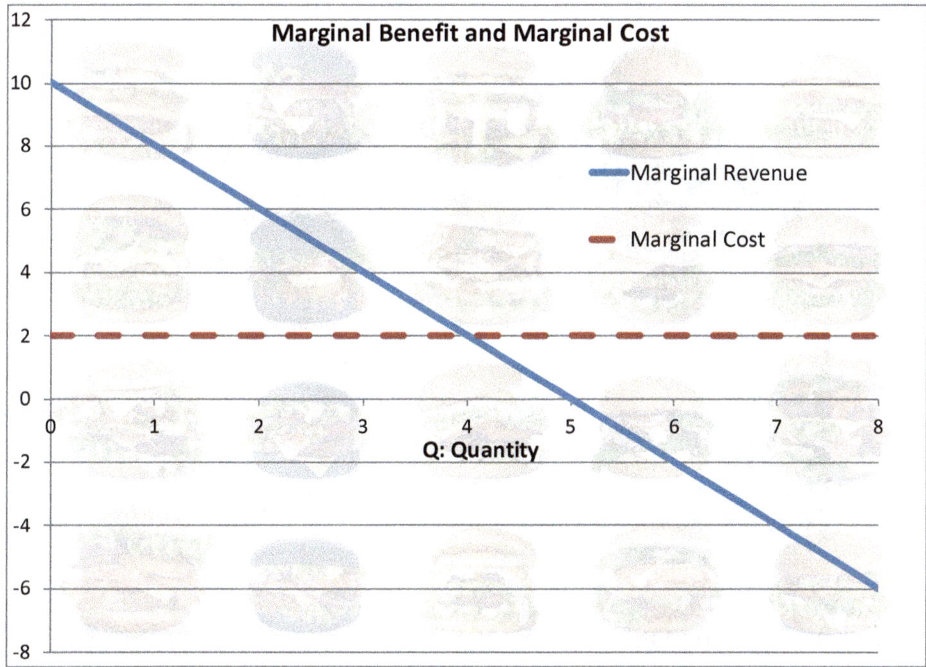

Figure 2.2: Luffy's burgers, marginal benefit and marginal cost.

To understand this point better, think about quantity choices other than at 4. Suppose the monopolist chooses $Q = 2$. Then the marginal revenue would be $10 - 2Q = 10 - 2(2) = 6$, and the marginal cost is still 2. Thus, there are extra profits (the difference between the marginal revenue and marginal costs) that the monopolist could get. The monopolist is producing too little at $Q = 2$ to maximize profits. What about $Q = 7$? The marginal benefits are now $10 - 2Q = 10 - 2(7) = -4$. A negative value here represents lost revenue and is clearly below the costs at 2. Thus, producing too large a quantity leads to lost profits as well. Compare these results for other levels of Q to the graph in Figure 2.1. Clearly $Q = 2$ or $Q = 7$ does not yield the most profits. The rule is that the function is maximized at the point exactly where the marginal benefits equal the marginal costs. Mathematically, that is the point in the objective function for profits where the slope of the curve is zero.

To finish the example, now that we know the profit maximizing quantity, we can figure out profits.

$$\Pi = 10\,(4) - (4)^2 - 2\,(4) = 16$$

We can see the result visually in Figure 2.1, but we can also derive it from Figure 2.2. The profits are the difference or the area between the marginal revenue curve and the marginal cost curve. In this case, to figure out the area we have a nice, simple triangle. The base has length of 4, and the height is $10 - 2$ or 8. So the area is $\frac{1}{2}(4)(8) = 16$.

2.2   Formal Definition of the Derivative

The derivative of a function tells us the slope of the function at any point. Consider the following graph of an arbitrary function, $y = f(x)$. Suppose we want to know the slope at the point x_0 where the function has the value $y = f(x_0)$. The true slope is drawn in the figure and is a line tangent to the actual curve at that point. The slope represents how the function itself changes. The slope tells us whether the function is increasing or decreasing and how fast it is changing as the value of x increases.

We can get an approximation of the slope by picking some other point away from x_0 and then using the point-slope formula from basic algebra. Suppose we add h_1 to x_0 to get $x_0 + h_1$. h_1 is just some arbitrary number. Now the value of the function at $x_0 + h_1$ is $y = f(x_0 + h_1)$. To find the slope we merely take the ratio of the change in y to the change in x by comparing the new point to the original point.

$$\text{Slope} = \frac{\Delta y}{\Delta x} = \frac{f(x_0 + h_1) - f(x_0)}{x_0 + h_1 - x_0} = \frac{f(x_0 + h_1) - f(x_0)}{h_1}$$

Look at Figure 2.3. Visually, h_1 is pretty large, and thus the slope from the formula will give us the line labeled "First Approximation." It's not a very good estimate. The true slope is much steeper. We can do better by choosing a point closer to our original x_0. Let our new try come from adding h_2 to x_0 to get $x_0 + h_2$. h_2 is deliberately chosen to be smaller than h_1 so that we are closer to the original point. Now the value of the function at $x_0 + h_2$ is $y = f(x_0 + h_2)$. Again, to find the slope we take the ratio of the change in y to the change in x.

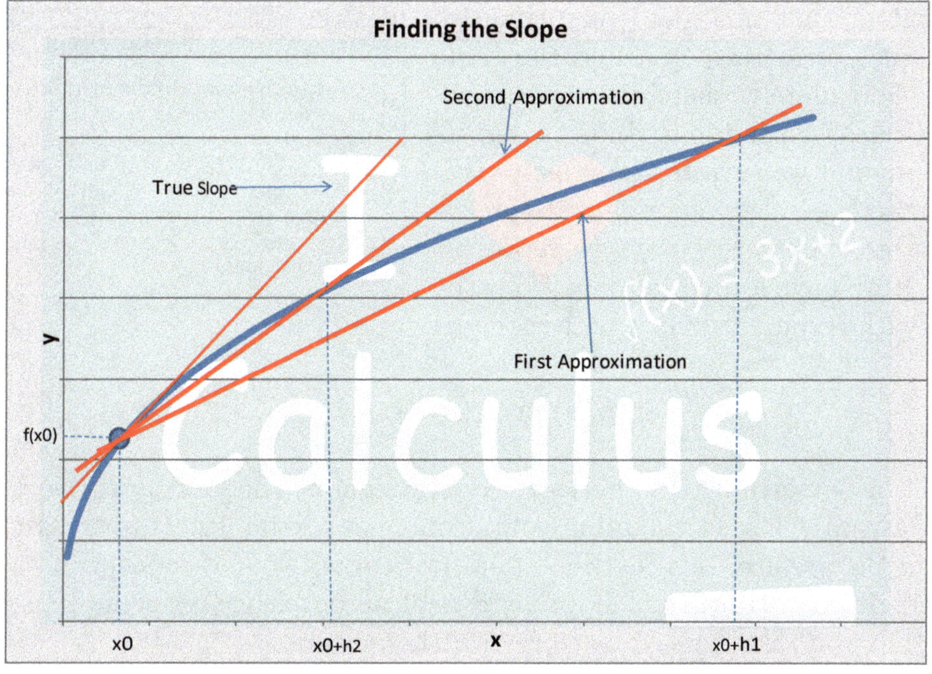

Figure 2.3: Finding the Derivative the Hard Way

$$\text{Slope} = \frac{\Delta y}{\Delta x} = \frac{f(x_0 + h_2) - f(x_0)}{x_0 + h_2 - x_0} = \frac{f(x_0 + h_2) - f(x_0)}{h_2}$$

We would now get the line labeled "Second Approximation," which is better than the first, but still not accurate. Now imagine choosing points closer and closer to x_0 such that our h value gets smaller and smaller, eventually approaching zero. If we look at the slope formula for any level of h we have

$$\text{Slope} = \frac{\Delta y}{\Delta x} = \frac{f(x_0 + h) - f(x_0)}{h},$$

which presents a problem because if h in the denominator really does go to zero the fraction is undefined. Thus, to get the slope we want we need to take the *limit* as $h \to 0$.

$$\text{Slope} = \frac{dy}{dx} = \lim_{h \to 0} \frac{f(x_0 + h) - f(x_0)}{h}$$

That is the formal definition of the derivative. It's really nothing different from the slope formula you used in basic algebra except that the change in x is infinitessimally small and effectively zero. In the preceding two equations, notice the transition from $\frac{\Delta y}{\Delta x}$ to $\frac{dy}{dx}$. They represent the same thing. The only difference is that in the second one, the change in x is marginal or infinitessimally small. The former is the algebra version; the latter is the calculus version.

It would, however, be a major pain if we had to take the limit of that function every time we wanted to find the derivative (i.e. , find the slope of a function). Fortunately, there are some fairly straightforward rules we can use to find the derivative of a function, and we turn to those next. But before going on to the rules, one point about notation: There are a number of ways to represent the derivative. If we have a function $y = f(x)$, the derivative of y with respect to x can be written as

> **Making the Connection**
>
> **Continuity and differentiability:** As a review and to aid your understanding, answer these questions using any resource: (1) Why is continuity necessary for taking derivatives? (2) Write down a function that is continuous but not differentiable? (3) What makes a function differentiable?

$$\frac{dy}{dx} \text{ or } f'(x) \text{ or } f_x(x) \text{ or } f_x.$$

They represent the same thing, the derivative. Throughout this textbook, we will use all these forms, as you will see them all in your economics textbooks. **Important note:** As mentioned in the preface, this textbook does not cover two key concept in calculus: **continuity** and **differentiability**. If you are unclear on either concept, it is highly recommended that you take some time to review these terms.

2.3 Value of the Slope: Interpretation

We just spent some time covering the fact that the derivative is the slope of a function at any point along the function. Therefore, it follows that a *positive* derivative means the slope is going up or the function is ***increasing*** at that point. Similarly, a *negative* derivative means the slope is going down or the function is ***decreasing*** at that point.

Definition

Increasing function: An increasing function has a positive slope, $\frac{dy}{dx} > 0$.

Decreasing function: A decreasing function has a negative slope, $\frac{dy}{dx} < 0$.

Whether a function is increasing or decreasing is really important when we build models, so the relationship between the first derivative and the direction of the function is important for interpretation and economic intuition.

For example, if you have a demand curve $D(P)$ that reads "*The quantity demanded is a function of price, P.*" We know from our most basic economics of supply and demand, that demand curves are downward sloping in price. That is, "*Demand is decreasing in price.*" Therefore, it would be reasonable to assume or expect that the derivative of the demand function with respect to P is negative, indicating a decreasing function. Formally, we would write that as

$$\frac{dD(P)}{dP} < 0,$$

$$\text{or}$$

$$D_P(P) < 0,$$

$$\text{or}$$

$$D'(P) < 0,$$

$$\text{or}$$

$$D_P < 0.$$

All of these versions of the notation represent the same thing, the first derivative of the demand function, $D(P)$, with respect to P is negative, which is the same as saying the demand function is decreasing in price, P. Analogously, we expect that a supply function, $S(P)$ is increasing in price. So, we would have $S_P(P) > 0$ or $\frac{dS(P)}{dP} > 0$.

Not all functions are *always* increasing or *always* decreasing. For example, $\Pi(Q)$, which represents a firm's profits as a function of quantity, similar to what we saw for Luffy's burgers, and illustrated in Figure 2.1, initially increases, hits the maximum point at $Q^* = 4$, then becomes decreasing. So, the slope of the function is positive between 0 and 4 and negative after 4. The slope, of course, is exactly zero at $Q^* = 4$, and therefore the function is neither increasing nor decreasing at that point.

2.4 Rules of Differentiation

This section goes through all the key rules for taking derivatives. I do not provide proofs for any of these but jump right to what is useful.

2.4.1 Constant Rule

The derivative of a constant term (i.e. , just a number) is zero. For example, if we consider the function $y = 5$, then

$$\frac{dy}{dx} = 0$$

because x does not enter the function at all and therefore does not change y. The graph of just a constant is a horizontal line, and therefore the slope is zero. Another way to say that is, with a constant function the value of y does not change with x.

2.4.2 Power Rule

The derivative of y with respect to x in a function of the form

$$y = x^n$$

is

$$\frac{dy}{dx} = nx^{n-1}.$$

Multiply by the exponent and subtract 1 from the exponent itself. If the function is linear (meaning its a straight line) such as

$$y = 5x,$$

then the exponent on x is 1 and the derivative is

$$\frac{dy}{dx} = 5$$

since the implied 1 in the exponent after subtracting 1 becomes 0 and anything raised to the power of 0 is 1 More generally if the x term is multiplied by any constant b, we have

$$y = bx$$
$$\frac{dy}{dx} = b.$$

If the exponent itself is a variable, the same rules apply. Multiply the expression by the exponent and then subtract 1 from the exponent.

$$y = bx^c$$
$$\frac{dy}{dx} = cbx^{c-1}$$

Here are some examples:

$$y = x^3 \rightarrow \frac{dy}{dx} = 3x^2$$

$$y = 4x^2 \rightarrow \frac{dy}{dx} = 8x$$

$$y = 3x^{1/2} \rightarrow \frac{dy}{dx} = \frac{3}{2}x^{-1/2}$$

$$y = 2x \rightarrow \frac{dy}{dx} = 2$$

$$y = 10x^{-8} \rightarrow \frac{dy}{dx} = -80x^{-9}$$

$$y = Ax^5 \rightarrow \frac{dy}{dx} = 5Ax^4, \text{ where } A \text{ is a parameter}$$

$$y = Bx^\alpha \rightarrow \frac{dy}{dx} = \alpha Bx^{\alpha-1}, \text{ where } B \text{ and } \alpha \text{ are parameters}$$

In the last two examples, when taking the derivative with respect to x, we treat the parameters (numbers represented by letters or Greek letters) as regular numbers. One of the main conceptual difficulties we note among students is moving from taking derivatives on specific functional forms with numbers, like the first example, to taking derivatives on functions with only variables and parameters, like the last example. Remember to treat the parameters and variables, the letter symbols, as numbers because that is what they are.

2.4.3 Sum-Difference Rule

When a function has two components that are additively separable (or subtracted), take the derivatives of each part separately. An additively separable function has the following form:

$$y = f(x) + g(x).$$

To find the derivative with respect to x, first take the derivative of $f(x)$ and add the derivative of $g(x)$. In the following example, $f(x) = 2x^2$ and $g(x) = 5x^3$.

$$y = 2x^2 + 5x^3 \rightarrow \frac{dy}{dx} = 4x + 15x^2$$

The derivative of $f(x)$ is $4x$, and the derivative of $g(x)$ is $15x^2$. If the components are linked by a minus sign, the minus sign stays.

$$y = 6x^4 - 2x^3 \rightarrow \frac{dy}{dx} = 24x^3 - 6x^2$$

If the exponent is negative, make sure the sign in front changes appropriately. Here are some examples.

$$y = 2x^2 + 5x^{-3} \rightarrow \frac{dy}{dx} = 4x - 15x^{-4}$$

$$y = 6x^4 - 2x^{-3} \rightarrow \frac{dy}{dx} = 24x^3 + 6x^{-4}$$

If the function contains an added constant term, then the constant rule applies to that part of the function. Here are some examples.

$$y = 3x^2 + 8 \rightarrow \frac{dy}{dx} = 6x$$

$$y = 6x^4 - 250 \rightarrow \frac{dy}{dx} = 24x^3$$

$$y = Ax^\alpha + Bx^\beta + Cx^\gamma + D \rightarrow \frac{dy}{dx} = \alpha Ax^{\alpha-1} + \beta Bx^{\beta-1} + \gamma Cx^{\gamma-1}$$

2.4.4 Product Rule

Sometimes the function is complicated and involves two smaller functions that multiply each other, $y = f(x)g(x)$. For example,

$$y = (x + 2)\left(x^2 + 3\right),$$

where we think of $f(x) = x + 2$ as the component in the first set of parentheses and $g(x)$ as the component in the second set of parentheses. We could algebraically multiply these out to get $y = x^3 + 3x + 2x^2 + 6$ and then apply the sum/difference rule. However, in many cases we will not be able to do that. For a general function of the form

$$y = f(x)g(x),$$

the derivative is

$$\frac{dy}{dx} = f'(x)g(x) + f(x)g'(x).$$

Note that here I have used the prime ($\prime$) symbol to indicate the derivative of the smaller functions with respect to x. In words, take the derivative of the $f(x)$ portion and multiply by $g(x)$, then add $f(x)$ times the derivative of $g(x)$. Let's return to the example.

$$\begin{aligned}
y &= (x+2)\left(x^2+3\right) \\
\frac{dy}{dx} &= (1)(x^2+3) + (x+2)(2x) \\
&= 3x^2 + 4x + 3.
\end{aligned}$$

You should verify that this result is the same if you took the derivative directly on $y = x^3 + 3x + 2x^2 + 6$, which is the result from carrying out the multiplication on $(x+2)\left(x^2+3\right)$.

Here's another example:

$$y = (3x^{-1} + 3)(4x^\alpha - 6).$$

Here, let $f(x) = 3x^{-1} + 3$ and $g(x) = 4x^\alpha - 6$.

$$\frac{dy}{dx} = (-3x^{-2})(4x^\alpha - 6) + (3x^{-1} + 3)(4\alpha x^{\alpha - 1})$$

Simplifying algebraically,

$$\begin{aligned}
\frac{dy}{dx} &= -12x^{\alpha-2} + 18x^{-2} + 12\alpha x^{\alpha-2} + 12\alpha x^{\alpha-1} \\
&= 12(\alpha - 1)x^{\alpha-2} + 12\alpha x^{\alpha-1} + 18x^{-2}.
\end{aligned}$$

Here are some more examples.

$$\begin{aligned}
y &= (10 - x)x^{1/2} \\
\frac{dy}{dx} &= (-1)x^{1/2} + (10 - x)\frac{1}{2}x^{-1/2} \\
&= -x^{1/2} + 5x^{-1/2} - \frac{1}{2}x^{1/2} \\
&= 5x^{-1/2} - \frac{3}{2}x^{1/2}
\end{aligned}$$

$$
\begin{aligned}
y &= x^{2/3}(6 - Bx^5) \\
\frac{dy}{dx} &= \frac{2}{3}x^{-1/3}\left(6 - Bx^5\right) + x^{2/3}(-5Bx^4) \\
&= 4x^{-1/3} - \frac{2}{3}Bx^{14/3} - 5Bx^{14/3} \\
&= 4x^{-1/3} - \frac{17}{3}Bx^{14/3}
\end{aligned}
$$

2.4.5 Quotient Rule

When a function is a fraction or a ratio of two smaller functions,

$$
y = \frac{f(x)}{g(x)},
$$

it gets a bit messier. The derivative is

$$
y = \frac{g\left(x\right) f'\left(x\right) - f(x)g'\left(x\right)}{\left[g(x)\right]^2}.
$$

In words, the derivative has the square of the $g(x)$ function in the denominator. In the numerator, we have the g function times the derivative of the f function minus the f function times the derivative of the g function. Let's do an example.

$$
\begin{aligned}
y &= \frac{3x + 2}{x^3 + 1} \\
\frac{dy}{dx} &= \frac{\left(x^3 + 1\right)(3) - (3x + 2)\left(3x^2\right)}{\left(x^3 + 1\right)^2}
\end{aligned}
$$

Simplifying using algebra,

$$
\begin{aligned}
\frac{dy}{dx} &= \frac{3x^3 + 3 - \left(9x^3 + 6x^2\right)}{\left(x^3 + 1\right)^2} \\
&= \frac{-6x^3 - 6x^2 + 3}{\left(x^3 + 1\right)^2}.
\end{aligned}
$$

Here are some more examples.

$$y = \frac{1-x}{1+x}$$

$$\frac{dy}{dx} = \frac{(1+x)(-1) - (1-x)1}{(1+x)^2}$$

$$= \frac{-1-x-1+x}{(1+x)^2}$$

$$= \frac{-2}{(1+x)^2}$$

$$y = \frac{x^2-3}{1+2x}$$

$$\frac{dy}{dx} = \frac{(1+2x)(2x) - (x^2-3)2}{(1+2x)^2}$$

$$= \frac{(2x+4x^2) - 2x^2 + 6}{(1+2x)^2}$$

$$= \frac{2x^2 + 2x + 6}{(1+2x)^2}$$

$$y = \frac{2x-1}{x^{-1/2}}$$

$$\frac{dy}{dx} = \frac{\left(x^{-1/2}\right)2 - (2x-1)\left(-\frac{1}{2}x^{-3/2}\right)}{x^{-1}}$$

$$= \frac{3x^{-1/2} - \frac{1}{2}x^{-3/2}}{x^{-1}}$$

$$= x\left(3x^{-1/2} - \frac{1}{2}x^{-3/2}\right)$$

$$= 3x^{1/2} - \frac{1}{2}x^{-1/2}$$

Note that in some cases you may find it easier to convert a function that is a fraction and requires quotient rule into a function that is multipilicative and would require product rule. Let's do an example to make that clear. In doing so, we take advantage of the following property of exponents:

$$\frac{a}{x^b} = ax^{-b}.$$

For example, going back to one of the earlier problems,

$$y = \frac{3x + 2}{x^3 + 1}$$
$$\frac{dy}{dx} = \frac{(x^3 + 1)(3) - (3x + 2)(3x^2)}{(x^3 + 1)^2}.$$

Let's do that problem again, but this time let's move the denominator into the numerator using that exponent rule.

$$y = \frac{3x + 2}{x^3 + 1}$$
$$y = (3x + 2)(x^3 + 1)^{-1}$$

Now apply the product rule to get

$$\frac{dy}{dx} = 3(x^3 + 1)^{-1} + (3x + 2)(-1)(x^3 + 1)^{-2}(3x^2).$$

Rearranging algebraically, we get back to

$$\frac{dy}{dx} = \frac{(x^3 + 1)(3) - (3x + 2)(3x^2)}{(x^3 + 1)^2}.$$

Try that method on the other quotient rule examples and verify you get the same answer. You might note that in the example we just did, when we applied the product rule to the $(x^3 + 1)^{-1}$ portion, the derivative was $(x^3 + 1)^{-2}(3x^2)$. That came from taking the derivative of the outer function, z^{-1} where $z = x^3 + 1$, and then multiplying by the derivative of the inner function, $z = x^3 + 1$. That is actually an application of the chain rule, which we turn to next.

Quotient rule

e) Find dy/dx from $y = \frac{x}{1+x}$.

f) Find dY/dG from $Q(L) = A\frac{L^\beta}{L^{1/2}+\Phi}$.

2.4.6 Chain Rule

This rule is probably the hardest and the one most frequently forgotten when trying to take derivatives. The chain rule applies when you can write one function inside another function. It sounds odd, but the general form looks like this:

$$y = f(g(x))$$
$$\frac{dy}{dx} = f'(g(x))\, g'(x).$$

In words, it says take the derivative of the outer function first and then multiply by the derivative of the inner function. This (hopefully) will become clearer with some examples. Consider the following:

$$y = \left(2x^2 + x\right)^3.$$

The inner function here is $g(x) = 2x^2 + x$. To represent the outer function let $z = 2x^2 + x$; then if we rewrite the function we let

$$y = z^3.$$

The outer function is $y = f(z) = z^3$. We know how to take the derivative of both of those parts from the basic rules of derivatives previously discussed. Applying the chain rule, we would first take the derivative of $y = z^3$, then multiply that by the derivative of the inner function, $2x^2 + x$.

$$\frac{dy}{dx} = 3z^2 \left(4x + 1\right)$$

Replacing the z with its original expression in terms of x we have

$$\frac{dy}{dx} = 3 \left(2x^2 + x\right)^2 \left(4x + 1\right).$$

Here are some more examples of the chain rule.

$$\begin{aligned}
y &= \left(x^2 + x + 1\right)^2 \\
\text{Let } z &= x^2 + x + 1 \\
y &= z^2 \\
\frac{dy}{dx} &= 2z(2x + 1) \\
\frac{dy}{dx} &= 2\left(x^2 + x + 1\right)(2x + 1) \\
&= 4x^3 + 6x^2 + 6x + 2
\end{aligned}$$

$$\begin{aligned}
y &= \left(x^2 + 1\right)^{1/2} \\
\frac{dy}{dx} &= \frac{1}{2}\left(x^2 + 1\right)^{-1/2} 2x \\
&= x\left(x^2 + 1\right)^{-1/2}
\end{aligned}$$

I skipped the step letting z equal the inner function $(x^2 + 1)$ and took the derivative of the outer function directly.

Here's another, more complex example:

$$y = \frac{x(2x + 1)^{1/2}}{x^2 - 1}.$$

In this case, we need to use product, quotient and chain rules. Since the entire function is one ratio, we start with the quotient rule. In doing that, when we take the derivative of the numerator, we apply the product rule because we have x times $(2x + 1)^{1/2}$. Furthermore, in taking the derivative of the second part of the numerator, $(2x + 1)^{1/2}$, we need to apply the chain rule.

$$\frac{dy}{dx} = \frac{\overbrace{\left(x^2 - 1\right) \overbrace{\left[(1)(2x + 1)^{1/2} + (x)\overbrace{\frac{1}{2}(2x + 1)^{-1/2}}^{\text{Chain Rule}}\right]}^{\text{Product Rule}} - x(2x + 1)^{1/2}(2x)}^{\text{Quotient Rule}}}{\left(x^2 - 1\right)^2}$$

After some algebra we can write it a little bit more compactly as

$$\frac{dy}{dx} = \frac{(x^2 - 1)\left[(2x + 1)^{1/2} + \frac{1}{2}x(2x + 1)^{-1/2}\right] - 2x^2(2x + 1)^{1/2}}{\left(x^2 - 1\right)^2}.$$

The following table summarizes the key rules presented here:

Table 2.1: Derivative Rules

Rule	Function	Derivative
Constant	$y = c$	$\frac{dy}{dx} = 0$
Linear	$y = bx$	$\frac{dy}{dx} = b$
Power	$y = x^n$	$\frac{dy}{dx} = nx^{n-1}$
Sum-Difference	$y = f(x) + g(x)$	$\frac{dy}{dx} = f'(x) + g'(x)$
Product	$y = f(x)g(x)$	$\frac{dy}{dx} = f'(x)g(x) + f(x)g'(x)$
Quotient	$y = \frac{f(x)}{g(x)}$	$\frac{dy}{dx} = \frac{g(x)f'(x) - f(x)g'(x)}{[g(x)]^2}$
Chain	$y = f(g(x))$	$\frac{dy}{dx} = f'(g(x))g'(x)$

2.5 Derivatives of Generalized Functions

Consider the following specific function and its derivative:

$$y = 5x^2$$
$$\frac{dy}{dx} = 10x.$$

We say the function is *specific* because we know exactly what the function is. Moreover, the rules for taking the derivative and finding the marginal value are clear. Now, recall the discussion from the end of Chapter 1 about general functions and how they are advantageous because they can account for a wider variety of possibilities. In contrast, consider

$$y = f(x).$$

That is very general. All it says is that y is a function of x. How would we take the derivative, $\frac{dy}{dx}$? None of the rules we went through directly apply. But, it is really just a matter of writing down the derivative notation as follows:

$$y = f(x)$$
$$\frac{dy}{dx} = f'(x).$$

All that says is that $f'(x)$ is the derivative of the function $f(x)$, which we label with the "prime" symbol, $'$, to indicate the first derivative. That's it.

Can we say anything more about $\frac{dy}{dx} = f'(x)$? No, not unless we make assumptions about the nature of the function. Suppose that we have good reason to believe that y is an increasing function of x, so we assume that is so. Therefore, $\frac{dy}{dx} = f'(x) > 0$. We are effectively assuming that the function is increasing. We do that by stating at the outset that we assume $f'(x) > 0$. If it is a decreasing function we would assume that $f'(x) < 0$.

What about when a function increases at first but then decreases, like the profit function in Figure 2.1? We could write that as follows: For $y = f(x)$, $f'(x) > 0$ for $x < \bar{x}$ and $f'(x) < 0$ for $x > \bar{x}$. In words that says that y is increasing with x until x reaches a particular point $\bar{x}$, and then y is a decreasing function after that when x greater than $\bar{x}$.

Let's use a more concrete example from economics. Suppose that the quantity demanded of straw hats is a function of the price of straw hats. We represent that idea in the function $Q = Q(P)$. Notice that the letter Q appears twice. However, the left-hand side represents the actual numerical value. The Q on the right-hand side is the function that determines the value of Q and that is a function of P, the price. So, a specific function might look like this: $Q(P) = 100 - P^2$. The actual function only has P in it on the right-hand side, but it determines the value of Q. Now go back to our general function $Q = Q(P)$. Consumer theory from microeconomics tells us that the higher the price the fewer of a good consumers wish to buy. Therefore, we expect a negative relationship between the price of straw hats and the quantity consumers want to buy. Q should decrease in P and therefore we would assume that

$$\frac{dQ}{dP} = Q_P(P) < 0.$$

The derivative of Q with respect to P is negative. Therefore, Q is a decreasing function of P, or, in economics terms, the higher the price, the lower the quantity the consumers wish to buy.

Suppose we have a function represented as $y = F(x, z)$. What does that mean? What's the comma for? That means that the variable y is a function of two variables, both x and z. For example, at Nami's ice cream stand, her quantity of sales, Q, depends on both the price she charges, P, for ice cream and her sales also depend on the temperature, T. We would write the function as $S = S(P, T)$. That is a *multivariate* function. Our dependent variable is now determined by more than one variable. While the concepts of slope, margin, increasing, decreasing, and so forth, do not change just because the function is multivariate, we do need to be precise about *which derivative* we are taking because there is more than one. We will expound on these types of functions and applying calculus in Chapter 6.

2.6 Exercises

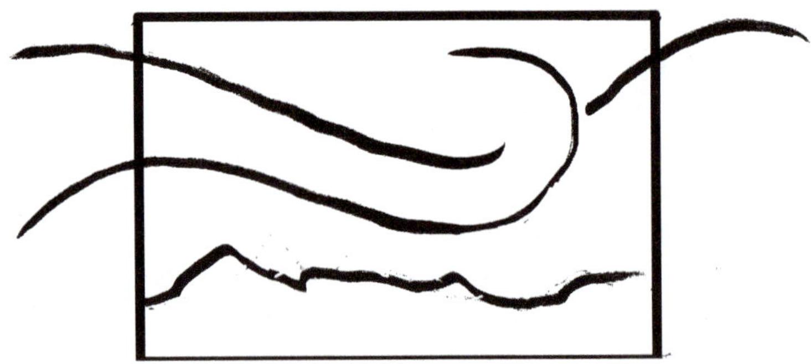

2.6.1 Quick Check Answers

a) $y = 8x^8 \quad \Rightarrow \quad \frac{dy}{dx} = 64x^7$

b) $Q(L) = AL^\rho \quad \Rightarrow \quad \frac{dQ(L)}{dL} = \rho AL^{\rho-1}$

c) $y = x(5 + 3x^2)^{1/2} \quad \Rightarrow$
$\frac{dy}{dx} = (1)(5 + 3x^2)^{1/2} + x\frac{1}{2}(5 + 3x^2)^{-1/2}(6x) = (5 + 3x^2)^{1/2} + 3x^2(5 + 3x^2)^{-1/2}$

d) $Q(L) = AL^\rho(1 + L^\phi) \quad \Rightarrow \quad \frac{dQ}{dL} = \rho AL^{\rho-1}(1 + L^\phi) + AL^\rho(\phi L^{\phi-1})$

e) $y = \frac{x}{1+x} \quad \Rightarrow \quad \frac{dy}{dx} = \frac{(1+x)(1)-x(1)}{(1+x)^2} = \frac{1}{(1+x)^2}$

f) $Q(L) = A\frac{L^\beta}{L^{1/2}+\Phi} \quad \Rightarrow \quad \frac{dQ}{dL} = A\frac{(L^{1/2}+\Phi)\beta L^{\beta-1}-L^\beta\left(\frac{1}{2}L^{-1/2}\right)}{(L^{1/2}+\Phi)^2}$
$= A\frac{(\beta-\frac{1}{2})L^{\beta-1/2}+\Phi\beta L^{\beta-1}}{(L^{1/2}+\Phi)^2}$

g) $y = (x^3 + 4)^2 \quad \Rightarrow \quad \frac{dy}{dx} = 2(x^3 + 4)^1(3x^2) = 6x^2(x^3 + 4)$

h) $U(C) = (C^\alpha - \rho)^3 \quad \Rightarrow \quad \frac{dU}{dC} = 3(C^\alpha - \rho)^2(\alpha C^{\alpha-1})$
$= 3\alpha(C^\alpha - \rho)^2 C^{\alpha-1}$

i) $S = S(P) \quad \Rightarrow \quad \frac{dS}{dP} = S_P(P) > 0$

j) $L^D = L^D(w) \quad \Rightarrow \quad \frac{dL^D}{dw} = L_w^D(w) < 0$

2.6.2 Practice Problems

1. Consider the following common functions from microeconomics and macroeconomics. From your knowledge of economics principles, determine whether the function is normally expected to be increasing, decreasing, or have ranges of both. Write out the formal derivative indicating your assumption textitand state the meaning in words. For example, if you believe the demand function, $D(P)$, is decreasing in price, you would write $D_P(P) < 0$: "Demand is decreasing in price."

   (a) Revenues as a function of quantity sold: $R(Q)$

   (b) Costs as a function of quantity produced: $C(Q)$

   (c) Utility as a function of cookies consumed: $U(C)$

   (d) Utility as a function of hours worked at a job: $U(H)$

   (e) Output produced as a function of physical capital available: $Y = F(K)$

   (f) Money demand as a function of income: $M^D = L(Y)$

   (g) Money demand as a function of nominal interest rates: $M^D = L(i)$

   (h) Demand for boats as a function of sales tax rates: $D(\tau)$

   (i) Firm profits as a function of the number of competitors: $\Pi(N)$

   (j) Labor supply as a function of wages: $L^S(w)$

   In the following two problems, find the profit maximizing quantity and graph the marginal revenue and marginal cost. Interpret your first-order condition in terms of marginal benefit and marginal cost.

2. $\max_Q \Pi = (20 - Q)Q - 4Q$

3. $\max_Q \Pi = (12 - Q)Q - 2Q^2$

In all the following problems, find the derivative of y with respect to x:

4. $y = 2x^2 + x$

5. $y = -x^2 + x - 5$

6. $y = x + \frac{1}{x} + 2$

7. $y = Ax^\alpha + Cx^{-\beta}$

8. $y = \left(x^2 - 5x + 2\right)\left(2x^{1/2} - 1\right)$

9. $y = \left(5x^3 + 1\right)\left(6 - x^{-2}\right)$

10. $y = \left(\beta x^2 - \alpha\right)\left(\alpha x^{\alpha-1} + \beta\right)$

11. $y = \frac{1}{x^2+1}$

12. $y = \frac{x^{1/2}}{3x^2+2}$

13. $y = \frac{Ax^\alpha}{x^2+\beta}$

14. $y = (2x - 1)^5$

15. $y = \left(x^2 - 9\right)^{-2/3}$

16. $y = \left(\frac{x+3}{x-4}\right)^2$

17. $y = \left(Ax^\alpha - \beta x\right)^{\alpha+1}$

2.7   Appendix

Here, we list some formal definitions and other items for your reference.

Continuous: A function $f(x)$ defined on an open interval, I, including the point $x = a$, is **continuous** if at $x = a$ the limit exists, that is $\lim_{x \to a} f(x)$ exists, and $\lim_{x \to a} f(x) = f(a)$. For $\lim_{x \to a} f(x)$ to exist, the limits from both the negative and positive sides of a are equal, $\lim_{x \to a-} f(x) = \lim_{x \to a+} f(x)$.

Decreasing function: If over an interval (a, b) the derivative of a function is $f'(x) \leq 0$ for all $x \in (a, b)$ and $f'(x) < 0$ for at least some $x \in (a, b)$, the function is said to be **decreasing over the interval**. If the function is such that $f'(x) \leq 0$ for all values of x in the domain, the function is said to be **decreasing** or **non-increasing**. If the function is such that $f'(x) < 0$ for all values of x in the domain, the function is said to be **strictly decreasing**.

Differentiable: A function $f(x)$ defined on an open interval including the point $x = x_0$ is **differentiable** at that point if

$$\lim_{\Delta x \to 0} \frac{f(x_0 + \Delta x) - f(x_0)}{\Delta x}$$

exists and is finite. Thus, if it exists, it is the derivative of the function

$$\frac{dy}{dx} \equiv \lim_{\Delta x \to 0} \frac{f(x_0 + \Delta x) - f(x_0)}{\Delta x}.$$

First-order condition for an extreme point: If the first derivative of a function, $y = f(x)$ at $x = x^*$, is such that $f'(x^*) = 0$, then a the point (x^*, y^*) on a function $y = f(x)$ is a **local maximum** if the derivative of the function at (x^*, y^*) is $f'(x^*) = 0$ and $f'(x) > 0$ for all x values within $x - \epsilon$ and $f'(x) < 0$ for all x values within $x + \epsilon$ for some $\epsilon > 0$.

b the point (x^*, y^*) on a function $y = f(x)$ is a **local minimum** if the derivative of the function at (x^*, y^*) is $f'(x^*) = 0$ and $f'(x) < 0$ for all x values within $x - \epsilon$ and $f'(x) > 0$ for all x values within $x + \epsilon$ for some $\epsilon > 0$.

c neither a minimum or maximum if the point (x^*, y^*) on a function $y = f(x)$ if the derivative of the function at (x^*, y^*) is $f'(x^*) = 0$ and the sign of $f'(x)$ for all x values within $x - \epsilon$ and for all x values within $x + \epsilon$ have the same sign.

Increasing function: If over an interval (a, b) the derivative of a function is $f'(x) \geq 0$ for all $x \in (a, b)$ and $f'(x) > 0$ for at least some $x \in (a, b)$, the function is said to be **increasing over the interval**. If the function is such that $f'(x) \geq 0$ for all values of x in the domain, the function is said to be **increasing** or **non-decreasing**. If the function is such that $f'(x) > 0$ for all values of x in the domain, the function is said to be **strictly increasing**.

Limit: A sequence has a **limit** if, for any $\epsilon > 0$, there exists a number X and for a_n, being the $n^t h$ number of the sequence, such that $|a_n - L| < \epsilon$ whenever $n > X$. A sequence that has a limit is said to be **convergent**.

3   Natural Logs and the Exponential Function

We return to further analysis of derivatives in Chapters 4 and 5 and, in particular, discuss optimization and the *second* derivative then. However, before doing so, it is highly advantageous to become familiar with two widely used functions in economics and many other sciences for modeling purposes: the *exponential function* and the *natural log*.

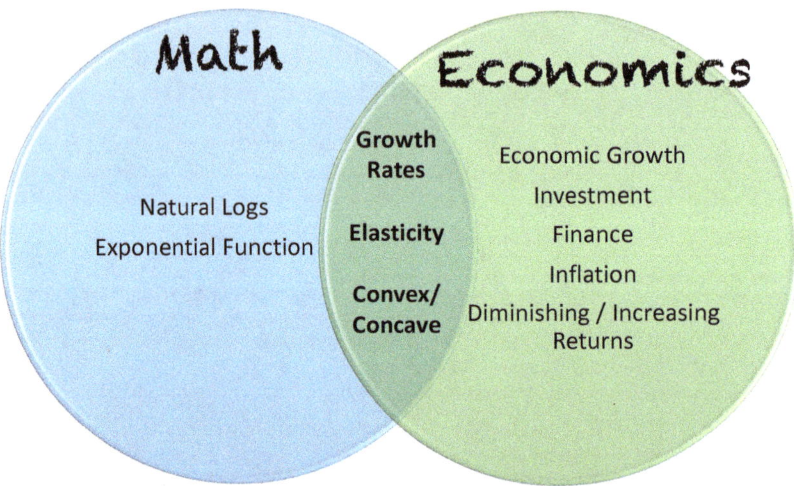

Many economic models involve assumptions or analyses of growth or changes over time in variables of interest. This feature is particularly true for macroeconomics and for values in finance (a subfield of economics). The exponential function and the natural logarithm are highly convenient for describing, for example, the growth rate of an economy, the rate of inflation over the long run, or the expected increase in the value of an asset. As we are often interested in the rate of change, in other words the growth rate of economic variables, the derivatives of these functions are highly informative. The good news is that, relatively speaking, the derivatives of these functions are easy to obtain. Furthermore, we also frequently use these functions in models because the shape of the natural log function is *concave* while the shape of the exponential function is *convex*. We define those terms more formally in Chapter 5 because they are intimately connected to the second derivative, but please pay attention to these shapes in the graphs that follow.

3.1 The Functions

We start by introducing the exponential and natural log functions and explain their relationship. We then move on to explore their properties and some applications to give a taste of their usefulness.

3.1.1 The Exponential Function

We have worked with and taken derivatives of function of the form $y = x^a$, where x is the base number, and we saw the power rule in the previous chapter, which yields $\frac{dy}{dx} = ax^{a-1}$. Now we look at functions in which the x variable is in the exponent itself. In general, if we have a function of the form

$$y = a^x$$

we say that x is the logarithm of y in base a, where a is some number greater than 1. If $0 < a < 1$, we can always rewrite that in terms of the inverse of a, meaning $1/a$, which would be greater than 1.

Rephrasing, the logic of the equation means that x is the power one needs to raise a to, to get the value y. The choice of base depends on the situation. Engineering and the sciences frequently use $a = 10$ to make expressing extremely large or small values more manageable. Computer science uses base 2 in which the digits are either 0 (open circuit or low voltate) or 1 (closed circuit or high voltage) because that is how computers were built to process and store information.

In economics, we usually stick with a very special base called the natural log, denoted by e. e is approximately equal to 2.71828... and is an irrational number, like $\pi = 3.14159...$ Most conveniently, e happens to be the base that satisfies the following property. Given the **exponential function**

$$y = e^x$$

the derivative of the exponential function is

$$\frac{dy}{dx} = e^x.$$

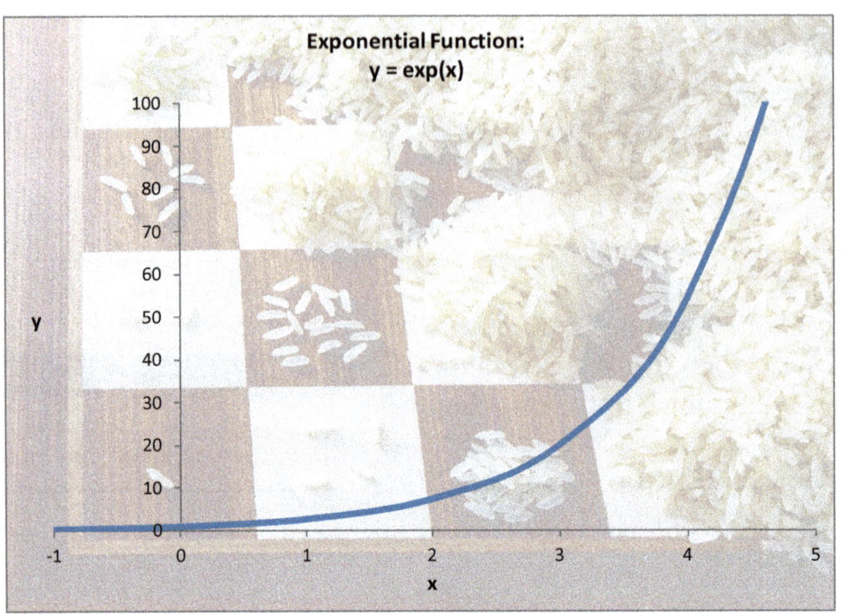

Figure 3.1: The exponential function.

The derivative is the same as the function!! So cool. That makes taking the derivative easy. Moreover, this result means that the slope of the function at any point is the value of the function itself. See Figure 3.1.

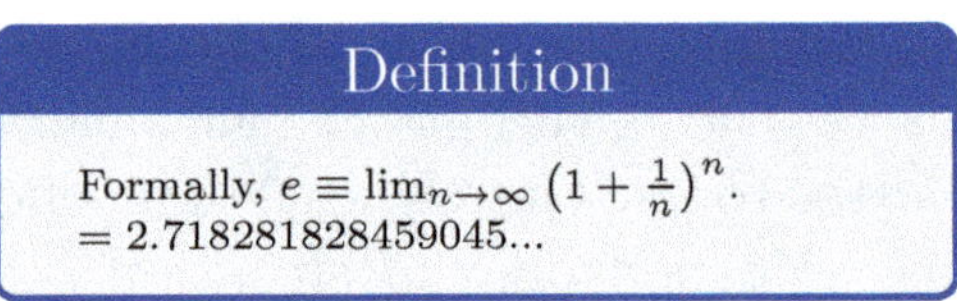

We can see that the exponential function is increasing at an increasing rate, and for reasons we will see later, the percentage of the rate increase is constant. That is, the growth rate of the function is constant. That makes the function relatively easy to work with when we are analyzing something that grows over time.

Sometimes the exponential function is written as

$$y = \exp(x)$$

instead of as $y = e^x$, but they mean exactly the same thing: If we raise e to the power x we get y.

3.1.2 The Natural Log

The following is the **natural log function**:

$$x = \ln(y) \text{ for } y > 0.$$

That is effectively the inverse of the exponential function. The natural log is only defined for positive numbers. It is not defined for zero or any negative number. That is why it states "for $y > 0$." Since the base, e, is a positive number, there is no value we can raise it to in order to generate zero or a negative number. To help fix the idea, suppose your base is 10. What power can you raise 10 to in order to get zero? What power can you raise 10 to in order to get a negative number, say -5? Neither are possible. The same applies to e. Thus, the natural log function is not defined for values less than or equal to zero.

Reading the above equation in words, you could say that in order to get y, with a base of e, raise it to the power of x, or x is the natural log of y. As the inverse of the exponential function we have

$$y = e^x.$$

Taking the natural log, ln, of both sides

$$\begin{aligned} \ln y &= \ln(e^x) \\ &= x \ln e \\ &= x. \end{aligned}$$

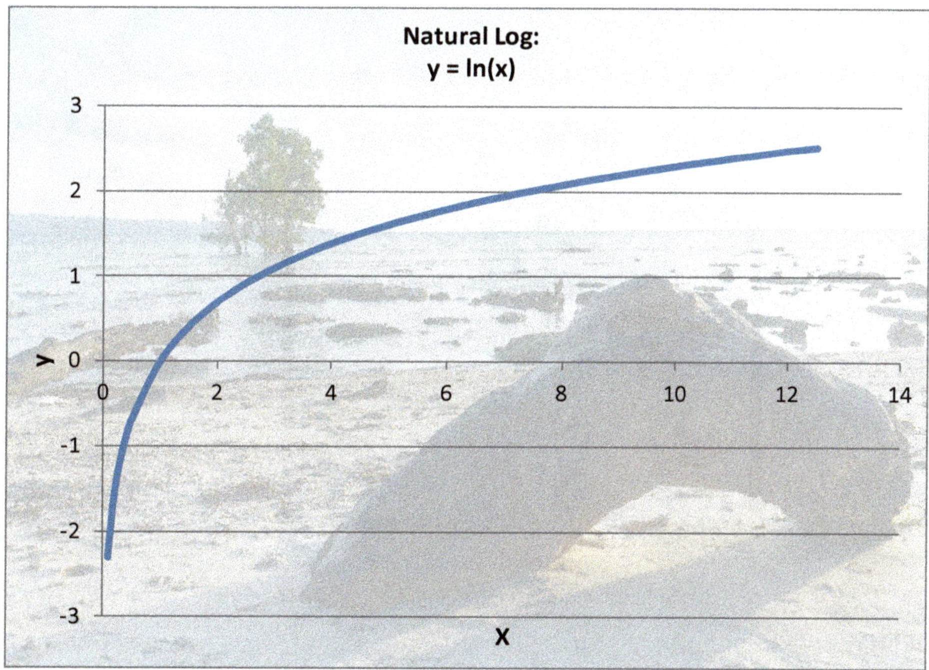

Figure 3.2: The natural log function.

The steps follow from a few of the properties of the natural log, which are listed with examples in Section 3.2. One of the properties is that $\ln(a^b) = b \ln(a)$. That allows us to move the exponent to a coefficient that multiplies (and that feature can often help us simplify complex math expressions). Furthermore, since the ln function is the inverse of e^x, then $\ln(e) = 1$ because $\ln(e)$ asks to what power do you need to raise e to get e. Figure 3.2 is a graph of the natural log.

Observe that the natural log function is also increasing, but unlike the exponential function, it is increasing at a *decreasing* rate. That feature is often used to represent the idea of diminishing returns. For example, suppose Edward Elric's utility of consuming quiche is given by $U = ln(q)$, where q is number of quiche slices he eats. Then, the more quiche he has, the higher the value of q, and the higher his utility. However, each *additional* slice of quiche will add a smaller and smaller amount to his total utility. Look at Figure 3.2 and think about how that graph shows this property of diminishing marginal utility.

The natural log also has an easy derivative to work with in models.

$$\begin{aligned} y &= \ln(x) \\ \frac{dy}{dx} &= \frac{1}{x} \end{aligned}$$

This derivative could be written as $\frac{dy}{dx} = x^{-1}$. So the natural log function, like the exponential function, has a simple derivative to obtain. Since the ln function exhibits diminishing returns and the derivative of the function is easy (which comes in handy in optimization problems for which we have to take derivatives), it is a commonly employed function our objective functions such as utliity functions.

3.2 Properties and Application

3.2.1 Properties

The following are properties and derivatives of the natural log and exponential function that you should become comfortable with using in economics models.

$$\ln(e) = 1$$

$$\ln(1) = 0$$

$$\ln e^x = x$$

$$\ln(ab) = \ln a + \ln b$$

$$\ln\left(\tfrac{a}{b}\right) = \ln a - \ln b$$

$$\ln a^b = b \ln a$$

$$y = \ln x \rightarrow \frac{dy}{dx} = \frac{1}{x}$$

$$y = \ln[f(x)] \rightarrow \frac{dy}{dx} = \frac{1}{f(x)} f'(x)$$

$$y = e^x \rightarrow \frac{dy}{dx} = e^x$$

$$y = e^{f(x)} \rightarrow \frac{dy}{dx} = f'(x) e^{f(x)}$$

Note that in the derivatives, when the natural log contains a function of $f(x)$ we need to apply the chain rule, where ln is the outer function and $f(x)$ is the inner function. For example:

$$
\begin{aligned}
y &= \ln(3x) \\
\frac{dy}{dx} &= \frac{1}{3x}(3) = \frac{1}{x}
\end{aligned}
$$

and

$$
\begin{aligned}
y &= \ln(3x^2 + x) \\
\frac{dy}{dx} &= \frac{1}{3x^2 + x}(6x + 1) = \frac{6x + 1}{3x^2 + x}.
\end{aligned}
$$

When e is raised to a power that is a function of x, we again need to apply the chain rule, where the exponential function, $e^{f(x)}$, is the outer function and $f(x)$ is the inner function. For example:

$$
\begin{aligned}
y &= e^{2x} \\
\frac{dy}{dx} &= e^{2x}(2) = 2e^{2x}
\end{aligned}
$$

and

$$
\begin{aligned}
y &= e^{3x^2 + x} \\
\frac{dy}{dx} &= e^{3x^2 + x}(6x + 1).
\end{aligned}
$$

3.2.2 Applied Example

A simple example may help illustrate the usefulness of those functions. Consider a share of stock in Winry's Automail Company, WAC for short. You can purchase a share of that stock today for A dollars. The value of the stock is expected to grow at a constant rate of 3% per year. How much will that share of WAC be worth in 5 years? The exponential function provides a way to find out. Let the value at some time t be $V(t)$, where t is the length of time from the purchase of the share of WAC. Let the initial value of the share be denoted as $V(0)$. $V(0)$ is equal to the A dollars you need to spend on it now in order to purchase it. Then,

> **Definition**
>
> **Discrete compounding:** Adding interest to a value at the end of each specific regular time interval we have
>
> $$V(t) = V(0)(1+r)^t,$$
>
> where the value after t time periods, $V(t)$, is calculated as the initial value $V(0)$, the value at $t = 0$, times the quantity 1 plus the interest rate, r, raised to the power of the number of time periods, t.

$$V(t) = V(0)e^{0.03t}.$$

This formula gives the value of the share after t years if it grows at a rate of 3% per year. Suppose the initial value is $A = \$100$; how much is it worth after 5 years?

$$V(t) = (100)e^{0.03(5)}$$

When you plug that into your calculator (make sure you know how to do that) you get

$$V(t) = (100)e^{0.03(5)} = \$116.18.$$

The exponential function computes the value over time for *continuous compounding* (finance majors take note!), where interest is added at all points in time. If we want to calculate growth with discrete compounding we use a different formula.

$$V(t) = V(0)(1+g)^t,$$

where g is the growth rate and t is again the length of time. The difference here is that interest is only added at the end of the period. Interest can be added annually, semi-annually, monthly, weekly, and so forth. The discrete compounding formula represents the growth in value when the interest is added at the end of each period. The continuous compounding formula can be thought of as the

> **Definition**
>
> **Continuous compounding:** Adding interest to a value continuouly through time we have
>
> $$V(t) = V(0)e^{rt},$$
>
> where the value after a length of time t, $V(t)$, is calculated as the initial value $V(0)$, the value at $t = 0$, times e to the power of the interest rate, r, times the length of time, t.

limit when the length of the time between points in time to add interest goes infinitessimally small so as to become continuous in the sense that there is no time between the points when interest is added.

For most calculations, the two formulas give very similar results unless the numbers are quite large. Under the continuous compounding formula using the exponential function, the value will be larger. To illustrate, suppose you are choosing between two banks where you will place your savings of $1,000. Bank A offers an interest rate of 2% and continuous compounding: interest is added to your account every fraction of every second. Bank B offers an interest rate of 2.01% with annual compounding: you get your interest added to your account at the end of each year. If you plan on holding your savings in the account for 3 years, which bank has the better offer? If you choose bank A, the value of your savings account after 3 years will be

$$V(3) = (\$1000)e^{0.02(3)} = \$1,061.84.$$

At bank B you would have

$$V(3) = (\$1000)(1 + 0.0201)^3 = \$1,061.52.$$

Bank A pays slightly more, even though the interest rate offered by bank B is slightly higher. The reason is that bank A adds interest throughout the entire year, and thus you are also earning interest on the interest immediately after it is added to your account. With bank B, you do not get the interest payment until the end of the year, so up until then you would not earn interest on the interest.

Reconsider the growth rate formula again, but let's think of it as representing the growth of an entire economy. Let's call our economy Amestris. In the formula, $y(t)$ is GDP per capita and the average growth rate of the economy of Amestris is g. To find GDP per capita after t years we would write

$$y(t) = y(0)e^{gt}.$$

If Amestris started with a GDP per capita of $5,000 and grew at an average rate of 2.5%, after 30 years the GDP per capita would be

$$y(t) = (\$5,000)\, e^{(0.025)(30)} = \$10,585.$$

According to the general growth formula, $y(t) = y(0)e^{gt}$, take the derivative of y with respect to time. What do we get?

$$\begin{aligned} y(t) &= y(0)e^{gt} \\ \frac{dy(t)}{dt} &= y(0)e^{gt}(g) \end{aligned}$$

Can we make any sense of that? Well, $\frac{dy(t)}{dt}$ is the change in GDP per capita with time. It is not the growth *rate*. It is analogous to the change from time period $t - 1$ to t as in $y_t - y_{t-1} = \Delta y$. If we divided by $y(t)$, we have an expression very, very close to the basic growth rate formula. That is, when you calculate the growth rate of GDP between 2 years you use the following:

$$\text{growth rate} = \frac{y_t - y_{t-1}}{y_{t-1}},$$

which gives you the percentage change between year $t - 1$ and year t. $\frac{dy(t)}{dt}$ is effectively equivalent to $y_t - y_{t-1}$, how much y changed over time, but for a marginally small amount of time. Thus, if we divide $\frac{dy(t)}{dt}$ by $y(t)$ we get a percentage back. That percentage is the growth rate of $y(t)$.

$$\begin{aligned} \frac{dy(t)}{dt} &= y(0)e^{gt}(g) \\ \frac{dy(t)/dt}{y(t)} &= \frac{y(0)e^{gt}(g)}{y(t)} \end{aligned}$$

But since $y(t) = y(0)e^{gt}$ we can replace the $y(t)$ in the denominator on the right-hand side and then almost everything cancels out.

$$\frac{dy(t)/dt}{y(t)} = \frac{y(0)e^{gt}(g)}{y(0)e^{gt}} = g,$$

which just says that the growth rate of the economy is g. Another way to see this relationship is by taking the natural log of both sides of the growth equation.

$$\begin{aligned} y(t) &= y(0)e^{gt} \\ \ln(y(t)) &= \ln\left[y(0)e^{gt}\right] \\ \ln(y(t)) &= \ln(y(0)) + \ln[e^{gt}] \\ \ln(y(t)) &= \ln(y(0)) + gt \\ g &= \frac{\ln(y(t)) - \ln(y(0))}{t} \end{aligned}$$

The growth rate is the difference in the natural logs divided by the length of time. If the length of time is 1 (e.g. , 1 year, 1 month), then the growth rate is just the difference in the natural logs, or effectively the slope of the natural log of y. Consider the second to last line: $\ln(y(t)) = \ln(y(0)) + gt$. How does $\ln(y(t))$ change with t? When taking the derivative observe that the right-hand side is a simple linear function.

$$\ln(y(t)) = \ln(y(0)) + gt$$
$$\frac{d\ln(y(t))}{dt} = g,$$

and this point is perhaps easiest to see graphically. Figure 3.3 shows the function

$$y(t) = y(0)e^{gt},$$

where the values are on the left-side y-axis, $y(0)$ is 100, and g is 5% (or 0.05). The straight dashed line is the graph of

$$\ln(y(t)) = \ln(y(0)) + gt$$

for the same numbers but uses the axis on the right side so that both functions can be seen easily on the same graph.

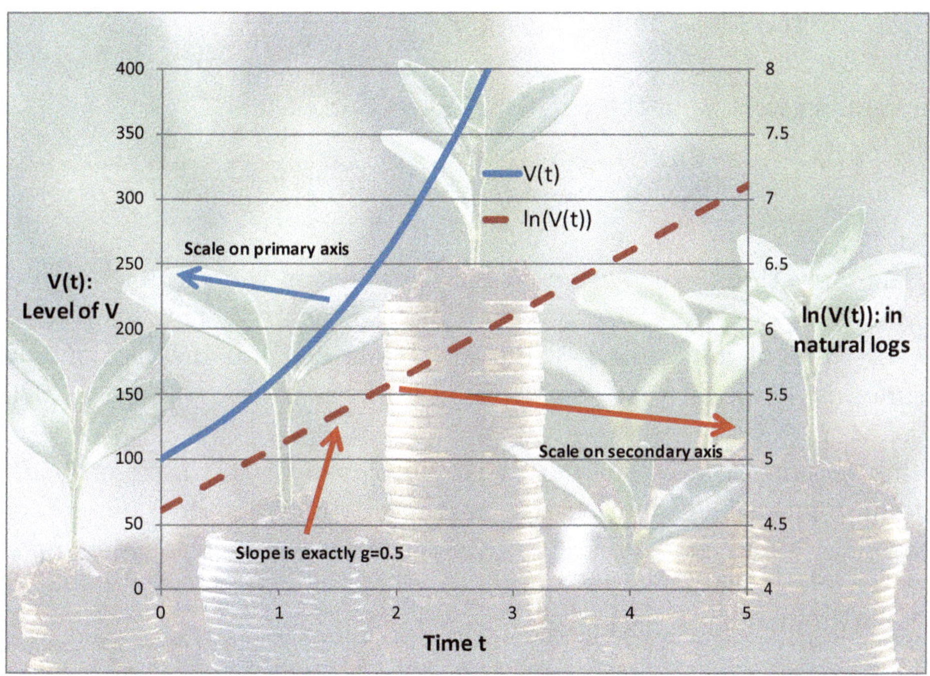

Figure 3.3: Growth in levels and natural logs

3.3   More on Growth Rates

3.3.1   Natural Logs and Growth Rates

One of the very convenient features about the natural log function, as discussed in the previous example, is the connection to growth rates. Growth rates are a percentage change in some variable over time. Normally, we represent percentage change as

$$\text{Percentage change in } y \;=\; \frac{\Delta y}{y} = \frac{y_t - y_{t-1}}{y_{t-1}}.$$

The calculus version, as we saw in the previous subsection, is

$$\text{Growth rate of } y \;=\; \frac{dy/dt}{y}.$$

Both represent the same thing except that in the first version there is a discrete unit of time (an hour, a day, a year, etc.) between the points measured. Thus, the subscript t means we have the value of y at time t, while the subscript $t-1$ means the value of y at one unit of time (day, month, year) before time t. If t is in years and $t = 2020$, then $t - 1 = 2019$.

In the calculus version we take the derivative with respect to time, so dy/dt is the marginal change in y with a marginal change in t, in other words, how much y changes with an infinitessimally small change in t. – or, in yet other words, the slope of y against t.

Now, consider the following function of z:

$$z \;=\; \ln(y(t)),$$

where y is a function of t. That means z is also a function of t. How does z change with t? Let's start by taking the derivative with respect to t and being careful to apply the chain rule correctly.

$$\frac{dz}{dt} \;=\; \frac{1}{y(t)}(dy/dt)$$

The result comes from applying the rule for taking the derivative of a function inside a natural log. The $\frac{1}{y(t)}$ is one over the function then multiplied by the derivative inside the function, dy/dt. In this case we only have $y(t)$. We know that y is a function of t but we do not have an explicit relationship. Therefore, we represent how y changes with t by simply writing the derivative itself, dy/dt.

Look again at the result and then look back at our calculus version of the growth rate. They are identical. What does that mean? If we have the natural log of a variable, which is a function of time, the derivative with respect to time gives us the growth rate. That feature gets exploited frequently in macroeconomics, when many variables may be growing and keeping track of their growth rates is made convenient by taking the natural log of all variables.

For example, consider the following equation:

$$Y \;=\; AK^{1/3}L^{2/3},$$

where Y is output or GDP, A is technology, K is physical capital, and L is labor. That function, which we use in multiple examples and problems in this textbook, is widely used in macroeconomics to represent economy-wide production. It tells us how much the economy will produce given the level of technology, the amount of physical capital (machinery, tools, equipment, factories, etc.), and the size of the labor force. However, in macroeconomics we are nearly always interested in how the economy changes over time. Is the economy growing? Is technology improving? Has the amount of capital grown or shrunk? Is the labor force getting bigger or smaller over the years? All these questions have to do with changes over time. Thus, all the variables in the equation can be thought of a functions of time: $Y(t)$, $A(t)$, $K(t)$, and $L(t)$.

If we rewrite our function to explicitly represent the fact that they are functions of time, we have

$$Y(t) \;=\; A(t)K(t)^{1/3}L(t)^{2/3}.$$

Now, if we asked how output, Y, changed over time, we could answer that by taking the derivative with respect to time. It will be a little messy because time, t, shows up in three places on the right-hand side, so we have a large product rule application.

$$\frac{dY(t)}{dt} \;=\; K(t)^{1/3}L(t)^{2/3}\frac{dA(t)}{dt} + \frac{1}{3}A(t)K(t)^{-2/3}L(t)^{2/3}\frac{dK(t)}{dt}$$
$$+\frac{2}{3}A(t)K(t)^{1/3}L(t)^{-1/3}\frac{dL}{dt}$$

Note the chain rule applications, where, for example, we took the derivative of $K(t)^{1/3}$ to get $\frac{1}{3}K(t)^{-2/3}$, but then for the chain rule also multiplied that by the inner function $K(t)$, which generates the $\frac{dK(t)}{dt}$. The same occured for labor, $L(t)$. If we want that as a growth rate of Y, we need to divide by $Y(t)$, as we did in previous subsection.

$$\frac{dY(t)/dt}{Y(t)} = \frac{K(t)^{1/3}L(t)^{2/3}\frac{dA(t)}{dt} + \frac{1}{3}A(t)K(t)^{-2/3}L(t)^{2/3}\frac{dK(t)}{dt} + \frac{2}{3}A(t)K(t)^{1/3}L(t)^{-1/3}\frac{dL}{dt}}{Y(t)}$$

$$= \frac{K(t)^{1/3}L(t)^{2/3}}{Y(t)}\frac{dA(t)}{dt} + \frac{\frac{1}{3}A(t)K(t)^{-2/3}L(t)^{2/3}}{Y(t)}\frac{dK(t)}{dt} + \frac{\frac{2}{3}A(t)K(t)^{1/3}L(t)^{-1/3}}{Y(t)}\frac{dL}{dt}$$

We can simplify that by replacing the $Y(t)$ in the three denominators with the orginal production function, $Y(t) = Y(t) = A(t)K(t)^{1/3}L(t)^{2/3}$, so is that many objects will cancel out, leaving an easier equation to work with and interpret.

$$= \frac{K(t)^{1/3}L(t)^{2/3}}{A(t)K(t)^{1/3}L(t)^{2/3}}\frac{dA(t)}{dt} + \frac{\frac{1}{3}A(t)K(t)^{-2/3}L(t)^{2/3}}{A(t)K(t)^{1/3}L(t)^{2/3}}\frac{dK(t)}{dt} + \frac{\frac{2}{3}A(t)K(t)^{1/3}L(t)^{-1/3}}{A(t)K(t)^{1/3}L(t)^{2/3}}\frac{dL}{dt}$$

$$= \frac{1}{A(t)}\frac{dA(t)}{dt} + \frac{1}{3}\frac{1}{K(t)}\frac{dK(t)}{dt} + \frac{2}{3}\frac{1}{L(t)}\frac{dL(t)}{dt}$$

$$= \frac{dA(t)/dt}{At} + \frac{1}{3}\frac{dK(t)/dt}{K(t)} + \frac{2}{3}\frac{dL(t)/dt}{L(t)}$$

In the last line, all three terms contain a growth rate. To make it easier to read, let's label each growth rate as g_x where the subscript x will indicate which variable: $g_A = \frac{dA(t)/dt}{At}$, $g_K = \frac{dK(t)/dt}{K(t)}$, $g_L = \frac{dL(t)/dt}{L(t)}$, and $g_Y = \frac{dY(t)/dt}{Yt}$. Then we can write this line as

$$g_Y = g_A + \frac{1}{3}g_K + \frac{2}{3}g_L.$$

This last line states that the growth rate of output, Y, is the sum of the growth rates of technology, A, one third the growth rate of physical capital, K, and two thirds the growth rate of the labor force, L.

More importantly, for our purposes, let's do that again, but this time, we'll take the natural logs fo the equation first. Starting again from the production function, take the natural log of both sides as follows:

$$Y(t) = A(t)K(t)^{1/3}L(t)^{2/3}$$

$$\ln Y(t) = \ln(A(t)) + \frac{2}{3}\ln(K(t)) + \frac{1}{3}\ln(L(t)).$$

Now, when we take the derivative with respect to time, notice how much less notation we have to deal with to get the expression into growth rates.

$$\frac{1}{Y(t)}\frac{dY(t)}{dt} = \frac{1}{A(t)}\frac{dA(t)}{dt} + \frac{1}{3}\frac{1}{K(t)}\frac{dK(t)}{dt} + \frac{2}{3}\frac{1}{L(t)}\frac{dL(t)}{dt}$$

$$g_Y = g_A + \frac{1}{3}g_K + \frac{2}{3}g_L$$

By taking the natural logs of the expression before taking the derivative with respect to t, we made our math life easier. Less notation means less chance of a sloppy error, and the overall expression after taking the derivative is easier to understand so that we can simplify it into the final expression.

3.4 Exercises

3.4.1 Quick Check Answers

a) $\ln\left(e^{\alpha+\beta x}\right) = \alpha + \beta x$

$\qquad \ln\left(e^{\alpha+\beta x}\right) = (\alpha + \beta x)\ln(e) = (\alpha + \beta x)(1) = \alpha + \beta x$

b) $3\ln(a) - 5\ln(a) = \ln\left(a^{-2}\right)$

$\qquad 3\ln(a) - 5\ln(a) = \ln a^3 - \ln a^5 = \ln\left(\frac{a^3}{a^5}\right) = \ln\left(a^{-2}\right)$

$\qquad$ or $3\ln(a) - 5\ln(a) = -2\ln a = \ln\left(a^{-2}\right)$

c) $y = \ln(3x)$

$\qquad \frac{dy}{dx} = \frac{1}{3x}3 = \frac{1}{3x}$

d) $y = 4\ln(x^2 + 1)$

$\qquad y = 4\ln(x^2 + 1) = \frac{dy}{dx} = \frac{4}{x^2+1}2x = \frac{8x}{x^2+1}$

e) $y = ae^{bx^\alpha}$

$\qquad y = ae^{bx^\alpha} = \frac{dy}{dx} = ae^{bx^\alpha}\left(\alpha bx^{\alpha-1}\right) = ab\alpha x^{\alpha-1}e^{bx^\alpha}$

f) At 3% interest over 50 years, \$100 at bank A would be \$438.39 and at bank B would be \$448.17. If the interest rate was 10%, then after 50 years \$100 at bank A would be \$11,739.09 and at bank B would be \$14,841.32.

3.4.2 Practice Problems

Find the derivatives of y with respect to x for the following:

1. $y = e^{5x}$

2. $y = 4e^{3x}$

3. $y = 6e^{-2x}$

4. $y = e^{ax^2+bx+c}$

5. $y = \ln(7x^5)$

6. $y = \ln(ax^c)$

7. $y = 5\ln(x+1)^2$

8. $y = 5x^4 \ln x$

9. Suppose that the price of a share of WAC after t years is given by

$$V(t) = 100e^{(1/2)(t)}.$$

Compute that price of a share after 2 years, 5 years, and 10 years.

10. From the WAC share problem in #9, take the derivative of V with respect to time, t. What is the growth rate of the value?

11. Take the natural lns of the value equation in problem #9. Graph the ln of the value against time.

12. Repeat #9–11 for the following:

$$V(t) = 100e^{t^{1/2}}.$$

13. Consider the data taken from the World Bank's World Development Indicators (https://databank.worldbank.org/source/world-development-indicators).

Economy	1970 GDP per Capita	2020 GDP per Capita	Average Annual Growth Rate
Argentina	$9,243	$11,344	0.41%
India	$363	??	3.27%
Japan	$14,114	??	1.81%
South Africa	$5,184	$5,659	??%
South Korea	$1,977	$31,327	??%
Sweden	??	$51,542	1.51%

(a) Using the formula for continuous compounding, confirm that if Argentina had a GDP per capita of $9,243 in 1970 and grew at an average annual rate of 0.41%, Argentina's GDP per capita would be $11,344 in 2020.

(b) Using the data in the table for India and the formula for continuous compounding, what would India's GDP per capita be in 2020?

(c) Using the data in the table for Japan and the formula for continuous compounding,, what would Japan's GDP per capita be in 2020?

(d) Using the data in the table and the formula for continuous compounding, what was the average annual growth rate of South Africa?

(e) Using the data in the table and the formula for continuous compounding, what was the average annual growth rate of South Korea?

(f) Using the data in the table and the formula for continuous compounding, what was the GDP per capita of Sweden in 1970?

3.5 Appendix

Here, we list some formal definitions and other items for your reference.

Definition of e: Take the limit of the function

$$e \equiv \lim_{x \to \infty} \left(1 + \frac{1}{x}\right)^x = 2.71828...$$

Changing a log base to base e: We can always convert a logarithm value in another base to base e via the following:

$$\log_b x = (\log_b e)(\log_e x) = (\log_b e)(\ln x).$$

Identify the log of e in base b (the value you need to raise e to in order to get the base b.) and multiply that by the natural log of the original value x. To identify the log of e in base b, apply the following:

$$\log_b e = \frac{1}{\log_e b} = \frac{1}{\ln b}.$$

Derivative of $y = a^x$: In the derivative rules presented in this chapter, we did not look at the case when the variable we are taking the derivative with respect to is in the exponent *except in the case where the base is e itself, as in e^x.* For the general case, that derivative is as follows:

$$y = a^x$$
$$\frac{dy}{dx} = a^x \ln(a),$$

where if $a = e$ then $\ln(a) = \ln(e) = 1$.

4 Unconstrained Optimization

We now get to the primary use of calculus in economics, and that is optimization. That is, calculus is our favorite tool for maximizing or minimizing things (objective functions). In your principles classes, you certainly have come across this notion, though probably without the calculus representation. Consumers maximize utility, firms maximize profits, or firms minimize costs, governments try to

> **Definition**
>
> **Unconstrained optimization**: Solving for an extreme point or points of a function without additional requirements or limitations on the dependent variables.

minimize dead weight losses, investors try to maximize returns while minimizing risk, the central bank tries to minimize the costs of inflation and unemployment, and so on. In many cases, our economic agents are trying to optimize some objective with a constraint of some form. Consumers have a budget constraint; firms have a technological constraint; governments are constrained by the resources available. In essence, constrained optimization is the formal expression of the basic definition of economics: How do people make decisions with unlimited wants with limited resources available?

However, we will start with the mathematical tools for **unconstrained optimization**. That means we will either do problems that do not have an explicit constraint and are thus unconstrained or we will manipulate a constrained optimization problem by merging the constraint with the objective function itself to form an unconstrained optimization problem. We will continue with more advanced tools for **constrained optimization** in

> **Definition**
>
> **Constrained optimization**: Solving for an extreme point or points of a function while also satisfying additional requirements or limitations on the dependent variables not expressed in the objective function.

Chapter 7 after covering some necessary topics for better understanding optimization generally.

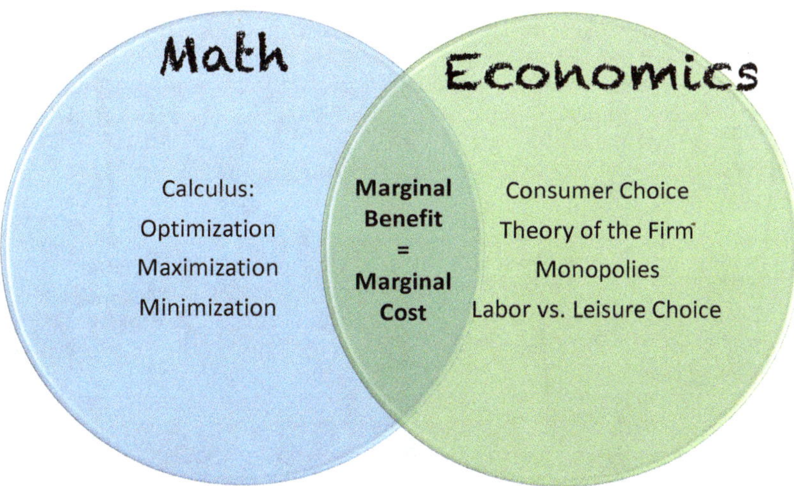

4.1 Finding Extreme Points

In optimization problems we are looking for an extreme point or an *extremum* collectively, meaning a minimum and/or maximum. Note that functions may have more than one extreme point. They can have many minima or maxima and can have both minima and maxima.

When a function has, for example, two maximum points, that does not mean those two points are equal. Consider the Figure 4.1 with several maxima. In the diagram there are three points that would satisfy the first-order condition, as the slope at each point is zero. Two of those points sit at a peak and thus both are maxima. However, the maximum to the left is a *local maximum*

> **Definition**
>
> **Stationary point**: A point on a function where the function is neither increasing nor decreasing, making it "stationary." That means the slope and the value of the first derivative is zero.

because while it is the highest point compared to the other points nearby, it is not the highest point for the entire function. The highest point for the entire function is the extremum marked to the right. That point is the *global maximum*, "global" indicates it is the highest point for the entire function. The extreme point in between the two maxima is a *local minimum* because it is the lowest point when compared with the points immediately around it, but it is not the lowest point of the entire function.

This chapter looks at examples of how to find these points, but we will need the tools explored in the next chapter to correctly identify the points as minima or maxima. So, stay tuned. We will use three common microeconomics models to illustrate: profit maximization, cost minimization, and utility maximization.

> **Definition**
>
> **Global maximum (minimum)**: A global maximum (minimum) is the point on the function that yields the largest (smallest) value over the entire domain.

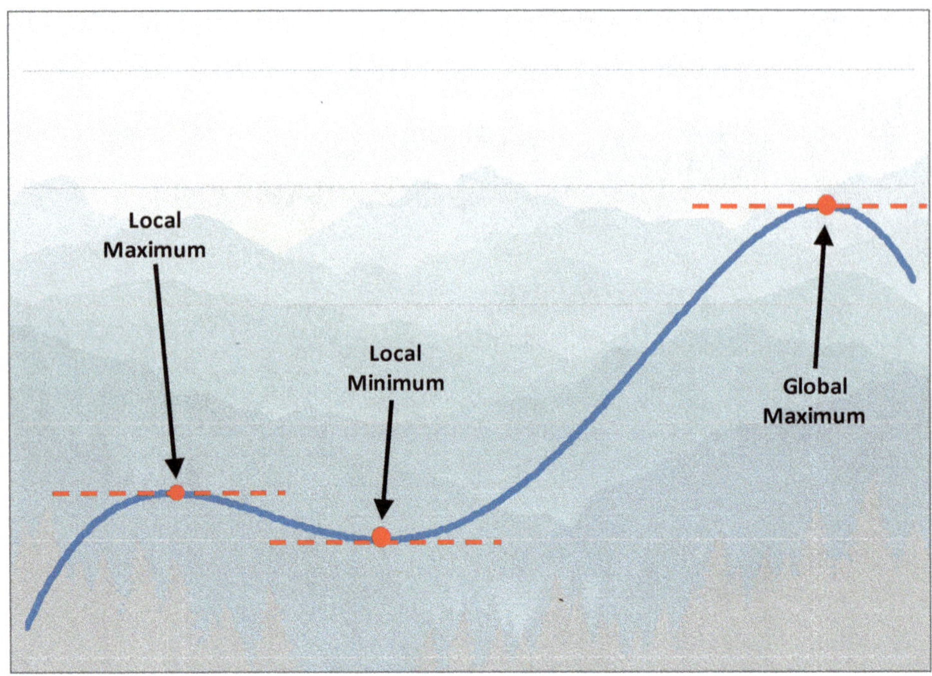

Figure 4.1: Example of function with multiple extrema.

4.2 Profit Maximization

To begin, let us return to the monopolist example from Chapter 2. The monopolist's optimization problem is

$$\max_{Q} \Pi = (10 - Q)Q - 2Q,$$

where Π represents profits and the price P is given by $P = 10 - Q$ and thus represents a downward sloping demand curve. Q is the quantity that the monopolist will choose to maximize its profits, and the cost per unit of Q produced is 2. In this problem, the monopolist has one choice, Q. Q can be chosen and is therefore an endogenous variable. Recall that it is written below *max* in the expression, indicating it is the variable chosen to maximize profits.

Notice that this expression is the entire problem. There is no "subject to" followed by another math expression representing a constraint of any kind. For a firm, there is no explicit constraint, though their production technology or cost function represents an implicit constraint. The firm would like to produce an infinite amount at zero

cost, just like a consumer would like to have an infinite amount of goods at zero price. However, the firm is constrained by its production technology, which means it is costly to produce. The consumer is bound by a limited budget and the fact that goods are costly.

Since $P = 10 - Q$, then $(10 - Q)Q$ is really price times quantity, or the total revenue of the firm. $2Q$ is the total cost to the firm. The difference between total revenues and total costs is profits. To find the level of Q that maximizes profits, we take the derivative with respect to Q. Rewriting slightly by multiplying the Q through the expression for price, we have

$$\max_{Q} \Pi = 10Q - Q^2 - 2Q.$$

Now, taking the derivative we get

$$\frac{d\Pi}{dQ} = 10 - 2Q - 2.$$

Recall Figure 2.1 from Chapter 2 and repeated here as Figure 4.2. The profits reach a maximum at 4. At exactly that point, if we draw a tangent line to the function, the tangent line will be a flat horizontal line with a slope equal to zero. Therefore, the derivative must be equal to zero for a maximum of profits. Relative to the maximum point, the slope is positive to the left, but negative to the right, which is another way of saying the function is *concave* - a concept we will define in the next chapter. Therefore, we set the derivative equal to zero to find the point where the slope is exactly 0.

$$\frac{d\Pi}{dQ} = 10 - 2Q - 2 = 0$$

This type of expression is referred to as a **first-order condition** (FOC) for an optimal solution. The FOC requires that the derivative of the function be equal to zero for a minimum **or** a maximum. Thus, the condition is *necessary* for a maximum or a minimum, but it is not *sufficient.* Sufficiency involves the second-order condition and, again, we save that for the next Chapter.

Let us now look at the monopolist's profit maximization problem, but using a more general function rather than the specific functional form. The monopolist's objective function is

$$\max_{Q} \Pi = P(Q)Q - C(Q).$$

In the expression, $P(Q)$ indicates that price, P, is an unspecified function of Q. Similarly, costs are an unspecified function of Q, $C(Q)$. Recalling our discussion from Section 2.4, we can make assumptions about the shape of these functions. That is, we will assume a standard downward sloping demand curve. That means that P is declining in Q or $\frac{dP(Q)}{dQ} < 0$ (or, equivalently, $P_Q(Q) < 0$). We also assume that production costs are increasing in quantity produced, $\frac{dC(Q)}{dQ} > 0$ (or equivalently, $C_Q(Q) > 0$).

Notice how, using calculus terms in the first derivative, we can succinctly state common assumptions about these functions. In introducing the problem, one could write

> *The firm chooses the quantity $Q \geq 0$ to maximize its profits*
> $\Pi = P(Q)Q - C(Q)$ *where* $P_Q(Q) < 0$ *and* $C_Q(Q) > 0$.

This statement encapsulates the information needed to proceed with analyzing the problem.

Let's move on to taking the derivative. Note that there is an application of product rule because Q appears *twice* in the revenue portion of the profit function. Q is an argument in the price function, $P(Q)$, *and* it appears multiplying price when it represents the quantity sold, Q. In words when we take the derivative of the revenue portion of the objective function, $\Pi = P(Q)Q$, we look at the marginal change in the price with respect to Q by taking the derivative of $P(Q)$ while holding the quantity sold (the second Q) fixed. Then, in the second part of the product rule, hold $P(Q)$ fixed and take the derivative with respect to the quantity sold, Q. Finally, we take the derivative of the cost function, $C(Q)$. Thus, we have

$$\frac{d\Pi}{dQ} = \underbrace{P_Q(Q)Q + P(Q)}_{\text{Marginal Revenue}} - \underbrace{C_Q(Q)}_{\text{Marginal Cost}}.$$

Let's break that expression down and interpret each component. The first two terms come from applying the product rule to the expression for revenues, $P(Q)Q$, which becomes $P_Q(Q)Q + P(Q)$ after taking the derivative. The first part, $P_Q(Q)Q$, shows the effect on profits with how the price changes (negatively because we assumed it is a decreasing function) as the quantity is increased. The second part, $P(Q)$, shows how the revenue increases, holding the price fixed but increasing the quantity sold. Together, they are the marginal increase in profits resulting from (1) the decrease in price holding the quantity sold fixed; plus (2) the marginal increase in profits from increasing the quantity sold while holding the price fixed. Taken together they make up the marginal revenue or marginal benefit to the monopolist from a marginal increase in quantity sold, Q. The last term in the first derivative is the derivative of the cost function with respect to Q and represents the marginal costs of producing more.

To find the profit-maximizing level of Q, set the first derivative equal to zero to form the FOC.

$$P_Q(Q)Q + P(Q) - C_Q(Q) = 0$$

Setting it equal to zero means that the slope of the function is zero. Recall our discussions in Chapter 2. I rearrange that FOC slightly and add the labels back to get:

$$\underbrace{P_Q(Q)Q + P(Q)}_{\text{Marginal Revenue}} = \underbrace{C_Q(Q)}_{\text{Marginal Cost}} \ .$$

That equation says that at the optimum, in other words, at the profit-maximizing level of output, marginal revenue must equal marginal cost. You learned that relationship in principles of microeconomics, but here you have the same idea formally expressed in the language of logic.

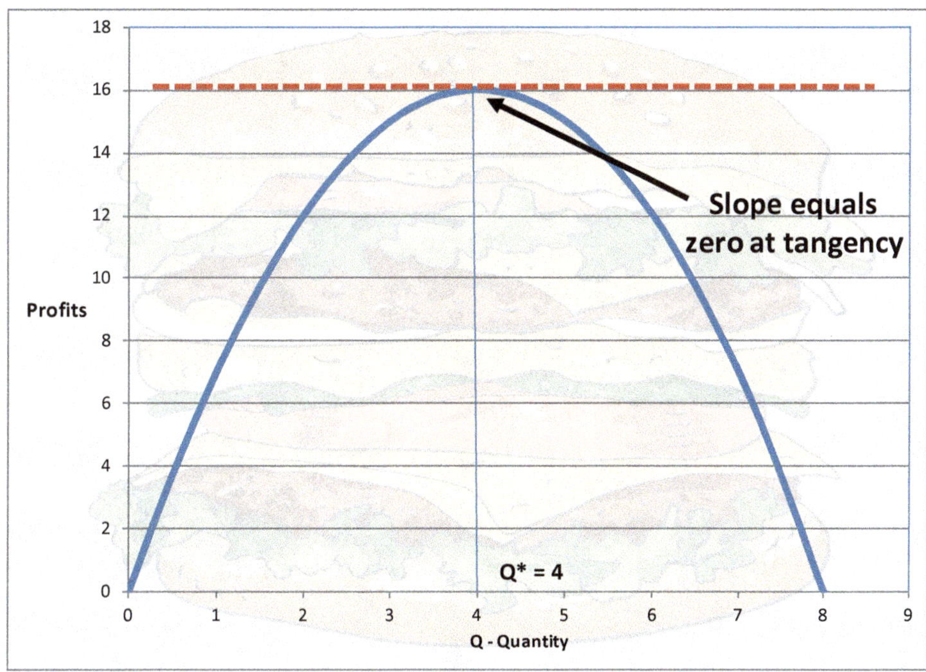

Figure 4.2: Slope is zero at extreme point.

Now, you might be asking yourself, how do I solve that for Q^*? The answer is you cannot solve it for an *explicit* solution for Q^*. However, the FOC provides you with all the information needed. We would simply state at this stage that Q^* is the value(s) of Q that solve(s) the FOC That is, the FOC *alone* defines the optimal solution.[6]

> **Definition**
>
> **First-Order Condition (FOC)**: For a single variable function to achieve a minimum or a maximum, the first derivative of the function must equal zero. This condition is necessary but not sufficient.

---

[6]Note that it is possible there is more than one solution, and with some further work and/or assumptions we could determine whether the solution is unique or not. However, we do not address that topic here as it would take us too far astray for our purposes.

In the definition for the FOC, it states the condition is *necessary* but not *sufficient*. What does that mean? In order for a point to be a minimum or maximum (an extremum), the first derivative has to be zero. So the condition the first derivative is zero is *necessary*. But even when the first derivative is zero, the point may not be a minimum or maximum. It could be an **inflection point**. So, the condition that the first derivative is zero is not *sufficient*.

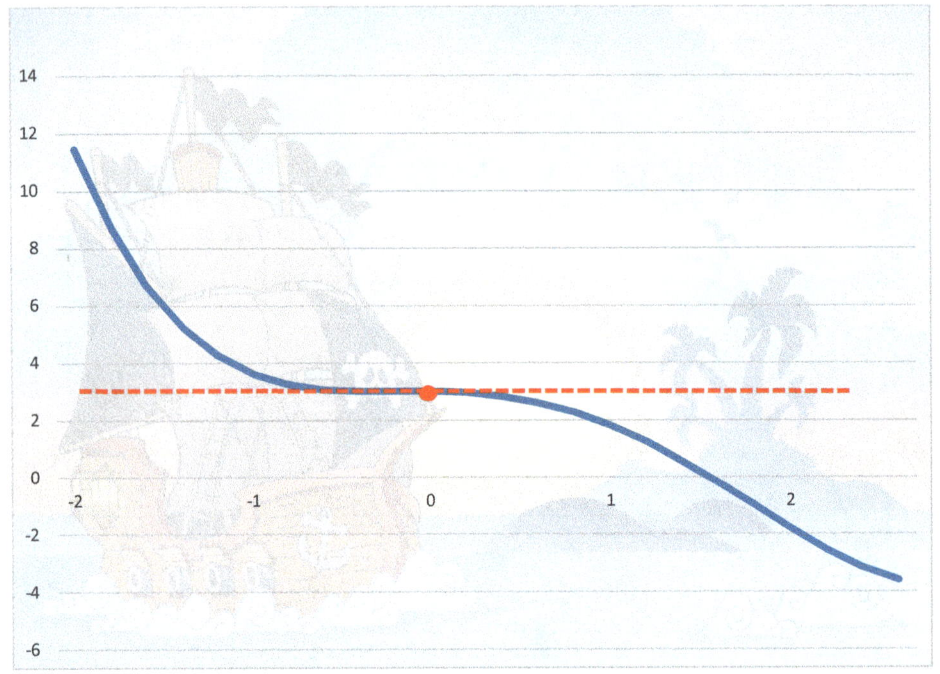

Figure 4.3: Inflection point example.

Figure 4.3 is an example of an inflection point. At that point the first derivative is zero. Notice how the curvature of the function changes. Right at that point indicated, the slope is zero, but it is obviously neither a minimum nor a maximum. For a maximum, the necessary conditions are the first derivative is zero and the second derivative is less than or equal to zero. For a minimum, the necessary conditions are the first derivative is zero and the second derivative is greater than or equal to zero for a minimum. The sufficient conditions are that the first derivative is zero and the second derivative must be strictly negative for a maximum or strictly positive for a minimum. However, those second-order conditions with strict inequalities are *sufficient* but not *necessary*, meaning a maximum or minimum could exist even if the second derivative is equal to zero at that point. We look at the second derivatives and why that is the case in the next chapter.

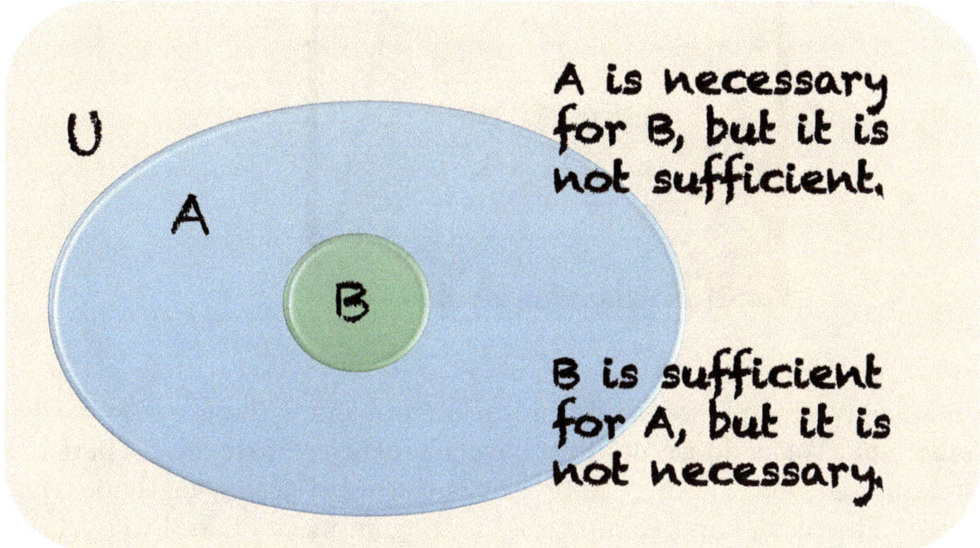

Figure 4.4: Venn diagram of necessary and sufficient conditions

4.2.1 Necessary and Sufficient Conditions: A Quick Aside

Having used the terms *necessary* and *sufficient* in the context of finding an extreme point, perhaps it would be helpful to illustrate their meanings. Figure 4.4 provides a visualization of the concepts. Imagine the full set of possibilities is the large rectangularish shape U. Inside U there are two more sets represented by the circles, A and B. Every point on the outer edge and within circle A is considered part of set A. Similarly, every point on the outer edge and within circle B is considered part

> **Making the Connection**
>
> **Necessary and sufficient conditions**: Necessary and sufficient conditions are not uncommon in math and economics. Think of a nonmath example when a condition is necessary but not sufficient and another which is sufficient but not necessary. For example, to get a passing grade in a class it is *necessary* to take the course, but that is not sufficient. If one gets a perfect score on all exams and assignments, that is *sufficient* to pass the course, but it is not necessary. Can you think of more examples?

of set B. Both, of course, are part of the universal set, U. As drawn, set B lies entirely within set A. Therefore we can state that being in set A is *necessary* for being in set B, but being in set A is not *sufficient* to be in set B. Conversely, if a point is in set B, that is *sufficient* to determine that it is in set A; however, being in set B is not *necessary* to also be in set A.

With that illustration in mind, where the first derivative is equal to zero is a *necessary* condition for a point to be an extreme point of a function. Therefore the FOC is necessary. However, it is not *sufficient* to be sure that it is an extreme point because the point could also be an inflection point.

4.3 Substituting a Constraint Into the Objective Function

The previous monopoly example was an optimization problem without a second equation
acting as a constraint. What do we do if we have a separate constraint? There are two
commonly used approaches. The first approach is the substitution method, which we
will describe here. The second approach is the Lagrangian method, and we will develop
that approach in Chapter 7.

4.3.1 Utility Maximization: Specific Utility Example

To illustrate the substitution method for constrained optimization, we will use a simple
example from microeconomics. Consider a single consumer, Jiji, who gets utility, U,
(happiness) from consuming two goods: x (donuts) and y (fish). The utility of this
consumer is given by

$$\max_{x,y} U(x,y) = x^{1/2} + y^{1/2}.$$

Jiji wants to be as happy as possible (i.e. , get the highest value of U possible). Right
now there is no constraint. If Jiji could choose the amount of donuts and fish to consume,
he would simply choose an infinite amount of both donuts and fish, and that would give
him an infinite level of utility. No calculus is needed. However, suppose our consumer
has a budget of \$32 dollars and donuts cost \$4 and fish costs \$2 (it's just a silly example
to illustrate key points, so don't worry about the unrealistic prices). Thus, the constraint
is

$$4x + 2y = 32,$$

which simply states that the amount spent on donuts, $4x$, plus the amount spent on fish,
$2y$, must be equal to the income level our consumer has available, 32. Now the problem
faced by our consumer would be written as

$$\max_{x,y} U(x,y) = x^{1/2} + y^{1/2}$$

$$s.t. \;\; 4x + 2y = 32,$$

which reads as the consumer maximizes utility by choosing the amounts x and y, subject
to ($s.t.$) the constraint. Now Jiji can't choose an infinite amount of donuts and fish

because he has to live within his budget. But how do we solve this problem for the utility-maximizing choices of x and y?

There are two issues here. First, we need to account for the constraint, that much is obvious, but also our consumer is choosing two things simultaneously, which actually makes this problem a multivariate calculus problem. Jiji is maximizing with respect to *two* variables. That requires different math than we have encountered thus far. We are going to hold off on addressing the multivariate calculus approach until the next section. However, we can utilize the budget constraint to change this problem from a multivariate constrained optimization problem into a single variable unconstrained optimization problem, just like those we have been solving all along.

If we use the budget constraint and solve it algebraically for either x or y, we can put it into the utility function. Suppose we solve it for y. We get the following:

$$\begin{aligned} 4x + 2y &= 32 \\ 2y &= 32 - 4x \\ y &= 16 - 2x, \end{aligned}$$

which says that the amount of y Jiji will buy is equal to 16 minus 2 times the amount of x he buys. With that we can replace the y in the objective function.

$$\max_{x} U(x, y) = x^{1/2} + (16 - 2x)^{1/2}$$

Now the constraint is accounted for because in the objective function as x increases y will decrease. Moreover, notice that our original problem, the multivariate problem (when we would have needed two derivatives: one for x and another for y), has been reduced to choosing just x and has become a single-variable problem. The y variable does not appear in the optimization problem. By first solving for the optimal x^* we can then use $y = 16 - 2x$ to find y^*. To maximize utility we know that the derivative with respect to x must be equal to zero; that is the necessary FOC. Taking the derivative we have

$$\frac{dU}{dx} = \frac{1}{2}x^{-1/2} + \frac{1}{2}(16 - 2x)^{-1/2}(-2) = 0.$$

Let's simplify a bit.

$$\frac{dU}{dx} = \frac{1}{2}x^{-1/2} - (16 - 2x)^{-1/2} = 0$$

Recall from Chapter 2 that every FOC has two parts: a marginal benefit and a marginal cost. In this example that decomposition would be

$$\frac{dU}{dx} = \overbrace{\frac{1}{2}x^{-1/2}}^{\text{Marginal Benefit}} - \overbrace{(16 - 2x)^{-1/2}}^{\text{Marginal Cost}} = 0.$$

The marginal benefit portion is $\frac{1}{2}x^{-1/2}$, which is the marginal utility of x. That is, it is the marginal increase in utility that comes from an infinitessimally small increase in consuming more x (donuts). The marginal cost portion (notice it has a negative sign) represents a marginal reduction in utility. $-(16-2x)^{-1/2}$ shows how utility decreases by choosing more x, because consuming more x means that Jiji must consume less y (fish). The second term is the marginal utility of fish, y, but it enters negatively because we are taking the derivative with respect to x, and as x increases y decreases, as dictated by the budget constraint.

To "solve" the problem and find the optimal choices of x and y, we solve the FOC for x algebraically.

$$\begin{aligned}
\frac{1}{2}x^{-1/2} - (16-2x)^{-1/2} &= 0 \\
\frac{1}{2}x^{-1/2} &= (16-2x)^{-1/2} \\
2x^{1/2} &= (16-2x)^{1/2} \\
4x &= 16-2x \\
6x &= 16 \\
x^* &= 8/3
\end{aligned}$$

Since the optimal choice for x is $8/3$, we can now use the budget constraint to find the optimal choice of y.

$$\begin{aligned}
y &= 16-2x \\
y &= 16-2\left(\frac{8}{3}\right) \\
y^* &= \frac{32}{3}
\end{aligned}$$

Therefore, when Jiji consumes $8/3$ donuts and $32/3$ fish, his utility is maximized given the budget constraint.

4.3.2 General Utility Problem

Now, let's set up the general two–good utility maximization problem. Suppose our consumer values two goods and the consumer's preferences are represented by the utility function $U(x,y)$. We assume $U_x > 0$ and $U_y > 0$, which means the utility function is increasing in both goods x and y. That is, the more of either good the consumer gets the happier they are (or utility is an increasing function of both x and y). The consumer also faces a standard budget constraint given by: $W = p_x x + p_y y$. In that budget constraint note the variables W, p_x, and p_y are all *exogenous*. The consumer does not get to choose the prices of the goods. Moreover, the consumer, in this context, does not get to choose W, their wealth level, either. Although in reality people can and do take actions to affect their income and wealth, in this model, we treat the acquisition of wealth as happening outside the model. Thus, the only things the consumer chooses are how much x and y to consume.

Our general, two–good utility maximization problem can be written as

$$\max_{x,y} U(x,y)$$

subject to:

$$W = p_x x + p_y y.$$

That is still a multivariate problem, but again we can solve the budget constraint for y and substitute that expression into the objective function.

$$y = \frac{W - p_x x}{p_y}$$

Then substituting we have

$$\max_{x} U\left(x, \frac{W - p_x x}{p_y}\right).$$

Taking the derivative with respect to x and forming the first-order condition (FOC), we get

$$\frac{dU}{dx} = \overbrace{U_x\left(x, \frac{W - p_x x}{p_y}\right)}^{\text{Marginal Benefit}} - \overbrace{U_y\left(x, \frac{W - p_x x}{p_y}\right)\frac{p_x}{p_y}}^{\text{Marginal Cost}} = 0.$$

Whoa. That looks complex. The interpretation is exactly the same as we had before when we did this problem: $\max_x U = x^{1/2} + (16 - 2x)^{1/2}$. The marginal benefit portion is the entire first term, which is just the derivative with respect to the first x, which is the marginal utility of x. That is, it is the marginal increase in utility that comes from an infinitessimally small increase in consuming more x. The marginal cost portion, notice it again has a negative sign on it, represents a reduction in utility. $U_y\left(x, \frac{W - p_x x}{p_y}\right)\frac{p_x}{p_y}$ shows how utility decreases by choosing more x, because consuming more x means consuming less y. The second term is the marginal utility of y, but it enters negatively because we are taking the derivative with respect to x, and as x increases y decreases. Because the x in the second term is multiplied by $-p_x$ and divided by p_y, we have to use the chain rule, which generates the p_x/p_y on the right.

Notice that we cannot solve this problem by using algebra to isolate x^*. Instead, we say the solution is the x^* that solves this first-order condition, and we would write it as $x^*(p_x, p_y, W)$ which means that the solution is a function of all the exogenous variables. Although we cannot get an explicit expression for x^* in terms of p_x, p_y, and W, we have the FOC which *implicitly* defines x^*. We can even find the derivatives of x^* with respect to p_x, p_y, and W by *implicit* differentiation, a topic we will cover in Chapter 6.

Let's go back and look at that derivative a little more carefully and how we got there. First, note that we are taking the derivative with respect to x on a function that has the form $U(x, f(x))$. Note that in this utility function, because it is general, we are listing, separated by commas, those items that affect utility. Each item in the list is referred to as an "argument." We would say "x and $f(x)$ are arguments of the utility function." Originally, that was $U(x, y)$. We replaced the y with a function of x, which we will call $f(x)$ for the moment, so let $y = f(x)$.

When we take the derivative of a function written in a general form, like $U(x, f(x))$, we take the derivative of each argument separately and add them together. We do need to note which argument we took the derivative of, and we need to apply the chain rule to the second argument because it is a function of x. If we were to take the derivative of $U(x, f(x))$ with respect to x we would have

$$\frac{dU}{dx} = U_x(x, f(x)) + U_{f(x)}(x, f(x))f_x(x).$$

The first term, $U_x(x, f(x))$, represents taking the derivative of U with respect to the first argument, which is x. Note that the subscript following the U is an x, as in U_x, which is standard notation for a derivative. Here, it carries the extra meaning that we took the derivative with respect to the first argument in the function, which is x. Alternatively, sometimes it is written as $U_1(x, f(x))$ where the subscript 1 indicates the derivative with respect to the first argument.

The second term is a little more complicated because we had to apply the chain rule. The result is $U_{f(x)}(x, f(x))f_x(x)$. First, notice the subscript following the U is $f(x)$, indicating the derivative with respect to that argument. We could have alternatively represented that by writing $U_2(x, f(x))f_x(x)$, where the subscript 2 indicates the derivative with respect to the second argument. At the end of the expression we find $f_x x$. That results from our application of the chain rule. We need to take the derivative with respect to the outer function and multiply it by the derivative of the inner function. The derivative of $f(x)$ is $f_x(x)$.

Let's now return to the FOC we started with and break it down again.

$$\frac{dU}{dx} = U_x \overbrace{\left(x, \frac{W - p_x x}{p_y}\right)}^{\text{Marginal Benefit}} - U_y \overbrace{\left(x, \frac{W - p_x x}{p_y}\right)\frac{p_x}{p_y}}^{\text{Marginal Cost}} = 0$$

The first term, $U_x\left(x, \frac{W - p_x x}{p_y}\right)$, represents the marginal utility of consuming more x. Note the subscript x after the U to indicate the derivative with respect to the first term inside the utility function, which is x. The second term is the negative of the marginal utility of consuming more y. First, note the subscript following the U here is y to indicate the derivative was taken with respect to the second argument, which is y. Though we replaced the y with the budget constraint expression in terms of x, it is still y. The fraction at the end, $\frac{p_x}{p_y}$, results from the application of the chain rule, and, in fact, the negative sign joining the two terms also comes from the chain rule application.

Note in the numerator the $-p_x x$ expression. When we take the derivative with respect to the x in $\frac{W - p_x x}{p_y}$, we get $-\frac{p_x}{p_y}$.

4.4 Exercises

Woof! Was section 4.3 tedious to read? It was tedious to write. Nevertheless, it is important and the best way to get a feel for what is being said is to do practice problems. Enjoy!

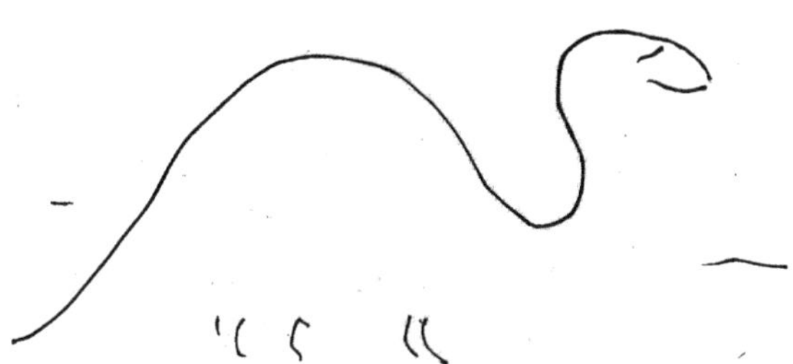

4.4.1 Quick Check Answers

a) $\Pi = (24 - Q^2)Q - Q^3 \Rightarrow \frac{d\Pi}{dQ} = 24 - 3Q^2 - 3Q^2 = 0 \Rightarrow \quad Q^* = 2$

b) Total costs $= TC = 4(Q - 5)^2 - 3Q^2 \Rightarrow \frac{dTC}{dQ} = 8(Q - 5) - 6Q = 0 \Rightarrow \quad Q^* = 20$

c) $y = \frac{1}{4}x^4 - x^3 + 3 \Rightarrow \frac{dy}{dx} = x^3 - 3x^2 = 0 \Rightarrow \quad x_1^* = 0$, and $x_2^* = 3$. There are two solutions to the FOC which we label with subscripts 1 and 2. Taking the second derivative we get: $\frac{d^2y}{dx^2} = 3x^2 - 6x$. Evaluating at $x_1^* = 0$ and $x_2^* = 3$, we get $\frac{d^2y}{dx^2} = 3(0)^2 - 6(0) = 0$ and $\frac{d^2y}{dx^2} = 3(3)^2 - 6(3) = 9 > 0$. The first solution, for reasons we will see in the next chapter, is an inflection point, but the first hint that it may be a inflection point is that the second derivative is zero at $x_1^* = 0$. The other solution generates a positive second derivative and is a minimum point.

d) $U = xy$ s.t. $x + y = 1 \Rightarrow \quad \max_x U = x(1 - x)$
$\frac{dU}{dx} = 1 - x - x = 0 \Rightarrow \quad x^* = 1/2$, and $y^* = 1/2$

e) $U = \ln(x) + \ln(y)$ s.t. $x + y = 1 \Rightarrow \quad \max_x U = \ln(x) + \ln(1 - x)$
$\frac{dU}{dx} = \frac{1}{x} - \frac{1}{1-x} = 0 \Rightarrow \quad x^* = 1/2$, and $y^* = 1/2$

4.4.2 Practice Problems

Maximize or minimize the following functions:

1. $\max_Q \Pi = (20 - Q)Q - 4Q$

2. $\max_Q \Pi = (12 - Q)Q - 2Q^2$

3. $\max_L \Pi = AL^{1/2} - 4L$

4. $\max_L \Pi = AL^{\alpha} - wL$ (where A and α are exogenous parameters and w is the wage rate on labor, L)

5. $\min_L$ Total Costs $= L^2 - 8L + 100$

6. $\min_Q$ Total Costs $= \alpha Q^2 - \beta Q + \Gamma$. What conditions must hold on α, β, and Γ for your answer to be a minimum?

In each of the following problems, do the following:

a) Write down the problem formally as in this example:

$$\max_{x,y} U = x^{1/2} + y^{1/2}$$
$$s.t. \quad 4x + 2y = 32.$$

b) Solve the budget constraint for one good.

c) Substitute your solution into the objective function to form a single variable unconstrained problem and write it formally like this example:
$\max_x U = x^{1/2} + (16 - 2x)^{1/2}$.

d) Take the derivative and form the first-order condition.

e) Interpret the first-order condition in words using the appropriate *marginal* terms.

f) Solve for the optimal values that maximize the consumer's utility.

Utility Maximization Problems

7. Solve Tombo's utility maximization problem. Tombo likes sushi, S, and takoyaki, T. His utility is given by U(S,T)=ST. He goes to the town fair with $100 in his wallet. At the fair, the price of sushi is $10, and the price of takoyaki is $5.

8. Solve Ursula's utility maximization problem. Ursula enjoys paintings, P, and ramen, R. Her utility is given by $U(P, R) = P^{1/3} R^{2/3}$. She has a budget of ¥5000 while a painting cost ¥2000 and a ramen costs ¥800.

9. Edward Elric needs to think about his future. He wants to consume now and in the future. His preferences are described by the following utility function: $U(c_1, c_2) = \ln(c_1) + 0.75 \ln(c_2)$, where c_1 is his consumption today and c_2 is his future consumption. He has an endowment from his father of $1,000,000. Not bad. If he saves any money for the future he gets a real interest rate return of 5%. His budget constraint is: $1,000,000 = c_1 + \frac{c_2}{1.05}$. Given his preferences, what is his utility maximizing consumption plan?

10. Go back to Section 4.4.1 and redo Jiji's specific problem. However, this time solve for x in terms of y from the budget constraint. Substitute that expression into the objective function and then proceed to solve the problem. Be sure to interpret the first-order condition in marginal benefit/marginal cost words.

11. Remember Margo from Chapter 1? Let's solve her problem now. Margo likes eating fish, F, and leisure time, N, that she spends with her cat. Her utility is given by $U(F, N) = F^{1/2}N^{1/2}$. She actually has a constraint and a production function. Her time constraint is $24 = H + N$, where 24 is her total time endowment of hours in a day and H is the hours she spends working. Her production function for fish is $F = 3H$. *Hint: First combine the constraint and the production function, then solve for one variable and substitute into the objective function.*

12. Suppose that King Bradley has the following preferences over gold, x, and power y, $U = Min(x, y)$. His budget contraint is $x + y = 1$. If you get stuck following steps (a)–(f), draw the indifference curves and budget constraint for this problem. It may help.

4.5 Appendix

Here, we list some formal definitions and other items for your reference.

First-order condition for an extreme point: If the first derivative of a function, $y = f(x)$ at $x = x^*$ is such that $f'(x^*) = 0$ then:

a) the point (x^*, y^*) on a function $y = f(x)$ is a **local maximum** if the derivative of the function at (x^*, y^*) is $f'(x^*) = 0$ and $f'(x) > 0$ for all x values within $x - \epsilon$ and $f'(x) < 0$ for all x values within $x + \epsilon$ for some $\epsilon > 0$.

b) the point (x^*, y^*) on a function $y = f(x)$ is a **local minimum** if the derivative of the function at (x^*, y^*) is $f'(x^*) = 0$ and $f'(x) < 0$ for all x values within $x - \epsilon$ and $f'(x) > 0$ for all x values within $x + \epsilon$ for some $\epsilon > 0$.

c) neither a minimum or maximum if the point (x^*, y^*) on a function $y = f(x)$ if the derivative of the function at (x^*, y^*) is $f'(x^*) = 0$ and the sign of $f'(x)$ for all x values within $x - \epsilon$ and for all x values within $x + \epsilon$ have the same sign.

Necessary and Sufficient: The concepts and examples further illustrate the ideas of necessary and sufficient conditions.

Proposition A statement that can be unambiguouly declared true or false.

A proposition normally takes the form $B \Rightarrow A$, where B represents a set of assumptions or initial conditions and A is the results or set of results that follow from B being true. In that structure we say that B is sufficient for A. Similarly A is necessary for B. It also follows that if NOT A then NOT B. That meas that if A is not true, then B is also not true. Refer back to the Venn diagram in Figure 4.4.

Consider the following example:

Example 1. Proposition: If $x \geq 4$ then $x + 3 \geq 0$.

The B set of assumptions is $x \geq 4$. That is *sufficient* to determine the result A that $x + 3 \geq 0$. However, B is not *necessary* for $x + 3 \geq 0$ because if $x = 1$, then B is not true, but A still holds. Conversely, A is *necessary* for B but not *sufficient*. That is, it must be the case that $x + 3 \geq 0$ for it to be possible for $x \geq 4$, but B is not enough information determine if $x \geq 4$ is true or not.

If B is *both* necessary and sufficient for A, we say that A and B are **equivalent** statements. It follows that if B is both necessary and sufficient for A, then A is both necessary and sufficient for B. Notationally we would write $A \Leftrightarrow B$.

Example 2. Proposition: If $x \geq 4$ then $x + 3 \geq 7$.

Those two statements are equivalent because one implies the other and the order does not matter. We could equivalently have written the proposition as, "If $x + 3 \geq 7$ then $x \geq 4$."

5 Second Order Conditions, Concavity, Convexity, and the Nth Derivative Test

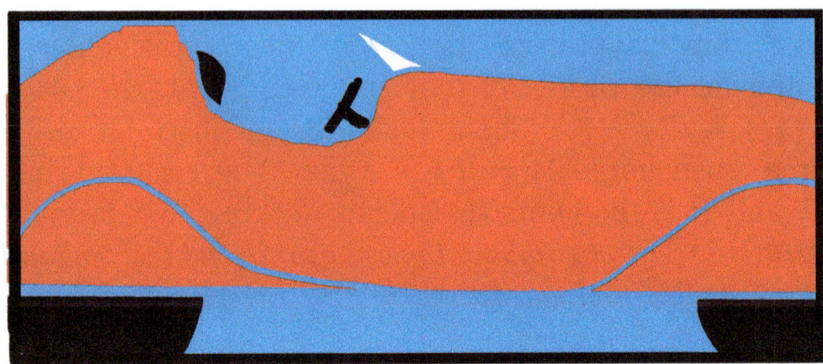

The previous chapter repeatedly talked about the first-order condition and how that is necessary but not sufficient to know if we have correctly identified a minimum or a maximum. The *sufficient* conditions include the second derivative. The first derivative tells us whether the slope is positive, zero, or negative, which corresponds to an increasing, flat, or decreasing function. The second derivative gives us the same information about the slope of the first derivative itself. Is the slope (first derivative) increasing, flat, or decreasing? In turn, that means if the original function is increasing, is it increasing at an increasing rate, increasing at a constant rate, or increasing at a decreasing rate? Or if the function is decreasing, is it decreasing at an increasing rate, decreasing at a constant rate, or decreasing at a decreasing rate? If that phrasing is confusing, hopefully the diagrams in this chapter will make all that a good bit clearer. Other related questions the second derivative helps answer include these: Does the function ever switch from increasing to decreasing or vice versa? If so, where? Those are the things that the second derivative tells us mathematically.

In economics terms, we usually use the phrase "diminishing returns," meaning a function is increasing at a decreasing rate. For example, you may have come across diminishing returns to labor. That means that hiring more workers yields more production, but the amount of increase for every worker hired gets smaller and smaller. To describe that type of pattern, we need both the first and second derivatives. As you may have inferred by now, the second derivative provides key information about the shape of functions. We will be very interested in functions that are **concave** or **convex**, as these shapes have many counterparts in economic intuition. Moreover, when we can identify the shape of the function, we can also tell whether we have indeed found a minimum or maximum and in some cases, even though the first derivative is zero, the point is neither a minimum or maximum! In that case, we may need to take more derivatives beyond the second derivative to determine the shape of the function.

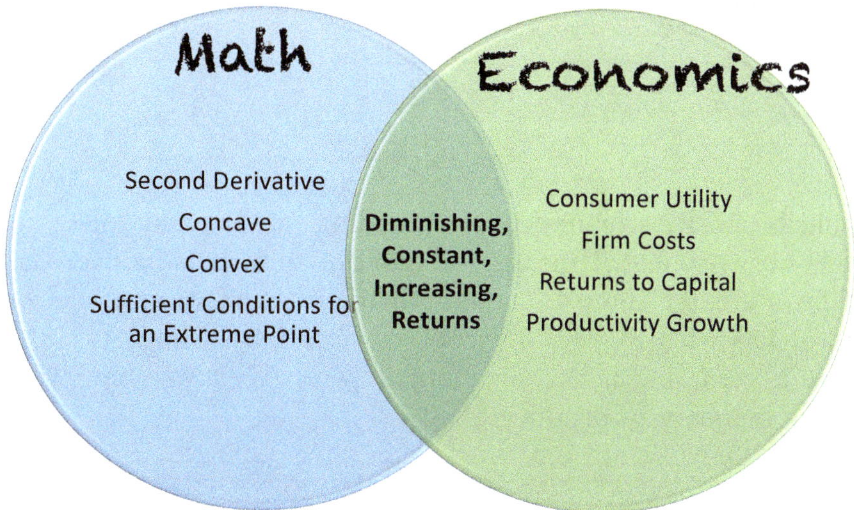

5.1 First Derivative Interpretation: A Brief Review

Now may be a good time to briefly review a few key points from the prior chapters. As we have seen, when we take a derivative, it reveals important information about a function that may not be obvious from looking at the mathematical expression. The sign of the derivative tells us whether the function is *increasing* or *decreasing*. For example, look at the following function and its derivative.

$$
\begin{aligned}
y &= 6x^3 \\
\frac{dy}{dx} &= 18x^2 > 0
\end{aligned}
$$

For any value of x other than zero, the first derivative is positive. That means the slope goes upward, which also means the function is **increasing**; larger values of x give larger values of y. If the derivative of a function is negative, the slope is downward and the function is **decreasing** as in the following example:

> **Definition**
>
> **Increasing function**: An increasing function has a positive slope, $\frac{dy}{dx} > 0$.
>
> **Decreasing function**: A decreasing function has a negative slope, $\frac{dy}{dx} < 0$.

$$
\begin{aligned}
y &= -6x^3 \\
\frac{dy}{dx} &= -18x^2 < 0.
\end{aligned}
$$

In the monopolist problem from Chapter 2, Luffy's burgers, we had the following:

$$
\max_{Q} \Pi = 10Q - Q^2 - 2Q
$$

and the derivative we got was

$$\frac{d\Pi}{dQ} = 10 - 2Q - 2$$
$$= 8 - 2Q \lesseqgtr 0.$$

That derivative could be positive or negative depending on the value of Q. If $Q < 4$, then the derivative is positive, and it means the function is increasing over the interval $[0, 4)$. However, if Q is greater than 4, the derivative is negative and the function is decreasing. Compare that with Figure 5.1 repeated here from Chapter 2. At exactly $Q = 4$, the derivative is zero, and hence the slope is zero. That was how we identified the profit– maximizing quantity to produce.

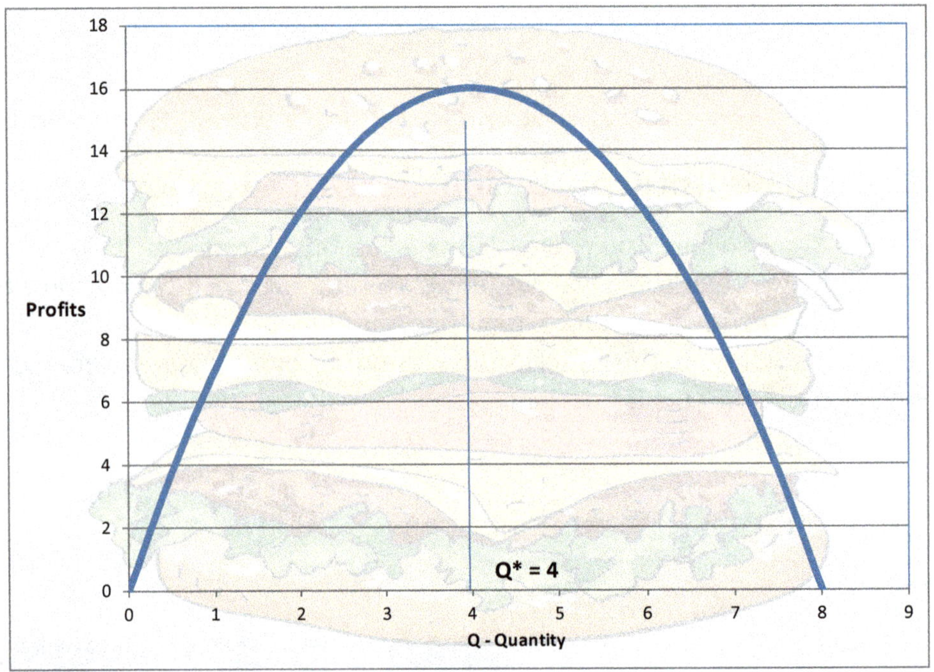

Figure 5.1: Luffy's monopoly profit function.

Similarly, consider the following production function, where Y represents output, α is a positive parameter, and K is physical capital. Taking the following function and the derivative with respect to K we have

$$Y = \alpha K^{1/2}$$
$$\frac{dY}{dK} = \frac{\alpha}{2} K^{-1/2}.$$

Logically, K can only take on positive values or be zero. α is assumed to be a positive parameter. Thus, the entire derivative is always positive. That means that output, Y, is always increasing with capital, K.

However, what we do not see directly from the first derivative is how the increase in output from adding more capital changes depending on how much capital the firm already has. We could plug in numbers for varying levels of K into the first derivative and figure it out by trial and error. However, the easier and more accurate approach is to take another derivative, the second derivative. We turn to that concept now.

5.2   The Meaning of the Second Derivative

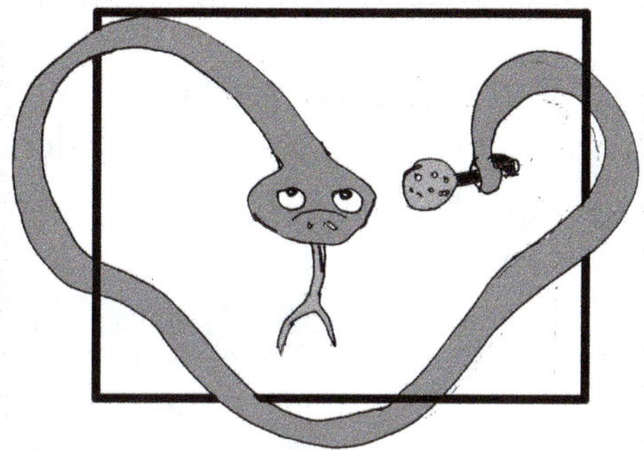

Before jumping into the meaning of second derivatives, let's quickly review how to get second derivatives. The second derivative is the derivative of the first derivative. So, for example if a function and its first derivative are

$$y = 5x^3, \text{ and}$$
$$\frac{dy}{dx} = 15x^2,$$

then to get the second derivative we would now take the derivative with respect to x again to get

$$\frac{d^2y}{dx^2} = 30x.$$

Notice the notation on the left-hand side. The superscript 2's indicate we have taken the derivative twice. The first 2 in d^2y is written between the d and y, indicating a second derivative of y. In the denominator we write the 2 to the right of dx, just like a squared term and that is the way to think of it. Note that the 2 applies to dx as one whole term and *not* to just the x. That is, $dx^2 = (dx)^2$, but $dx^2 \neq d(x^2)$. We could equivalently write it as

Quick Check

Find the second derivatives, $\frac{d^2y}{dx^2}$, of the following functions:
a) $y = 7x^4 + 2x^2$
b) $y = ln(x) + 5$

$$\frac{d^2y}{dx^2} = \frac{d^2y}{dxdx}.$$

The denominator shows we took the derivative with respect to x two times. Specifically, dx is a complete term that refers to a marginal change in the value of x as we explored in Chapter 2. We will see similar notation when we get to multivariate calculus in the next chapter, when we might take the derivative with

> **Definition**
>
> **Concave function**: A function, $y = f(x)$, is concave if the derivative for all possible values of x is such that $\frac{dy}{dx} \leq 0$. If it is such that $\frac{dy}{dx} < 0$, then we say the function is *strictly concave*. If the derivative is $\frac{dy}{dx} \leq 0$ for x values in the interval I, then we say the function is concave over the interval I.

respect to x and then some other variable (e.g. w), and we might have something like $\frac{d^2y}{dxdw}$. However, we are still in the single-variable calculus world for now, so we only have one right-hand side variable, x. Thus, we can focus on $\frac{d^2y}{dx^2}$ for now.

In addition to the first derivative telling us whether a function is increasing or decreasing, the second derivative also carries important information about the shape of the function. It tells us how the slope itself is changing over the function. Specifically, when the second derivative is negative, it means the slope is decreasing. We call functions of that type **concave**.

> **Definition**
>
> **Convex Function**: A function, $y = f(x)$, is convex if the derivative for all possible values of x is such that $\frac{dy}{dx} \geq 0$. If it is such that $\frac{dy}{dx} > 0$, then we say the function is *strictly convex*. If the derivative is $\frac{dy}{dx} \geq 0$ for x values in the interval I, then we say the function is convex over the interval I.

If the second derivative is positive, the slope is increasing and the function is **convex**.[7] Figures 5.2 and 5.3 show examples of concave and convex functions.

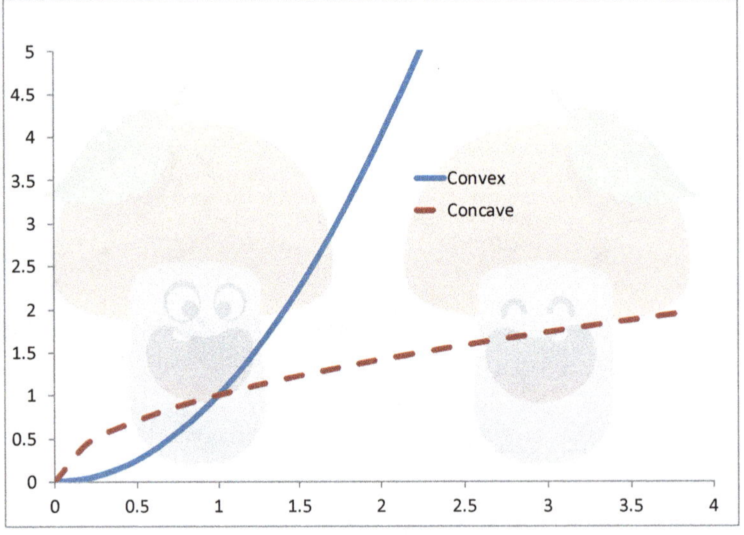

Figure 5.2: Convex and concave: Example 1.

---

[7]Some math textbooks use the terms *concave down* and *concave up* for concave and convex, respectively.

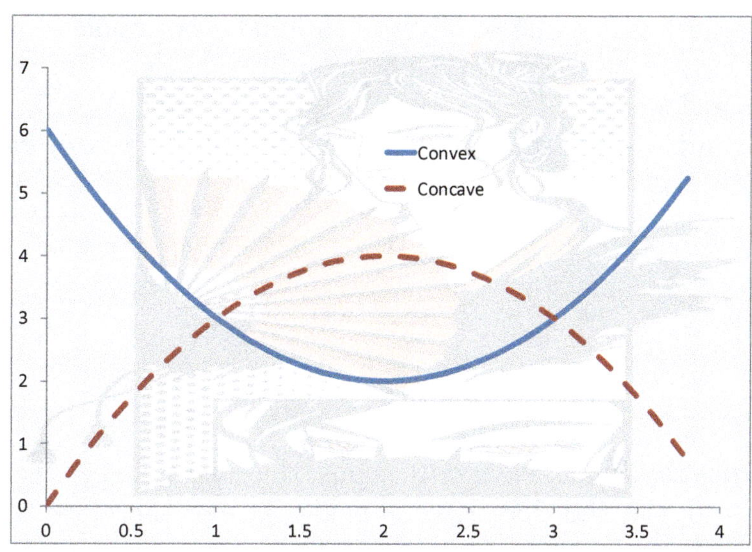

Figure 5.3: Convex and concave: Example 2

In both diagrams, notice that, for the convex functions (solid line), if you draw in several tangent lines for slopes, the slope is always increasing, as you go from left to right (as x increases), whereas for the concave functions (dashed line), the slope is always decreasing from left to right. From the monopolist problem we examined in the previous section, the derivative we got was

$$\frac{d\Pi}{dQ} = 10 - 2Q - 2.$$

Taking the second derivative tells us whether we have a concave function or a convex function:

$$\frac{d^2\Pi}{dQ^2} = -2.$$

The 2nd derivative is -2, and therefore we have a concave function (see Figure 5.1). The negative value means the slope is always decreasing. In fact, since the 2nd derivative is always -2 for any value of x,

> ### Making the Connection
>
> **Graphing** To help the terminology like *decreasing slope* or *increasing slope*, graph the following two functions by hand or on a spreadsheet, then print them out.
> a) $y = 4x^{1/2} - x$
> b) $x^2 - 8x$
> Draw in the tangencies (the slopes) to the two functions at the following values for x: 1, 2, 3, 4, 5, 6, and 7. Look at how the slope changes as x increases. Which change in the tangencies is convex? Which is concave?

it means the slope is always decreasing, and therefore the function is strictly concave.

Similarly, consider the production function from Section 5.1 and the derivative with respect to K:

$$\frac{dY}{dK} = \frac{1}{2}\alpha K^{-1/2}$$
$$\frac{d^2Y}{dK^2} = -\frac{1}{4}\alpha K^{-3/2} < 0.$$

If we assume $K > 0$ (negative amounts of physical capital do not make economic sense), then the 2nd derivative is always negative. Therefore, the function is strictly concave for $K > 0$.

Let's look at an example when the function is not always convex or concave. Take the second derivatives of the following to get

$$
\begin{aligned}
y &= 6x^3 \\
\frac{dy}{dx} &= 18x^2 > 0 \\
\frac{d^2y}{dx^2} &= 36x.
\end{aligned}
$$

The 2nd derivative is positive whenever $x > 0$. Thus, the function is convex for positive values of x. When $x < 0$, the 2nd derivative is negative, and therefore the function is concave for the interval from $-\infty$ to 0. Now consider the following function where we just added a minus sign:

$$
\begin{aligned}
y &= -6x^3 \\
\frac{dy}{dx} &= -18x^2 < 0 \\
\frac{d^2y}{dx^2} &= -36x.
\end{aligned}
$$

This function has the opposite result. The 2nd derivative is negative whenever $x > 0$, and therefore is concave, but is convex when $x < 0$. Figure 5.4 shows graphs of both $y = 6x^3$ and $y = -6x^3$. Note the ranges that are convex and concave.

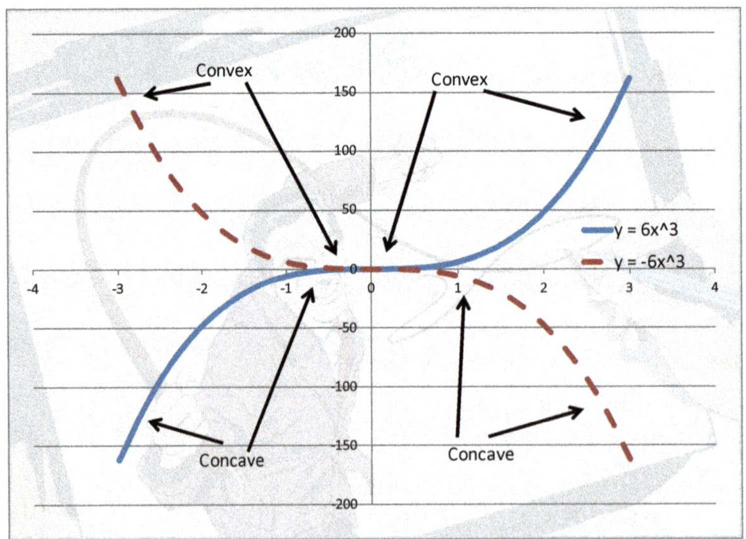

Figure 5.4: Convex and concave: Example 3.

Some useful features to note about concave and convex functions includ the following:

- If a function $f(x)$ is concave, then $-f(x)$ is convex, and vice versa.

- A straight line is both concave and convex.

- A straight line is not *strictly* concave nor *strictly* convex.

When we say strictly, the function is constantly "curving," or the slope is always changing. A function with flat portions that make straight lines or segments can still be concave or convex, but then we say it is not *strictly* concave or *strictly* convex. Refer to the definitions Also note that functions, like those in Figures 5.3 and 5.4 have regions that are convex and regions that are concave. In those cases we refer the function as being concave or convex over an *interval* of x values.

Quick Check

Determine whether these functions are concave or convex by taking the second derivatives, $\frac{d^2 y}{dx^2}$:

c) $y = 3e^{3x}$

d) $y = -\frac{1}{x+1}$

Concavity has a straightforward economics interpretation: **diminishing marginal returns**. In the context of the production function example, it means that capital exhibits diminishing marginal returns. In other words, output rises with more capital (as we found from the first derivative), but the amount of the increase gets smaller and smaller with more capital; that is what the second derivative tells us.

Utility functions are also typically concave for each good the consumer wishes to have. In one of our previous examples, Jiji liked donuts. If we were to express that in a utility function we would want Jiji's utility to increase with donuts consumed. That means the first derivative of utility with respect to donuts should be positive.

Definition

Diminishing marginal returns: The concept that additional units may increase the value of a function, but the amount of the increase gets smaller the larger the initial amount (e.g. diminishing marginal utility or diminshing marginal product).

Thus, if $U(D)$ is Jiji's utility based on donuts, then $\frac{dU}{dD} > 0$ means utility is increasing with donuts. However, if we also believe that Jiji has diminishing marginal utility with respect to donuts, then the second derivative should be negative. That is, if more donuts make Jiji happier, the extra amount of happiness or utility should be getting smaller and smaller with each donut. Thus, $\frac{d^2 U}{dD^2} < 0$.

Convexity is frequently associated with cost functions. Consider the following cost function and it derivatives:

$$\begin{aligned}
TC &= 8Q^2 - 4Q + 10 \\
\frac{dTC}{dQ} &= 16Q - 4 \\
\frac{d^2 TC}{dQ^2} &= 16
\end{aligned}$$

The second derivative is always 16 and therefore always positive. Thus, the marginal costs (which are given by the first derivative $16Q - 4$) are always increasing in quantity. Try graphing that cost function and convince yourself that it is indeed convex.

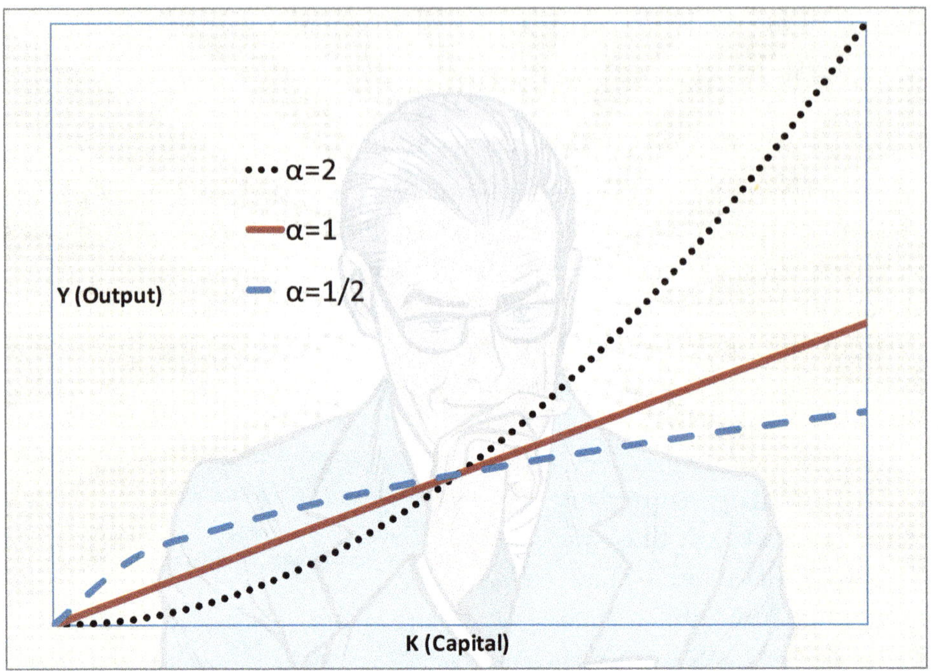

Figure 5.5: Concave, convex, and linear production functions.

Whether a function is convex or concave is really important for the behavior of economic variables. To see this point clearly, let's look at a variation f the production function again. This function is often used to represent the production of an entire economy:

$$Y = AK^\alpha$$

Now, output is given by productivity, A, which must be positive, and capital, K. The exponent is the parameter α. The value of α determines the shape of the function and how output responds to adding physical capital. Take the first derivative of Y with respect to K and we get

$$\frac{dY}{dK} = \alpha A K^{\alpha-1}.$$

Since we expect more capital to lead to increased output, it makes sense that $\alpha > 0$. That means that the first derivative is always positive, and we have a positive *marginal product of capital*. However, the function could still be concave or convex. Taking the second derivative we have:

> **Definition**
>
> **Marginal product of capital (MPK)**: The additional amount of output generated by a marginal increase in the amount of physical capital used in production.

$$\frac{d^2Y}{dK^2} = (\alpha - 1)\,\alpha A K^{\alpha-2}$$

where the sign depends on whether α is greater than, less than, or equal to 1. The graph in Figure 5.5 shows the difference.

The curve labeled $\alpha = 2$ is convex. The slope is increasing. In this case, it says that we are getting increasing returns for each additional unit of capital. The straight line ($\alpha = 1$) shows constant returns; each additional unit of capital yields the same increase in output. The final line ($\alpha = 1/2$) is concave and shows diminishing returns to capital.

5.3 Example: Cost Minimization and the Second Derivative

Let us briefly work with a very specific example to illustrate a minimization problem and the role of the second derivative. Mr. Sato sells cabbages. The costs to Mr. Sato of selling in the city of Omashu depends on the number of cabbages, Q, that he sells. He would like to know where his costs are minimized.

His cabbage growing cost function is $TC = 8Q^2 - 4Q + 10$. So, Mr. Sato's problem is

> ### Making the Connection
>
> **Concave and convex:** There are many examples of objects and designs around you that have these fundamental shapes we tie to powerful economic intuition. What shape is a bowl? How about the shape of the glass in eyeglasses? How many everyday objects can you identify that are concave? Convex? Can you find strictly convex objects? Or convex, but not strictly convex? What about concave and strictly concave? What shape is a horse's saddle?

$$\min_{Q} TC = 8Q^2 - 4Q + 10,$$

where TC represents total costs and the quantity of cabbages is Q. Q is the endogenous choice variable. Taking the derivative and setting it equal to zero we get

$$\frac{dTC}{dQ} = 16Q - 4 = 0.$$

Setting the derivative equal to zero forms the first-order condition for an extreme point. Recall, that means the slope of the TC function will be zero at the level of Q that satisfies this equation. Solving, we get $Q^* = 1/4$.

Although we got $Q* = 1/4$, we cannot be sure if we have a minimum or a maximum or neither until we check the second-order condition. For a minimum we need a positive second derivative. Taking the second derivative we get:

$$\frac{d^2TC}{dQ^2} = 16,$$

which is clearly positive. Therefore, the function is convex, and we found the minimum point.

Now, suppose Mr. Sato, our cabbage merchant, moves to the city of Ba Sing Se to sell cabbages. His cost function changes to $TC = Q^3 - 8Q^2 + 16Q + 25$. Again, he wants to know where his costs are minimized.

$$\min_Q TC = Q^3 - 8Q^2 + 16Q + 25$$

$$\frac{dTC}{dQ} = 3Q^2 - 16Q + 16 = 0$$

Solving the first-order condition we get:

$$(3Q - 4)(Q - 4) = 0.$$

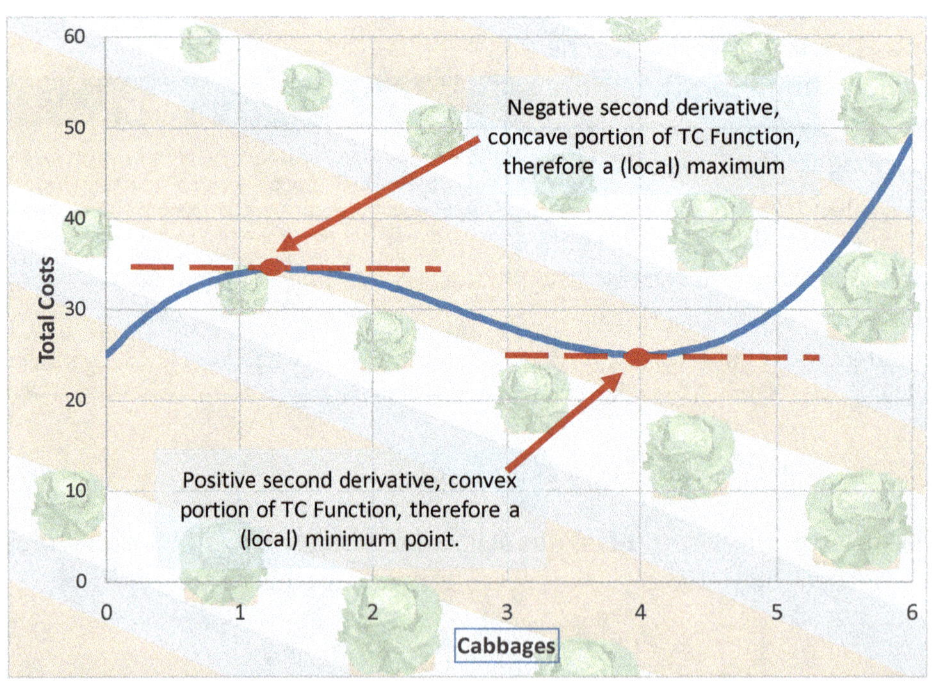

Figure 5.6: Mr. Sato's total costs of cabbage production in Ba Sing Se.

That yields two solutions, $Q_1^* = 4/3$ and $Q_2^* = 4$. The subscripts 1 and 2 on Q simply indicate solution 1 and solution 2. All we know from the first-order condition is that the slope of the function at these two points is zero. They meet the necessary condition for an extreme point. However, they could be minima, maxima, or neither. To emphasize that point, the first-order condition is necessary but not sufficient. We will see an example of how we can get a zero first-order condition but find that the solution is neither a minimum nor a maximum. We need to include conditions on the second derivative to get *sufficient* conditions. So, to determine the nature of Q_1 and Q_2, take the second derivative with respect to Q:

$$\frac{dTC}{dQ} = 3Q^2 - 16Q + 16$$

$$\frac{d^2TC}{dQ^2} = 6Q - 16.$$

So, to check our two candidate points, put those values into the second derivative. Doing so yields the following values:

Definition

Second order condition: The second order condition for a maximum point, x^*, of $y = f(x)$ is that the function is concave, $\frac{d^2y}{dx^2} < 0$ at x^*. The second-order condition for a minimum point, x^*, of $y = f(x)$ is that the function is convex, $\frac{d^2y}{dx^2} > 0$ at x^*.

$$Q_1^* = 4/3 \quad : \quad 6(4/3) - 16 = -8 < 0$$
$$Q_2^* = 4 \quad : \quad 6(4) - 16 = 8 > 0.$$

What do we conclude? At the first solution, $Q_1^* = 4/3$, the second derivative is negative, indicating a concave portion of the function and a maximum point. However, at the second solution, $Q_2^* = 4$, the second derivative is positive, indicating a convex portion of the function and a minimum. So, the solution to Mr. Sato's problem is $Q_2^* = 4$.

A quick summary of our conditions so far for minima and maxima can be found in Table 5.1. However, as we shall see in the next section, we are not yet done. There are cases where the 2nd derivative is zero, and we turn to that situation next.

Table 5.1: First and second order conditions for an extremum

	First-order condition	Second-order condition
Maximum	$\frac{dy}{dx} = 0$	$\frac{d^2y}{dx^2} < 0$, concave
Minimum	$\frac{dy}{dx} = 0$	$\frac{d^2y}{dx^2} > 0$, convex

5.4 N^{th} Derivative Test

As noted at the very end of the last section, what if we are trying to verify a minimum or maximum but find that the second derivative is zero? If the second derivative is zero, it tells us the slope itself is neither increasing nor decreasing at that point. That means we do not have enough information to tell whether the function is convex, concave, or neither.

So what do we do? If we get a second derivaitive that is equal to zero at our extreme point candidate x^*, *keep taking derivatives until you get a derivative that is not zero.* Essentially, take the third derivative. If that is not zero, stop. If it is zero, take the fourth derivative. Is that zero? If it is still zero, keep taking derivatives until you get a nonzero. Once you have something that is not zero at the point you are evaluating, stop. Determine how many derivatives you had to take to get to this point (was it the third derivative? The fourth? The tenth?), let N to be equal to the number of derivatives you took to get to a nonzero result. Then check Table 5.2.

N^{th} Derivative Test

Let x^* be the x value in question and $f(x^*)$ be the candidate extreme point satisfying the first-order condtion, $f'(x^*) = 0$. Then, depending on whether N is an even or odd number, we have

Table 5.2: The Nth derivative test

N	$f^N(x^*)$	Result
Even	$f^N(x^*) < 0$	Maximum, Concave
Even	$f^N(x^*) > 0$	Minimum, Convex
Odd	$f^N(x^*) \gtrless 0$	Neither, Inflection

Related to the Nth derivative test, here is some useful notation. Let $f^N(x)$ be the Nth derivative. For derivatives of three or less we often use the prime ($\prime$) mark instead, as in the first derivative is $f'(x)$, the second is $f''(x)$, and the third is $f'''(x)$.

What is an **inflection point**? Consider, for example, the following function: $y = x^7 - 1$. The first derivative yields

$$\frac{dy}{dx} = 7x^6$$
$$7x^6 = 0$$
$$x^* = 0.$$

So, at $x^* = 0$, the value of the function is $f(x^*) = f(0) = 0$. Is the point $(x^*, f(x^*)) = (0, 0)$ a minimum, maximum, or neither? Taking the second derivative we have

$$\frac{d^2y}{dx^2} = 42x^5 = 42(x^*)^5 = 42(0)^5 = 0.$$

> **Definition**
>
> **Inflection point:** An inflection point of a function is a point where the function is convex on one side and concave on the other. In other words, at the inflection point, the second derivative is zero, but to the immediate left and right of the inflection point the second derivative is positive on one side and negative on the other side. An inflection point of a function has second derivative equal to zero. However, a zero second derivative is not sufficient to know if the point is an inflection point.

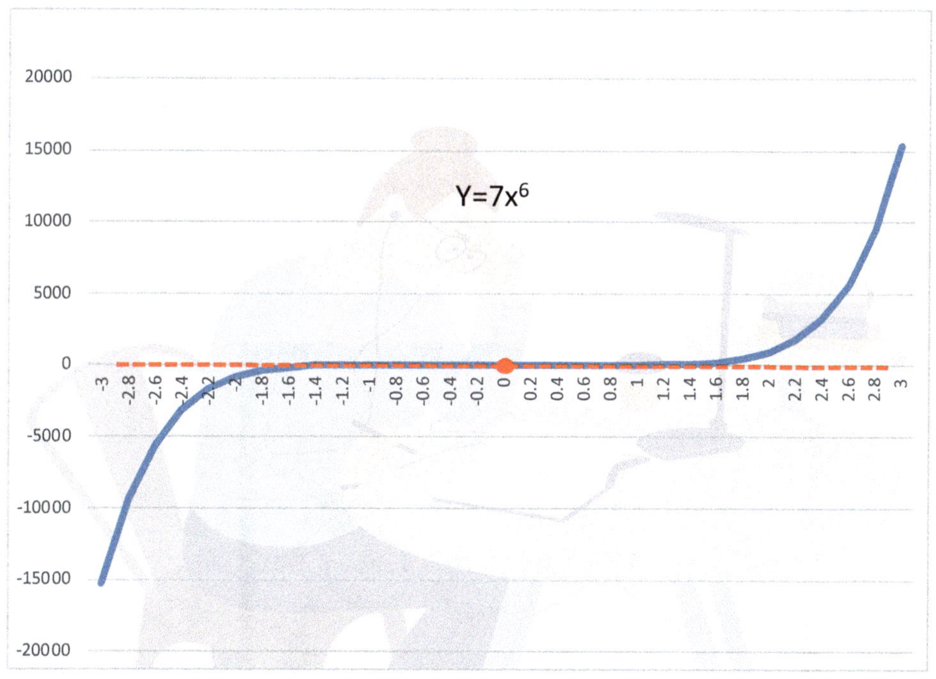

Figure 5.7: Inflection point example.

So, the second derivative is equal to zero at our candidate extreme point, thus the second order condition does not tell us whether the function is concave or convex. (Remember, the second order condition is sufficient, but not necessary). We keep taking derivatives until we get something that is not zero when evaluated at $x^* = 0$:

$$\frac{d^3y}{dx^3} = 210x^4 = 210(x^*)^4 = 210(0)^4 = 0$$

$$\frac{d^4y}{dx^4} = 840x^3 = 840(x^*)^3 = 840(0)^3 = 0$$

$$\frac{d^5y}{dx^5} = 2520x^2 = 2520(x^*)^2 = 2520(0)^2 = 0$$

$$\frac{d^6y}{dx^6} = 5040x = 5040(x^*) = 5040(0) = 0$$

$$\frac{d^7y}{dx^7} = 5040 > 0.$$

Our first nonzero value when evaluating at the candidate extreme value for x^* came on the seventh derivative. Seven is an odd number, and checking our table, that means the point is neither a minimum nor a maximum, but it is an **inflection point**. An inflection point is a point where a function switches from being concave to convex, or the other way around, from convex to concave.

In the graph of the function (see Figure 5.7) we just examined, one can see the point at zero does indeed have a slope of zero. However, the function moves from concave to the left of zero to convex to the right of zero. See Figure 5.4 in Section 5.2 for two more examples of functions with inflection points.

5.5 Exercises

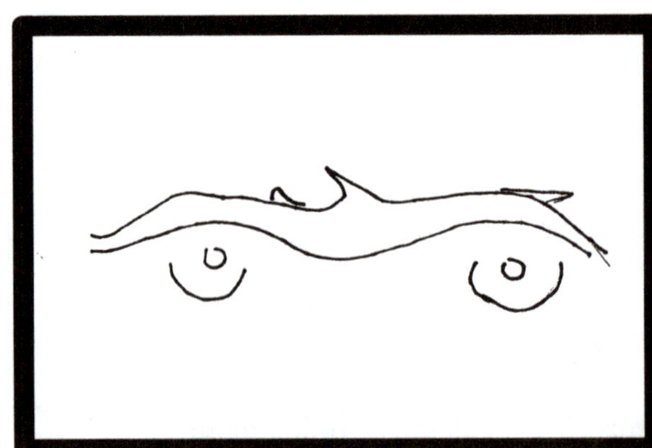

5.5.1 Quick Check Answers

a) $y = 7x^4 + 2x^2 \Rightarrow \qquad \frac{dy}{dx} = 28x^3 + 4x \Rightarrow \qquad \frac{d^2y}{dx^2} = 84x^2 + 4$

b) $y = ln(x) + 5 \Rightarrow \quad \frac{dy}{dx} = \frac{1}{x} \Rightarrow \quad \frac{d^2y}{dx^2} = -\frac{1}{x^2}$

c) $y = 3e^{3x} \Rightarrow \quad \frac{dy}{dx} = 9e^{3x} \Rightarrow \quad \frac{d^2y}{dx^2} = 27e^{3x} > 0$. The second derivative is always positive for any value of x. Therefore, the function is convex.

d) $y = -\frac{1}{x+1} \Rightarrow \quad \frac{dy}{dx} = \frac{1}{(x+1)^2} \Rightarrow \quad \frac{d^2y}{dx^2} = -\frac{2}{(x+1)^3}$. The second derivative is always negative for any value of $x > -1$. Therefore, the function is concave for $x > -1$. For values $x < -1$, the second derivative is always positive. Therefore, the function is convex for $x < -1$. The function has an inflection point at $x = -1$.

e) $y = 8x^{1/2} - x \Rightarrow \quad \frac{dy}{dx} = 4x - 1 = 0 \Rightarrow \quad x^* = 1/4 \Rightarrow \quad \frac{d^2y}{dx^2} = 4 > 0$. The second derivative is always positive, and thus the function is convex. Therefore the solution to the first-order condition, $x^* = 1/4$, is a minimum.

f) $x^3 - 7x^2 - 5x \Rightarrow \quad \frac{dy}{dx} = 3x^2 - 14x - 5 = (3x - 1)(x - 5) \Rightarrow \quad x_1^* = 1/3, \quad x_2^* = 5 \Rightarrow \quad \frac{d^2y}{dx^2} = 6x - 14 \Rightarrow \quad 6x - 14 = 6(x_1^*) - 14 = 6(1/3) - 14 = -12 < 0 \Rightarrow \quad = 6x - 14 = 6(x_2^*) - 14 = 6(5) - 14 = 16 > 0$. Therefore, the first solution $x_1 = 1/3$ is a maximum because the second derivative is negative, indicating the function is concave around that point. The second solution, $x_2 = 5$, is a minimum because the second derivative is positive, indicating the function is convex around that point.

g) $y = \frac{1}{5}x^5 \Rightarrow \quad \frac{dy}{dx} = x^4 \Rightarrow \quad x^* = 0 \Rightarrow \quad \frac{d^2y}{dx^2} = 4x^3 \Rightarrow \quad \frac{d^3y}{dx^3} = 12x^2 \Rightarrow \quad \frac{d^4y}{dx^4} = 24x \Rightarrow \quad \frac{d^5y}{dx^5} = 24 > 0$. Therefore, referring to the Nth derivative test table, the fifth derivative is the first nonzero derivative at $x^* = 0$. Since 5 is odd, the point is an inflection point, and $x^* = 0$ is neither a minimum nor a maximum.

h) $y = (x - 5)^4 \Rightarrow \quad \frac{dy}{dx} = 4(x - 5)^3 \Rightarrow \quad x^* = 5, \Rightarrow \quad \frac{d^2y}{dx^2} = 12(x - 5)^2 \Rightarrow \quad \frac{d^3y}{dx^3} = 24(x - 5) \Rightarrow \quad \frac{d^4y}{dx^4} = 24 > 0$. Therefore, referring to the Nth derivative test table, the fourth derivative is the first nonzero derivative at $x^* = 5$. Since 4 is even, the point is not an inflection point. Since $4 > 0$, the function is convex and the point is therefore a minimum.

5.5.2 Practice Problems

Find the second derivatives of the following and figure out whether they are concave or convex functions (or have ranges of both):

1. $y = 2x^2$

2. $y = -x^2 - 5$

3. $y = \ln x$

4. $y = 4e^x$

5. Go back to practice problems 1–6 from Chapter 4 and check the second-order conditions for each problem.

6. For the following functions, are they concave or convex, and what does that depend on? If the answer is possibly both, over what ranges are the functions convex or concave?

(a) $\Pi = 10Q - Q^2 - 2Q$

(b) $Y = AL^{1-\alpha}$

(c) $TC = 3Q^3 + 2Q^2 - Q + 10$

(d) $\frac{1}{2}x^{-1/2} - (16 - 2x)^{-1/2}$. From Section 4.4, we found the optimal solution was $(x^*, y^*) = (8/3, 32/3)$. Verify that solution is indeed a maximum.

7. Return to problem 1 from Chapter 2, where you were asked to write down the first derivatives for general functions. Now write down what you expect for the second derivatives for those same functions and explain your logic.

For example, for (a) $R(Q)$, revenues as a function of quantity sold, the first derivative would normally be $R_Q(Q) > 0$. That is, R is an increasing function in Q, meaning the higher the quantity sold, the higher the revenues. However, the 2nd derivative would be $R_{QQ}(Q) < 0$. Note how the 2nd derivative here is indicated with a pair of Q's as subscripts. We also could have written that as $\frac{d^2R}{dQ^2} < 0$. The logic is that revenues exhibit diminishing returns with quantity sold because the firm gets more in sales but has to lower the price in order to sell more.

5.6 Math Appendix

Here, we list some key math definitions, theorems, and formulae in formal math for your reference.

Concave: A function is concave on an interval if for any two points, x_1 and x_2 on the interval, if and only if $f(\alpha x_1 + (1 - \alpha)x_2) \geq \alpha f(x_1) + (1 - \alpha)f(x_2))$ for $0 < \alpha < 1$. If the relation holds with strict inequality, $>$, then we say the function is strictly concave on the interval.

Convex: A function is convex on an interval if for any two points, x_1 and x_2, on the interval, if and only if: $f(\alpha x_1 + (1 - \alpha)x_2) \leq \alpha f(x_1) + (1 - \alpha)f(x_2))$ for $0 < \alpha < 1$. If the relation holds with strict inequality, $<$, then we say the function is strictly concave on the interval.

Marginal product of captial (MPK): The additional amount of output generated by a marginal increase in the amount of physical capital used in production. For a general production function with physical capital, K, as an input, $Y = F(K, X)$, and X represents the set of other inputs (e.g. physical capital, raw materials, human capital, energy, etc.), the marginal product of capital is $MPK = \partial Y/\partial K = F(K, X)/\partial K$.

Necessary and sufficient conditions for an extremum: For the point x^* to be a maximum point, the necessary first-order condition is that $f'(x^*) = 0$. The necessary second-order condition is that $f''(x^*) \leq 0$. The sufficient second-order condition is that $f''(x^*) < 0$. For the point x^* to be a minimum point, the necessary first-order condition is that $f'(x^*) = 0$. The necessary second-order condition is that $f''(x^*) \geq 0$. The sufficient second-order condition is that $f''(x^*) > 0$. In all cases, the point x^* could be either a local or global extremum.

6　Multivariate Calculus

The previous chapters' discussions of derivatives were all for single-variable calculus. In economics, frequently, as in almost always, we are work with multivariate calculus. While that might sound daunting at first, especially if you have never taken multivariate calculus before, it turns out that if you can do single variable calculus, then doing multivariate calculus is a pretty simple step forward. You do not need to learn new derivative rules; the same ones apply, and the minimum and maximum rules are similar, though we have to add one condition.

For multivariate calculus, you do need to learn some new notation and some new terms, in particular the meaning of *partial* derivative and *total* derivative or *total* differential. In addition, we will use our new notation to think about how to take derivatives on general functions with many arguments, such as $U(x, y, z(x))$, where we need the chain rule. The chapter is heavy on some long examples in which we go back and forth between specific and general functions.

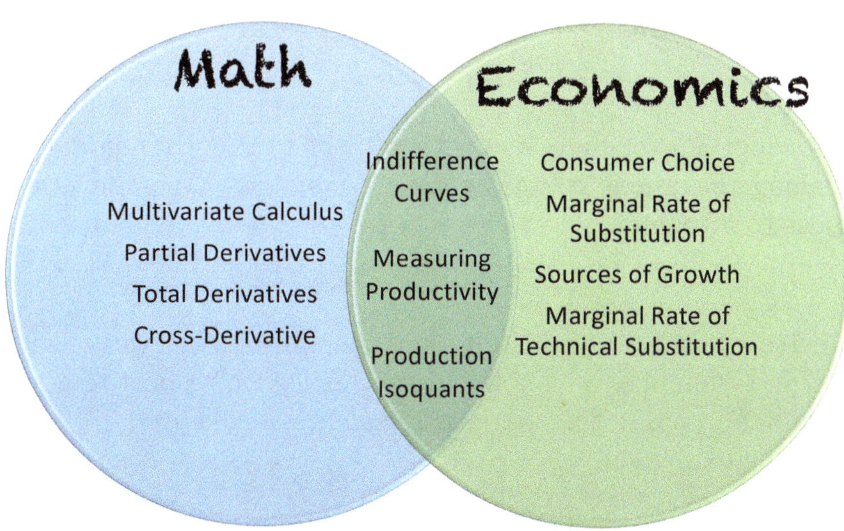

6.1 Partial Derivatives

When we say "single-variable calculus" we mean using calculus on a function like the following:

$$y = 3x^2 + 8$$

where y is a function of just one other variable, x. The derivative of y with respect to x is

$$\frac{dy}{dx} = 6x.$$

However, suppose we replace the 8 with z. z is also a variable:

$$y = 3x^2 + z.$$

Now we are in the world of multivariate calculus. y changes with both x and z, so y is a multivariate function. So how do we take derivatives? Pretty much the same way we did before. What we do, however, is start by using a "partial" derivative. That is, if we want to know how y changes with x, then we don't care about z, and we treat z just like any other number, as a constant term like 8. The only immediate difference is that instead of writing $\frac{dy}{dx}$ we write $\frac{\partial y}{\partial x}$, where the script ∂ sign indicates a partial derivative. For the previous example we have

$$\frac{\partial y}{\partial x} = 6x$$

which means the following: The change in y with respect to x is $6x$ *holding z constant* OR the slope of y along the x-axis is $6x$, *holding z constant*. We can also take the derivative with respect to z to get:

$$\frac{\partial y}{\partial z} = 1$$

where the term $3x^2$ is treated as just a constant, and therefore its derivative with respect to z is zero. The interpretation of $\frac{\partial y}{\partial z} = 1$ is the change in y with respect to z is 1 *holding x constant* OR the slope of y along the z-axis is 1, *holding x constant*.

Consider the following example:

$$y = 4x^2 + 2xz + z^{1/2}$$

Taking partial derivatives with respect to x and z we would get:

$$\frac{\partial y}{\partial x} = 8x + 2z$$

$$\frac{\partial y}{\partial z} = 2x + \frac{1}{2}z^{-1/2}.$$

The interpretation of each is as follows: $\frac{\partial y}{\partial x}$ shows how y changes with x holding z constant (i.e. it is the slope of the function along the x-axis.) $\frac{\partial y}{\partial z}$ shows how y changes with z holding x constant (i.e. it is the slope of the function along the z-axis).

Notice how the interpretation keeps repeating the phrase *holding $<$ variable $>$ constant*. That means we are looking at the slope along one dimension or axis only; all other variables are unchanging. That is the essence of how the derivative is *partial*. With a *total* derivative (discussed soon) we allow all the variables to change simultaneously and the really good news here is that a total derivative is easy to find because it is a sum of partial derivatives!

The same symbols and procedures apply to second derivatives as well. Using our example to illustrate:

$$\frac{\partial^2 y}{\partial x^2} = 8$$

$$\frac{\partial^2 y}{\partial z^2} = -\frac{1}{4}z^{-3/2}.$$

Notice again how the other variable, the one we are not taking a derivative on, is treated as a constant or just a number.

The second derivatives here, too, have the interpretation of showing us concavity or convexity with respect to that variable alone. That is, they show how the function is shaped because the second derivative shows how the slopes are changing. We must be careful here, though. Just because a function, like this one,

has a positive second derivative with respect to x, that does **not** mean the entire function is convex. Determining if a multivariate function is convex or concave takes some additional steps, which we will address in a few sections. Next, we will introduce one critical part of doing that here: finding the **cross-derivative**.

6.2 Young's Theorem and the Cross-Derivative

The cross-derivative is a second derivative and comes from taking the derivative of a function in one variable, then taking the derivative of the result with respect to another variable. That is probably easier understood by looking at an example. Using the same function from before, let's first take the derivative with respect to x:

$$\begin{aligned} y &= 4x^2 + 2xz + z^{1/2} \\ \frac{\partial y}{\partial x} &= 8x + 2z. \end{aligned}$$

Now take the derivative of $\frac{\partial y}{\partial x}$ with respect to z:

$$\frac{\partial^2 y}{\partial x \partial z} = 2.$$

That result, too, is a second derivative, as indicated by $\partial^2 y$, but is referred to as a "cross-partial derivative" or just "cross-derivative," as it shows how the slope with respect to one variable changes with respect to another variable.

You might ask, does the order in which we take the derivatives matter for the cross-partial derivative? Would it make a difference if we did $\frac{\partial y}{\partial z}$ first, then took the derivative with respect to x? The answer is, Nope! No, the order does not matter; you will always get the same result. This feature of cross-partial derivatives is referred to as **Young's Theorem**.

Let's redo the example to illustrate Young's theorem by changing the order we take the derivatives:

$$y = 4x^2 + 2xz + z^{1/2}$$
$$\frac{\partial y}{\partial z} = 2x + \frac{1}{2}z^{-1/2}.$$

Now take the derivative with respect to x:

$$\frac{\partial^2 y}{\partial x \partial y} = 2.$$

We get the same answer. That should always happen. In fact, a good way to check your work is to find the cross-derivative both ways. It is a particularly good practice when initially learning this material. We will return to the cross-derivative and how it is used in checking second-order conditions for an extreme point in Section 6.5.

<table>
<tr><td>

Definition

Young's theorem: For a twice differentiable function $y = f(x_1, x_2, ..., x_n)$, the order in which partial derivatives are taken does not affect the resulting cross-derivatives. That is, $\frac{\partial^2 y}{\partial x_i \partial x_j} = \frac{\partial^2 y}{\partial x_j \partial x_i}$.

</td></tr>
</table>

Let's go through the exercise of finding the cross-derivative again, but using general functions. Let's start with the most basic multivariate function, $z = f(x, y)$, where our dependent variable is z and the two independent variables are x and y. Starting with the first partial derivatives we get

<table>
<tr><td>

Quick Check

Find the cross-derivatives. Do them both ways. $\frac{\partial^2 z}{\partial x \partial y}$ and $\frac{\partial^2 z}{\partial y \partial x}$:

c) $z = 2x^{1/2} + y^4$

d) $z = 2x^{1/2}y^4$

</td></tr>
</table>

$$z = f(x, y)$$
$$\frac{\partial z}{\partial x} = f_x(x, y), \qquad \frac{\partial z}{\partial y} = f_y(x, y).$$

Notice that with the general functions, we are representing the partial derivatives with a subscript of the variable we took the partial derivative of in each case. We have that $\frac{\partial z}{\partial x}$ is $f_{\mathbf{x}}(x, y)$ with the subscript x indicating we took a partial derivative with respect to x. Analogously, $\frac{\partial z}{\partial y}$ is $f_{\mathbf{y}}(x, y)$, with the subscript y indicating we took a partial derivative with respect to y.

Moving to the second derivatives, we have:

$$\frac{\partial^2 z}{\partial x^2} = f_{xx}(x, y), \qquad \frac{\partial^2 z}{\partial y^2} = f_{yy}(x, y),$$
$$\frac{\partial^2 z}{\partial x \partial y} = f_{xy}(x, y).$$

Now note the subscripts again. For the second partial derivatives with respect to x, there are two x subscripts f_{xx} indicating we took the partial derivative of x twice. f_{yy} indicates we took the partial derivative of y twice. Finally, $f_{xy}(x,y)$ indicates a cross-partial derivative. The order of x then y in the subscript has no bearing on the result, and convention says we list them alphabetically. However, we could write it as $f_{yx}(x,y)$. That would still be "correct," but the ordering would be awkward and unconventional.

6.3 Examples

6.3.1 Cobb–Douglas Production Function

Here's a standard application from macroeconomics of multivariate calculus. Suppose that the total output (GDP) of an economy is given by the following production function:

$$Y = AK^{1/3}L^{2/3},$$

where Y is total GDP, A is productivity (or technology), K is the capital stock (the stock of machinery, factories, equipment, etc.), and L is the size of the labor force. The production function tells us how much output (Y) we get for the inputs (K and L) that we have and the level of productivity (the efficiency with which the economy transforms inputs into outputs). In this model we have output depending on three variables: A, K, and L. If we want to know how output changes with capital, we take the partial derivative of Y with respect to K, treating A and L as constant

Making the Connection

Cobb–Douglas production function: The functional form used here with two variables, K and L, in this case, multiplying each other with exponents on each input and a coefficient in front multiplying them all, is named after economists Charles Cobb and Paul Douglas, who used this form to study the US manufacturing sector a century ago. This mathematical function has become very common for modeling firm or economy-wide production and consumer preferences because the function has a large number of desirable properties that fit with empirical observations. You can read more about their contribution at the Inomics website, a highly useful site for searching for economics programs and jobs internationally.

numbers. We get

$$\frac{\partial Y}{\partial K} = \frac{1}{3}AK^{-2/3}L^{2/3}.$$

In words, $\frac{\partial Y}{\partial K}$ shows the change in output (GDP) with a marginal change in capital stock holding productivity and labor constant. That is more commonly referred to as the **marginal product of capital** (MPK), which is the marginal increase in output that comes from a marginal increase in capital. If we take the derivative with respect to labor we would have:

$$\frac{\partial Y}{\partial L} = \frac{2}{3}AK^{1/3}L^{-1/3}$$

which is the **marginal product of labor** (MPL), the additional amount of output we get with an additional amount of labor.

Let's now get the second partial derivatives, including the cross-derivative between K and L:

$$\frac{\partial^2 Y}{\partial K^2} = -\frac{2}{9}AK^{-5/3}L^{2/3}$$

$$\frac{\partial^2 Y}{\partial L^2} = -\frac{2}{9}AK^{1/3}L^{-4/3}$$

$$\frac{\partial^2 Y}{\partial K \partial L} = \frac{2}{9}AK^{-2/3}L^{-1/3}.$$

How do we interpret these results? Note again that we can safely assume A, K, and L are positive because negative values make no economic sense and zero values are uninteresting. So, we can say that the first two derivatives are strictly negative. That means that Y is concave in K, and Y is also concave in L. In economic terms, it means there are diminishing returns to physical capital, holding labor and productivity constant and diminishing returns to labor, holding capital and productivity constant. Similarly, from the result on the cross-derivative we see that it is positive. That means the marginal product of capital is increasing with labor or, conversely, the marginal product of labor is increasing with capital. More succinctly, we would say labor and capital are **complements** in production.

6.3.2 General Utility Example

Consider a general utility function for Toph Beifong over three items she enjoys consuming. Toph likes pet badgers, B, ginseng tea, G, and attending theatrical plays, P. We summarize her utility from these three items as $U = U(B, G, P)$. If we take the partial derivatives for each item, we would write

$$\frac{\partial U}{\partial B} = U_B(B, G, P)$$
$$\frac{\partial U}{\partial G} = U_G(B, G, P)$$
$$\frac{\partial U}{\partial P} = U_P(B, G, P).$$

Each of those is interpreted as a marginal utility holding the other two items constant. In all cases, economics would tell us that each partial derivative is positive, assuming that Toph likes all three items. So, the second term, $U_G(B, G, P)$, for example, represents her marginal utility of consuming ginseng tea.

The associated second derivatives would be written as:

$$\frac{\partial^2 U}{\partial B^2} = U_{BB}(B, G, P)$$
$$\frac{\partial^2 U}{\partial G^2} = U_{GG}(B, G, P)$$
$$\frac{\partial^2 U}{\partial P^2} = U_{PP}(B, G, P)$$
$$\frac{\partial^2 U}{\partial B \partial G} = U_{BG}(B, G, P)$$
$$\frac{\partial^2 U}{\partial B \partial P} = U_{BP}(B, G, P)$$
$$\frac{\partial^2 U}{\partial G \partial P} = U_{GP}(B, G, P).$$

The first three deriviatves are sometimes referred to as "own" second derivatives, meaning you take the deriviatve on the same variable twice. As with what we learned in the previous chapter, they represent the shape of the function, along the axis of that particular variable. That is, U_{BB}, shows whether U is concave or convex with respect to B. If $U_{BB} > 0$, then utility is convex in B. If $U_{BB} < 0$, then utility is concave in B. A function can be concave in one variable but convex in other variables. The question of whether the entire function across all variables is convex or concave can be addressed with the total derivative, and we look at that in the next subsection.

The second three derivatives are all cross-derivatives. Each one shows how the marginal utility of one good changes with the amount of the other good. For example, $U_{G,P}$ shows how Toph's marginal utilty of ginseng tea changes with the number of plays she attends. It *also* shows how Toph's marginal utility of plays changes with the amount of ginseng tea she consumes. If $U_{G,P}$ is positive, then G and P are **complements**. That means that Toph's utility, her enjoyment, of ginseng tea goes up with more plays she attends. It *also* means her enjoyment of plays goes up with the amount of ginseng tea she consumes. One can imagine she always has a cup of tea when attending a play and having tea makes the play even better than attending a play with no tea or makes the tea even better than having tea without attending a play. If the price of plays went down, Toph would increase her consumption of both plays and ginseng tea.

Suppose that $U_{BP} < 0$. That makes B and P **substitutes**. An increase in one good reduces the marginal utility of the other good. Perhaps, the draconian theater does not allow pets to attend. So, when Toph attends more plays she has less time for her pet badgers, for example. If the price of plays rises, Toph would go to less plays, which would increase her marginal utility from getting more pet badgers and her demand for pet badgers would increase.

6.4 Total Derivatives

6.4.1 Total Derivatives are Made Up of Partial Derivatives

Recall that the term *partial* indicates holding all other variables constant. There is another way to take the derivative, where we let all the variables change, and this is called the *total derivative* or the *total differential*, which is a closely related concept. Taking the total derivative follows the same rules for differentiation, but we now need to go through the function and take the partial derivative for every variable separately and then add these together. Return to our first example. The following is what we called the **total differential**:

$$y = 3x^2 + z.$$

The total differential would be

$$dy = 6xdx + 1dz,$$

where the $6x$ is the partial derivative from taking the derivative with respect to x, and 1 is the partial derivative from taking the derivative with respect to z as before. However, notice that the left-hand side is dy and not a ratio like $\frac{dy}{dx}$. dy represents the marginal change in y and is called the differential of y, but it is now dependent on changes in two variables: x and z. The changes in those variables are represented by dx and dz, respectively (the differentials in x and z). The total change in y is given by the slope of the function along the x-axis ($6x$) times the change in x, dx, plus the slope of the function along the z-axis (1) times the change in z, dz.

What does it mean? Let's use the following general function to help fix ideas:

$$H = f(x, y)$$

Definition

Derivatives and **Differentials**: *differentials* are the infinitessimally small change in a variable and represented as dx, for example. Derivatives, (e.g., $\frac{dy}{dx}$) are the ratio of differentials showing how an infinitessimally small change in one variable, x, alters the value of another variable, y. In the context here, total derivatives and total differentials are used interchangeably, and you will see both terms employed in economics textbooks. See the section on chain rule in this chapter for more on the distinction.

First, keep in mind that our function describes a three-dimensional object. If you draw x-y coordinates on the paper in front of you, imagine the H-axis going up vertically from the paper (and vertically down below the paper would be negative H values). H is used to represent height. Think of the x-y plane as being described by east, west, north, and south, where to the right is east, negative y values would be south, left (negative x values would be west, and directly away from you would be north. The value of the function would then be the height.

If you then put a three-dimensional object on the paper (imagine a ball or a bowl, for example), then further imagine the function, H=f(x,y), describes that object, where H is the height of the object. Put (or imagine putting) your finger on the object. Then, $\frac{\partial H}{\partial x}$ tells you how fast your finger goes up or down (along the H-axis) on the object as you move east to west (left to right) in the direction of the x-axis. $\frac{\partial H}{\partial y}$ tells you how fast your finger goes up or down on the object as you move in a south to north (from close to you to directly away from you) in the direction of y-axis. These are the two partial derivatives where you are moving your finger only along the x-axis (east to west) while keeping the same y level or you are moving your finger only along the y-axis (south to north) while keeping the same x level.

Now, move your finger in a northeast direction, along the 45° line. How fast your finger goes up or down now depends on moving *simultaneously* east to west *and* south to north. You are looking at how the height changes in both directions. That is the total differential (derivative), the marginal change in the height with marginal changes in both the x and y directions.

In our previous numerical example, when we got $dy = 6xdx + 1dz$, notice that if we hold z constant that means z doesn't change. If z does not change that means $dz = 0$.

The total differential with $dz = 0$ would become

$$\begin{aligned} dy &= 6x\,dx + 1\,dz \\ dy &= 6x\,dx + 1\,(0) \\ dy &= 6x\,dx, \end{aligned}$$

and we can divide by dx to get

$$\frac{dy}{dx} = 6x,$$

which is the same as the partial derivative of y with respect to x.

Let's use the Cobb–Douglas example from Section 6.3.1. Suppose we want the total differential to see how Y changes with A, L, and K simultaneously. Taking the total differential we have

$$\begin{aligned} Y &= AK^{1/3}L^{2/3} \\ dY &= \left(K^{1/3}L^{2/3}\right)dA + \left(\frac{1}{3}AK^{-2/3}L^{2/3}\right)dK + \left(\frac{2}{3}AK^{1/3}L^{-1/3}\right)dL. \end{aligned}$$

In words, the total change in GDP is given by the change in productivity (dA) times the marginal product of A, plus the marginal product of capital times the change in capital (dK), plus the marginal product of labor times the change in labor (dL). Another way of phrasing that is any change in dY must come from a change or changes in the other variables, A, K, or L, times the slope on each of those variables.

We can also take a total *second* differential (derivative). The procedure is the same: Go through the function, taking the partial derivative with respect to each variable, and multiply by the change in each variable. Let's go back to the numeric example we started this section:

$$\begin{aligned} y &= 3x^2 + z \\ dy &= 6x\,dx + 1\,dz. \end{aligned}$$

In taking the second derivative, we do it a longer way by writing in two zeros (0) to help illustrate the steps:

$$d^2y = (6dx + 0 * 1dz)dx + (0 * 6dx + 0dz)dz.$$

Here we first take the derivative of the first derivative with respect to x. That results in $6dx + 0 * 1dz$, where the second term has a derivative of zero because x does not appear. We again multiply by another dx, as we took the change with respect to x again, and we combine these in the following expression. We then add to that result the partial derivative with respect to z, where the first term, $6xdx$, drops out/gets mutiplied by zero because no z terms appears, and the same occurs with the second term, which is just a constant. Simplifying yields

$$d^2y = 6dx^2.$$

6.4.2 Application to Indifference Curves

We use partial and total derivatives to better understand and interpret commonly used functions. To illustrate and complete this discussion, let us do one more example with a general functional form for representing utility and **indifference curves** from microeconomics.

We start this application with a general utility function over two goods, x and y. To fix ideas and provide a visual interpretation, let us imagine an individual consumer, Uncle Iroh, who enjoys cake, x, and jasmine tea, y. His general utility function is given by

$$U = U(x, y),$$

where we will assume the following about the first partial derivatives: $U_x > 0$ and $U_y > 0$, which implies that utility is increasing with x and y (cake and tea). Furthermore, for the second partial derivatives we will assume that $U_{xx} < 0$ and $U_{yy} < 0$, which implies diminishing marginal utility for both cake and tea. Finally, let's assume $U_{xy} > 0$. What does that mean? Start with the first derivative of utility with respect to x, U_x. We know that derivative is positive because utility increases with x (cake). If the cross-derivative is positive, that means the marginal utility of x is increasing with y. Intuitively, the cake tastes better when consumed with more tea. Since, from Young's theorem, we know that we can get to the same result starting from either first derivative, we can also interpret the cross-derivative as starting from U_y. In that case, the cross-derivative also says that the marginal utility of y (tea) is increasing with the amount of cake consumed.

Hopefully, by now you have encountered the concept of indifference curves in your economics courses. Indifference curves are sets of combinations of goods for a consumer that all yield exactly the same level of utility. See Figure 6.1. The graph presents two indifference curves for a specific utility function. The lower dashed red curve is all the combinations of cake (x) and tea (y) that generate exactly a utility level of 3. They form a set that is sometimes referred to as a **level set**, which means equal values. From the

point of view of the consumer, Iroh, he is equally happy with any combination of cake and tea that falls on that indifference curve. Of course, Iroh would prefer more of both. The upper blue solid–line indifference curve represents a higher level of utilty, $U = 4$, to be exact. All the points on the upper curve are equally good from Iroh's point of view; however, all the points on the upper solid–blue line indifference curve are better than all the points on the lower red–dashed line indifference curve.

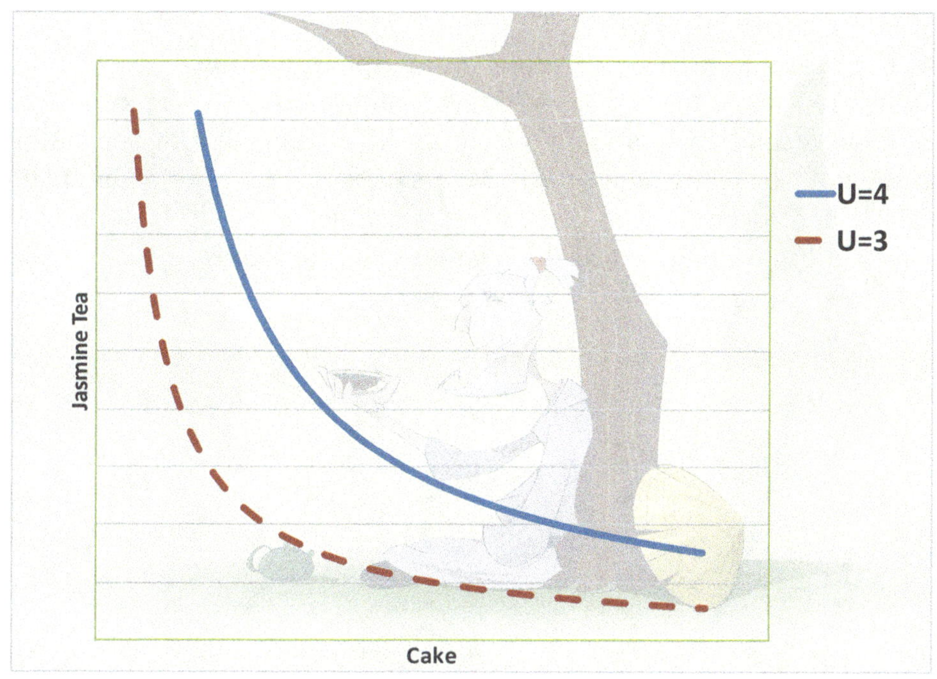

Figure 6.1: Indifference curves for $U = \ln(x) + \ln(y)$.

The indifference curve graph that is typically presented in textbooks, and the ones here, are two–dimensional graphs. But look at the utility function for a moment: $U = \ln(x) + \ln(y)$. That has three variables: x, y, and U. Shouldn't it be a three–dimensional graph? Well, yes, but drawing three–dimensional graphs is a lot

> **Definition**
>
> **Marginal Rate of Substitution**: The rate at which a consumer is willing to give away one good in exchange for another good such that they are just as happy before and after the trade, i.e. utility does not change.

more work, and if your art skills are as bad as mine, you would likely just make a big incomprehensible mess. If x and y are the axes you see, there is a third axis, which is perpendicular to the page, and the utility levels should be rising off the page. The two-dimensional representation here is akin to a topological map of mountains. Each line represents a particular height such that with a bit of imagination you can envision the mountains rising off the page.

OK. a concept directly related to indifference curves: the **marginal rate of substitution (MRS)**. The MRS is the rate at which a consumer is willing to trade one good for another while maintaining the same level of utility. That is, suppose Uncle Iroh has 2 cups of tea and 1 cake. His friend Zhao asks him for 1 cup of tea. Iroh declines, saying that would make him less happy because his utility would decline with the loss of a cup of tea. Zhao offers to give Iroh two cakes in exchange for 1 cup of tea. Iroh thinks it over and realizes he is indifferent between having 2 cups of tea and 1 cake or having 1 cup of tea and 3 cakes. He is equally happy with both options. Thus, his MRS between tea and cakes is 1 cup of tea per 2 cakes when he starts from 2 cups of tea and 1 cake. This idea of the trade-off, or willingness to trade one item for another, becomes precise with the help of multivariate calculus.

Think about the example in the preceding paragraph again for a second: 1 cup of tea for 2 cakes which is 1 less unit on the y-axis and 2 more units on the x-axis, $\Delta 1/\Delta 2$. It sounds like a slope *because* it is a slope of curve. It is the slope of the indifference curve at exactly the point where Iroh started (2 cups of tea and 1 cake). Great, but how do we find the slope here because it is measuring x against y, not the slope of U with respect to x or y? See Figure 6.2 for the slope we are seeking here.

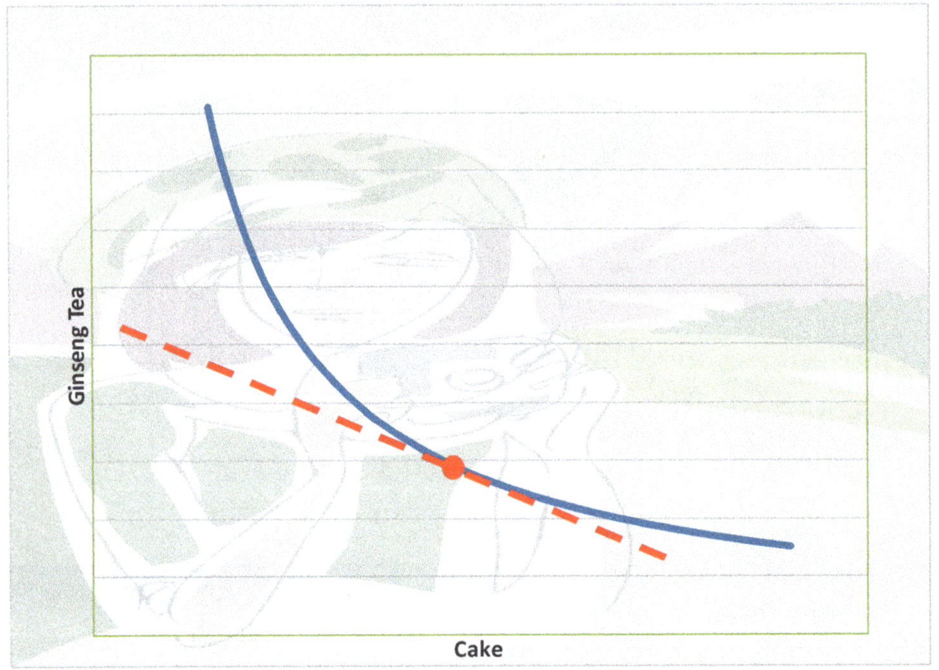

Figure 6.2: Indifference curves and the marginal rate of substitution.

What we want is the change in y (tea) for a change in x (cakes), holding U (Utility) constant. So, let's now compute the total differential of the utility function and see what that provides:

$$U(x, y)$$
$$dU(x, y) = U_x(x, y)dx + U_y(x, y)dy.$$

Reading that expression in words, the change in utilty, dU, is equal to the slope of utility with respect to x times the change in x, which is dx plus the slope of utility with respect to y times the change in y, which is dy. We have the components that make up the MRS, dy and dx, which are the changes in y and x. How can we solve it for dy/dx?

To get there, think about what it means to hold utility constant here. That is, think about what the definition of the MRS means when it says, "holding utility constant." That means we do not let U change, which, in turn, means we set $dU = 0$. So, if utility is held constant

$$dU = 0 = U_x(x, y)dx + U_y(x, y)dy.$$

Then if we change x (the amount of cake Uncle Iroh gets), that means increasing or decreasing dx, then in order for the equation to remain true and utility to remain unchanged such that $dU = 0$, dy has to change. In fact, after setting $dU = 0$ we can solve directly for dy/dx:

$$\frac{dy}{dx} = -\frac{U_x(x,y)}{U_y(x,y)},$$

which we can also write in the partial derivative notation:

$$MRS = -\frac{\partial U/\partial x}{\partial U/\partial y}.$$

That last line is often the representation you see for the MRS in microeconomics textbooks. It is the slope of the indifference curve.

6.5 Multivariate Sufficient Second Order Conditions

Let's quickly review the necessary and sufficient conditions for a solution to be a maximum or minimum in single– variable calculus without constraints. The first-order necessary condition requires that the first derivative in the choice variable be equal to zero, $f_x(x) = 0$. The second-order condition requires that at the optimal point, the 2nd derivative be negative for a maximum (concave) or the 2nd derivative be positive for a minimum.

Table 6.1: First and Second Order Conditions for an Extremum

	First–Order Condition	Second–Order Condition
Maximum	$\frac{dy}{dx} = 0$	$\frac{d^2y}{dx^2} < 0$, Concave
Minimum	$\frac{dy}{dx} = 0$	$\frac{d^2y}{dx^2} > 0$, Convex

6.5.1 Second Order Conditions

The same conditions for an extremum extend in a fairly straightfoward manner to the multivariate calculus world, but we need one additional condition to ensure we have an extreme point. Loosely speaking, to give the reader a sense of where this section is going, for a function with two choice variables (e.g., $z = f(x, y)$), we need first–order conditions from taking partial derivatives on each choice variables, $f_x(x, y) = 0$ and $f_y(x, y) = 0$. Both of those must be equal to zero. We will also need two second–order partial derivatives on each choice variable, $f_{xx}(x, y)$ and $f_{yy}(x, y)$, where, loosely, they are both less than zero for a maximum or greater than zero for a minimum. We say "loosely" here because we can find a minimum or maximum where one of the 2nd derivatives is equal to zero but the other is not. We will be more precise later.

Furthermore, we need a condition involving the cross-derivative to rule out the following possibility: Both the first-order conditions are equal to zero and both the second-order conditions are either positive or negative, but the function is actually concave in some directions and convex in another that is not revealed by $f_{xx}(x, y)$ and $f_{yy}(x, y)$ alone. We refer to such points as "saddle points" because the centerpoint on a horse's saddle has exactly this feature and is one example of a shape that will satisfy the first-order conditions but fail this new condition. See Figure 6.3 for an example.

With this brief introduction in mind, let's go through the second–order conditions for an extreme point carefully. Starting with a general two–variable function that we wish to maximize (or minimize), we have

$$max_{x,y} \quad z \quad = \quad f(x, y)$$

or

$$min_{x,y} \quad z \quad = \quad f(x, y).$$

The necessary first-order conditions are

$$\frac{\partial z}{\partial x} \quad = \quad f_x(x, y) = 0$$
and
$$\frac{\partial z}{\partial y} \quad = \quad f_y(x, y) = 0.$$

In words, each first-order condition states that the slope of the function with respect to the variable is zero at the point(s) (x^*, y^*) that solve(s) the equation. If the point(s) (x^*, y^*) solve(s) both first-order conditions, then the slope is zero in both the x and y directions.

However, as we encountered in single–variable calculus, the first-order condition alone is not enough to tell us if the points we found are minima, maxima, or neither. To establish what type of point we have found, we take the full set of second partial derivatives, including the cross-derivative:

$$\frac{\partial^2 z}{\partial x^2} = f_{xx}(x, y) \gtreqless 0,$$

$$\frac{\partial^2 z}{\partial y^2} = f_{yy}(x, y) \gtreqless 0$$

and

$$\frac{\partial^2 z}{\partial x \partial y} = f_{xy}(x, y).$$

From these, the same second-order conditions that correspond to concavity and convexity that we had before carry through. Both second partial derivatives in the same variable, f_{xx} and f_{yy}, need to be positive for convexity (a minimum), and both need to be negative for convcavity (a maximum). However, our condition involving the cross-derivative, sometimes referred to as the saddle point condition, is:

$$f_{xx} f_{yy} - (f_{xy})^2 > 0.$$

This condition must be positive for *either* a minimum or maximum. If it is not, we have a saddle point, which you can think of as a multidimensional inflection point. Table 6.2 summarizes the conditions introduced for a minimum or maximum at x^* and y^* if $f_x(x, y) = 0$ and $f_y(x, y) = 0$.

Table 6.2: Second Order Conditions for an Extremum in a Two Variable Function

2nd Derivatives	Sign	Result
$f_{xx}(x, y)$	> 0	
$f_{yy}(x, y)$	> 0	Minimum
$f_{xx} f_{yy} - (f_{xy})^2$	> 0	(Convex)
$f_{xx}(x, y)$	< 0	
$f_{yy}(x, y)$	< 0	Maximum
$f_{xx} f_{yy} - (f_{xy})^2$	> 0	(Concave)
$f_{xx} f_{yy} - (f_{xy})^2$	< 0	Neither

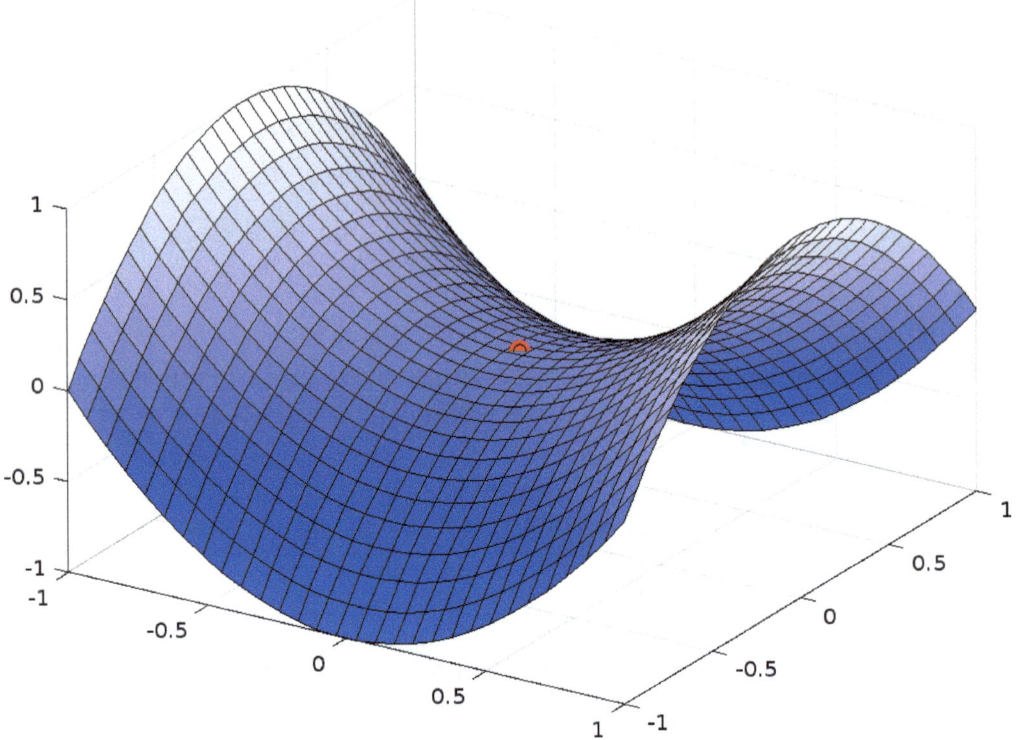

Figure 6.3: Saddle point.

But where does that saddle point condition come from? Actually, that condition and the conditions on f_{xx} and f_{yy} all come from the *total* derivative. Let's take the second total derivative of a general function, $z = f(x, y)$, and see how that is related to everything:

$$
\begin{aligned}
z &= f(x, y) \\
dz &= f_x(x, y)dx + f_y(x, y)dy.
\end{aligned}
$$

There's our first total derivative. Taking the second total derivative we have

$$
d^2 z = f_{xx}(x, y)dx^2 + 2f_{xy}(x, y)dxdy + f_{yy}(x, y)dy^2.
$$

To determine whether the function is convex or concave, we need to know the sign of the total derivative. Moreover, here it matches nicely with what we saw in the single variable case. A positive second total derivative means the function is convex and a negative value means concave. That is, if $d^2 z < 0$ at some (x^*, y^*), which solves the first-order conditions, we have a maximum; and if $d^2 z > 0$ at some (x^*, y^*), which solves the first-order conditions, we have a minimum. What we need to do is figure out the sign of the second total derivative. Doing so gives us those conditions for a minimum or maximum that we previously introduced.

We can see exactly where the conditions on the second partials listed in the table

come from through some algebra work. Starting with the second total derivative we add
and subtract the following term:

$$\frac{(f_{xy}(x,y))^2}{f_{xx}(x,y)}dy^2$$

to get

$$d^2z \;=\; f_{xx}(x,y)dx^2 + 2f_{xy}(x,y)dxdy + \frac{(f_{xy}(x,y))^2}{f_{xx}(x,y)}dy^2 + f_{yy}(x,y)dy^2 - \frac{(f_{xy}(x,y))^2}{f_{xx}(x,y)}dy^2.$$

Why do that? That will allow us to complete the square. First, factor out $f_{xx}(x,y)$ from
the first three terms:

$$d^2z \;=\; f_{xx}(x,y)\left(dx^2 + 2\frac{f_{xy}(x,y)}{f_{xx}(x,y)}dxdy + \frac{(f_{xy}(x,y))^2}{(f_{xx}(x,y))^2}dy^2\right) + f_{yy}(x,y)dy^2 - \frac{(f_{xy}(x,y))^2}{f_{xx}(x,y)}dy^2.$$

Factor dy^2 out from the last two terms and put dy^2 on the right-side:P

$$d^2z \;=\; f_{xx}(x,y)\left(dx^2 + 2\frac{f_{xy}(x,y)}{f_{xx}(x,y)}dxdy + \frac{(f_{xy}(x,y))^2}{(f_{xx}(x,y))^2}dy^2\right) + \left(f_{yy}(x,y) - \frac{(f_{xy}(x,y))^2}{f_{xx}(x,y)}\right)dy^2.$$

Now, the three terms inside the first set of parentheses are a squared term, and we can
write it as follows:

$$d^2z \;=\; f_{xx}(x,y)\left(dx + \frac{f_{xy}(x,y)}{f_{xx}(x,y)}dy\right)^2 + \left(f_{yy}(x,y) - \frac{(f_{xy}(x,y))^2}{f_{xx}(x,y)}\right)dy^2.$$

Finally, combine the two terms in the second set of parentheses by creating a common
denominator:

$$d^2z \;=\; f_{xx}(x,y)\left(dx + \frac{f_{xy}(x,y)}{f_{xx}(x,y)}dy\right)^2 + \left(\frac{f_{xx}(x,y)f_{yy}(x,y) - (f_{xy}(x,y))^2}{f_{xx}(x,y)}\right)dy^2.$$

There are two terms in the last line, and we were able to complete the square for the
first term that begins with $f_{xx}(x,y)$. Why do all that? We can now better identify the
sign of dz^2 based on the second partial derivatives. Let's break it down. Starting with
the first term

$$f_{xx}(x,y)\left(dx + \frac{f_{xy}(x,y)}{f_{xx}(x,y)}dy\right)^2 \gtrless 0.$$

Since the portion of the expression in parentheses is squared, we know that can never be
negative. Thus, the entire sign of the first term is completely dependent on the sign of
$f_{xx}(x,y)$, the second partial derivative on x. That was easy and the reason we did that
complete the square trick. The second term is a little more involved, but not much:

$$\left(\frac{f_{xx}(x,y)f_{yy}(x,y) - (f_{xy}(x,y))^2}{f_{xx}(x,y)}\right)dy^2 \gtrless 0.$$

dy^2 is also never negative, so look at the term inside the large parentheses. In order for $dz^2 > 0$, we know that $f_{xx}(x, y) > 0$ must hold. That term appears in the denominator, so for the entire second term in parentheses to be positive, the numerator must also be positive:

$$f_{xx}(x, y)f_{yy}(x, y) - (f_{xy}(x, y))^2 > 0.$$

Hey! That looks familiar. In order for that to be positive, we must have that $f_{xx}(x, y)f_{yy}(x, y) > (f_{xy}(x, y))^2$, which also requires that $f_{yy}(x, y) > 0$. So, for a convex function and $dz^2 > 0$, we must have that $f_{xx}(x, y) > 0$, $f_{yy}(x, y) > 0$, and $f_{xx}(x, y)f_{yy}(x, y) - (f_{xy}(x, y))^2 > 0$. We leave working out the analogous logic for a maximum as an exercise.

6.5.2   Examples

Let's start with a pair of numerical examples.

Example 1: Consider the function

$$z = 7x - 2x^2 + 8y - 5y^2 + 10.$$

Our task is to find the extreme points and determine whether they are minima, maxima, or neither. Taking the first-order conditions we have:

$$\frac{\partial z}{\partial x} = 7 - 4x = 0$$

$$\frac{\partial z}{\partial y} = 8 - 10y = 0.$$

The solution is $x^* = 7/4$ and $y^* = 4/5$, but we do not yet know if that is a minimum or maximum or neither. Taking the full set of second partial derivatives we have

$$\frac{\partial^2 z}{\partial x^2} = -4$$

$$\frac{\partial^2 z}{\partial y^2} = -10$$

$$\frac{\partial^2 z}{\partial x \partial y} = 0.$$

The two direct partial derivatives are both negative, suggesting a concave function. Checking the saddle point condition we have:

$$f_{xx}(x, y)f_{yy}(x, y) - (f_{xy}(x, y))^2 \gtrless 0$$
$$(-4)(-10) - (0)^2 \gtrless 0$$
$$40 > 0.$$

The condition is positive, which means we do indeed have an extreme point and not a saddle point, so our solution is a maximum and the function is concave.

Example 2: Consider the function:

$$z = \frac{x^2}{a^2} - \frac{y^2}{b^2}.$$

Again, find the extreme points and determine whether they are minima, maxima, or neither. Taking the first-order conditions we have

$$\frac{\partial z}{\partial x} = \frac{2x}{a^2} = 0,$$
$$\frac{\partial z}{\partial y} = -\frac{2y}{b^2} = 0.$$

The solution to the first-order conditions is $x^* = 0$ and $y^* = 0$. Taking the full set of second partial derivatives, we have

$$\frac{\partial^2 z}{\partial x^2} = \frac{2}{a^2} > 0,$$
$$\frac{\partial^2 z}{\partial y^2} = -\frac{2}{b^2} < 0,$$
$$\frac{\partial^2 z}{\partial x \partial y} = 0.$$

The two direct partial derivatives are contradictory, meaning we have neither a minimum nor a maximum. That is also confirmed via the saddle point condition:

$$f_{xx}(x, y) f_{yy}(x, y) - (f_{xy}(x, y))^2 \gtrless 0$$
$$\left(\frac{2}{a^2}\right)\left(\frac{-2}{b^2}\right) - (0)^2 \gtrless 0$$
$$\frac{-4}{a^2 b^2} < 0.$$

These results tell us is that the function is concave in some directions and convex in others. Indeed, that is the equation for a *hyperbolic parabola*, which produces the saddle shape.

Example 3: Profit maximization problem:

$$max_{K,L} \Pi = AK^\alpha L^\beta - rK - wL,$$

where α and β are parameters, assumed greater than zero but less than one, w is the wage rate for labor, L, and r is the rental price of capital, K. Let's find the solution to this problem and verify that we have a maximum by checking the second–order conditions:

$$\frac{\partial \Pi}{\partial K} = \alpha A K^{\alpha-1} L^{\beta} - r = 0$$

$$\frac{\partial \Pi}{\partial L} = \beta A K^{\alpha} L^{\beta-1} - w = 0.$$

After a bit of tedious algebra, the solutions are

$$K^* = \left[A \left(\frac{\alpha}{r}\right)^{1-\beta} \left(\frac{\beta}{w}\right)^{\beta} \right]^{1/(1-\alpha-\beta)}$$

$$L^* = \left[A \left(\frac{\alpha}{r}\right)^{\alpha} \left(\frac{\beta}{w}\right)^{1-\alpha} \right]^{1/(1-\alpha-\beta)}.$$

The second partial derivatives are

$$\frac{\partial^2 \Pi}{\partial K^2} = \alpha(\alpha-1) A K^{\alpha-2} L^{\beta}$$

$$\frac{\partial^2 \Pi}{\partial L^2} = \beta(\beta-1) A K^{\alpha} L^{\beta-2}$$

$$\frac{\partial^2 \Pi}{\partial K \partial L} = \alpha\beta A K^{\alpha-1} L^{\beta-1}.$$

The second partial derivatives on K and L are always negative. We can see that because K and L must be positive as negative values make no sense. We also know that α and β are between zero and one, so that must mean that $\alpha(\alpha-1) < 0$ and $\beta(\beta-1) < 0$, making them both negative. That will, of course, remain true when we evaluate at our optimal solutions, K^* or L^*, where we want to know if that they indeed maximize profits:

$$\frac{\partial^2 \Pi}{\partial K^2} = \alpha(\alpha-1) A \left\{ \left[A \left(\frac{\alpha}{r}\right)^{1-\beta} \left(\frac{\beta}{w}\right)^{\beta} \right]^{\frac{1}{1-\alpha-\beta}} \right\}^{\alpha-2} \left\{ \left[A \left(\frac{\alpha}{r}\right)^{\alpha} \left(\frac{\beta}{w}\right)^{1-\alpha} \right]^{\frac{1}{1-\alpha-\beta}} \right\}^{\beta} < 0$$

$$\frac{\partial^2 \Pi}{\partial L^2} = \beta(\beta-1) A \left\{ \left[A \left(\frac{\alpha}{r}\right)^{1-\beta} \left(\frac{\beta}{w}\right)^{\beta} \right]^{\frac{1}{1-\alpha-\beta}} \right\}^{\alpha} \left\{ \left[A \left(\frac{\alpha}{r}\right)^{\alpha} \left(\frac{\beta}{w}\right)^{1-\alpha} \right]^{\frac{1}{1-\alpha-\beta}} \right\}^{\beta-2} < 0$$

$$\frac{\partial^2 \Pi}{\partial K \partial L} = \alpha\beta A K^{\alpha-1} L^{\beta-1}.$$

Again, as complicated as these look, the two terms inside the braces, $\{\cdot\}$, are K^* and L^*, which have to be positive. That leaves the sign dependent on the terms in front, $\alpha(\alpha-1)A$ and $\beta(\beta-1)A$. Both terms are clearly negative so long as α and β are both less than one, as we assumed at the start. Those signs are consistent with concavity and a maximum point, but we need to check the final condition:

$$\left(\alpha(\alpha-1)AK^{\alpha-2}L^{\beta}\right)\left(\beta(\beta-1)AK^{\alpha}L^{\beta-2}\right)-\left(\alpha\beta AK^{\alpha-1}L^{\beta-1}\right)^2 \gtrless 0$$

$$\alpha\beta(1-\alpha-\beta)\left(AK^{\alpha-1}L^{\beta-1}\right)^2 \gtrless 0.$$

Some algebra gets us to that last line, which implies that the condition is indeed positive, as required for an extreme point, provided $\alpha+\beta<1$. Substituting in K^* and L^* does not change that because the term containing K and L is squared. The sign of the entire expression depends on $1-\alpha-\beta$.

6.6 Multivariate Calculus and the Chain Rule

6.6.1 Numerical Examples

Recall from Sections 2.4 and 2.5, when we first began working with derivatives and general functions, that the chain rule applies to situations in which we have one function nested inside another. The general representation we used then was

$$\begin{aligned} y &= f\left(g\left(x\right)\right) \\ \frac{dy}{dx} &= f'\left(g\left(x\right)\right)g'\left(x\right). \end{aligned}$$

We encountered this situation in the single–variable context in Section 4.3.2 when we substituted the budget constraint into a general utility function. Frequently, we need to take derivatives of multivariate general functions in which the same variable may appear in more than one argument of the function. For example, consider taking the derivative with respect to x of the following:

$$U \;=\; U(x, y, z(x))$$

Notice that x appears in the first argument *and* in the third argument, where z is a function of x. How do we take that derivative, or at least, how do we represent it? First, consider the function again, but where z is not a function of x such that we would have

$$U \;=\; (x, y, z)$$
$$\frac{\partial U}{\partial x} \;=\; U_x(x, y, z).$$

That's straightforward since the left– and right–hand sides are two different representations of the same thing. But now when z is a function of x, that means we also have to apply the chain rule such that we treat $U(x, y, z)$ as the outer function and $z(x)$ as the inner function, as follows:

$$U \;=\; U(x, y, z(x))$$
$$\frac{\partial U}{\partial x} \;=\; U_x(x, y, z(x)) + U_z(x, y, z(x))\frac{\partial z}{\partial x}.$$

x appears twice, and we need to account for how U changes with both components. The equation showing $\frac{\partial U}{\partial x}$ has the partial derivative with respect to the first argument x as before, but now we *add* the derivative with respect to the third argument, z, and apply the chain rule, multiplying by the derivative of the inner function $z(x)$. That line reads as the marginal change in U with a marginal change in x is equal to the partial derivative of U as it depends directly on x plus the slope of U with respect to x times the change in z that comes from a marginal change in x.

Let's do some examples to see how this general representation maps to more specific functions. Consider the following specific function, where $z(x) = (3x + 2)^{1/2}$:

$$U \;=\; U(x, y, z(x)) = x^2 + y^2 + (3x + 2)^{1/2}$$
$$U \;=\; U(x, y, z(x)) = x^2 + y^2 + z(x)$$
$$\frac{\partial U}{\partial x} \;=\; 2x + \frac{\partial z(x)}{\partial x}$$
$$\frac{\partial U}{\partial x} \;=\; \overbrace{2x}^{U_x} + \overbrace{\frac{3}{2}(3x + 2)}^{U_z \frac{\partial z}{\partial x}}.$$

We know how to take the partial derivative of the specific function, but notice how the end result is U_x plus $U_z(x)\frac{\partial z}{\partial x}$. We do not have a specific function, we can still represent that result by appropriately using the chain rule to represent that derivative.

What if the underlying specific function isn't nicely additive like the previous example? Suppose, instead we had:

$$U = U(x, y, z(x)) = x^2 y^2 (3x + 2)^{1/2}$$
$$U = U(x, y, z(x)) = x^2 y^2 z(x).$$

The same rule applies. We would still be able to represent that specific function as the same general function with the same result:

$$U = U(x, y, z(x))$$
$$\frac{\partial U}{\partial x} = U_x(x, y, z(x)) + U_z(x, y, z(x))\frac{\partial z}{\partial x}.$$

To see this, take the derivative of the specific function:

$$\frac{\partial U}{\partial x} = 2xy^2(3x + 2)^{1/2} + x^2 y^2 \left(\frac{3}{2}\right)(3x + 2)^{-1/2}$$

$$\frac{\partial U}{\partial x} = \overbrace{2xy^2(3x + 2)^{1/2}}^{U_x} + \overbrace{\left(\frac{3}{2}\right)x^2 y^2(3x + 2)^{-1/2}}^{U_z \frac{\partial z}{\partial x}}$$

$$= U_x(x, y, z(x)) + U_z(x, y, z(x))\frac{\partial z}{\partial x}.$$

6.6.2 Modeling Examples

Example 1

Let's do an example using a model we will introduce more fully in the next chapter. Suppose we have a person, Prince Zuko, who is considering how much to consume today versus how much to consume in the future and, in effect, how much to save or borrow to achieve his optimal consumption plan. Zuko's utility is given by the general function $U[c_1, c_2]$, where c_1 represents consumption today and c_2 is future consumption. Note the use of square brackets $[\cdot]$ here is merely to help distinguish between the arguments of the function $U[c_1, c_2]$ and actual quantities that we will see later. We could have used $U(c_1, c_2)$ as we usually do, but it may be harder to see the distinction in what follows.

Consumption now and in the future are related by the following constraint: $c_2 = y_2 + (1 + r)(y_1 - c_1)$. Zuko gets income in periods 1 and 2, as represented by y_1 and y_2. If Zuko chooses a consumption level today, c_1, that is less than his income today, he saves money and earns interest at the rate of r (r is the interest rate). If Zuko spends more today than he has current income such that $y_1 < c_1$, then he needs to borrow. His future consumption, c_2, will be reduced by repaying the loan with interest at the same rate, r.

If we substitute that constraint into our multivariate utility function, $U(c_1, c_2)$, we would have:

$$U[c_1, c_2] = U[c_1, y_2 + (1 + r)(y_1 - c_1)].$$

Zuko wants to maximize his utility, so the first-order condition would be:

$$\frac{dU[c_1, c_2]}{dc_1} = \overbrace{U_{c_1}[c_1, y_2 + (1+r)(y_1 - c_1)]}^{U_{c_1}} + \overbrace{U_{c_2}[c_1, y_2 + (1+r)(y_1 - c_1)](-1)(1+r)}^{U_{c_2}\frac{\partial c_2}{\partial c_1}}$$

$$\frac{dU[c_1, c_2]}{dc_1} = \overbrace{U_{c_1}[c_1, c_2)]}^{U_{c_1}} + \overbrace{U_{c_2}[c_1, c_2](-1)(1+r)}^{U_{c_2}\frac{\partial c_2}{\partial c_1}}.$$

The first term is the partial derivative of U with respect to the first argument, c_1. The second term is the partial derivative with respect to the second argument, c_2, then applying chain rule, the partial derivative of c_2 with respect to c_1, which is $-(1+r)$. Note the use of the square brackets $[\cdot]$ to help distinguish the function from values like $(y_1 - c1)$ and $(1+r)$. In the second line, we replaced $y_2 + (1+r)(y_1 - c_1)$ with c_2, returning to the original utility function before we substituted in the constraint. That just helps cut down on the notation, making it easier to read.

Example 2

Here's one more example using a general economy–wide production function given by:

$$Y(t) = A(t)F[K(t), L(t)].$$

Here the total production of the economy at time t is $Y(t)$. $Y(t)$ is a determined by productivity A times the general production function $F[\cdot]$, which has two arguments: (1) K which is physical capital, and (2) L which is labor. Note that Y, A, K, and L are all functions of time, t. Let's take the total derivative of this function with respect to t to see how Y changes over time:

$$\frac{dY(t)}{dt} = \frac{dA(t)}{dt}F[K(t), L(t)] + AF_K[K(t), L(t)]\frac{dK}{dt} + AF_L[K(t), L(t)]\frac{dL}{dt}.$$

In the first term on the right-hand side, A is outside the general function, so we just take that derivative of $A(t)$ and represent it with the standard derivative notation. However, when we get to the next item that is a function of t it is $K(t)$, which is inside the general, unspecified production funciton. We need to apply the chain rule by first accounting for how $F(\cdot)$ changes with K, then multiplying by the derivative of $K(t)$ with respect to t to account for how K changes with t. The same logic applies to $L(t)$. We apply the chain rule by first accounting for how $F(\cdot)$ changes with L, then multiplying by the derivative of $L(t)$ with respect to t to account for how L changes with t.

Note how, we add each component separately. We can do that because it is the same logic that makes the total derivative the sum of the partial derivatives. We add up the individual changes, the slopes times the differentials, of multiple variables to see how the overall total function changes.

Finally, the derivative is a total derivative as opposed to a differential as the right-hand side consists of all the derivatives of the input variables with respect to time, t.

The total differential would be:

$$dY(t) \;=\; F[K(t), L(t)]dA(t) + AF_K[K(t), L(t)]dK(t) + AF_L[K(t), L(t)]dL(t).$$

Now, we are looking at the change, the differential, in Y, with respect to changes, differentials, in all the variables. The two concepts, total derivative and total differential, are distinct but so close that terms are used interchangeably even when that is not technically correct.

6.7 Exercises

6.7.1 Quick Check Answers

a) $\frac{\partial z}{\partial x} = x^{-1/2}$, $\frac{\partial z}{\partial y} = 4y^3$

b) $\frac{\partial z}{\partial x} = x^{-1/2}y^4$, $\frac{\partial z}{\partial y} = 8x^{1/2}y^3$

c) $\frac{\partial^2 z}{\partial x \partial y} = 0$

d) $\frac{\partial^2 z}{\partial x \partial y} = 4x^{-1/2}y^3$

e) $\frac{\partial Y}{\partial A} = K^{1/3}L^{2/3}$, $\frac{\partial^2 Y}{\partial A^2} = 0$, $\frac{\partial^2 Y}{\partial A \partial K} = (1/3)K^{-2/3}L^{2/3}$, $\frac{\partial^2 Y}{\partial A \partial L} = (2/3)K^{1/3}L^{-1/3}$;

f) $\frac{\partial^2 U(x,y)}{\partial x \partial y} = -\frac{2x}{y^2} - \frac{2y}{2x} < 0 \Rightarrow$ Substitutes

g) $\frac{\partial^2 U(x,y)}{\partial x \partial y} = 1 > 0 \Rightarrow$ Complements;

h) $dz = x^{-1/2}dx + 4y^3 dy$

i) $dz = x^{-1/2}y^4 dx + 8x^{1/2}y^3 dy$

j) $x^* = 0$, $y^* = 0$. $\frac{\partial^2 z}{\partial x^2} = 2 > 0$, $\frac{\partial^2 z}{\partial y^2} = 2 > 0$, $\frac{\partial^2 z}{\partial x \partial y} = 1 \Rightarrow \frac{\partial^2 z}{\partial x^2}\frac{\partial^2 z}{\partial y^2} - (frac\partial^2 z \partial x \partial y) = (2)(2) - 1 > 0$. All second–order conditions for a minimum hold.

6.7.2 Practice Problems

1. **For the following find:**

    i. All the first partial derivatives.

    ii. All the second partial derivatives. **Do not forget to also get the cross-derivative.**

    iii. The first total differential (derivative).

iv. The second total differential (derivative).

(a) $z = x^2 + 2xy + y^2$

(b) $z = \ln(x_1 + x_2^2 + x_3^3)$

(c) $z = e^x \ln y$

(d) $z = f(x, y)$

(e) $z = \frac{x}{x^2 + y^2}$

(f) $Y = AK^{1/2}L^{1/2}$ (Take derivatives on K and L only, not A).

(g) $Y = AK^\alpha L^{1-\alpha}$ (Take derivatives on K and L only, not A. α is a parameter, and you do not need to take derivatives on it.)

(h) $Y = AF(K, L)$

(i) $U(x_1, x_2) = x_1 + x_2$

(j) $U(x_1, x_2) = \ln(x_1) + \ln(x_2)$

(k) $U(x_1, x_2) = x_1 + \alpha \ln(x_2)$

2. Find the second total derivative of the example from 6.3.1 $Y = AK^{1/3}L^{2/3}$.

3. Find the MRS for the utility functions in question 1: i, j, and k.

4. Starting from our quadratic form of the second total derivative,

$$d^2z \;=\; f_{xx}(x, y)\left(dx + \frac{f_{xy}(x, y)}{f_{xx}(x, y)}dy\right)^2 + \left(\frac{f_{xx}(x, y)f_{yy}(x, y) - (f_{xy}(x, y))^2}{f_{xx}(x, y)}\right)dy^2,$$

derive the conditions on the second partial derivatives for a concave function such that $dz^2 < 0$.

5. Solve the maximization problem and check the second order conditions: $\max_{x,y} z = (10 - x)x + (10 - y)y + xy$.

6. Solve the maximization problem and check the second order conditions: $\max_{x,y} z = (10 - x)x + (y - 10)y + xy$.

7. Solve the utility maximizaiton $U(x_1, x_2) = \ln(x_1) + \ln(x_2)$ s.t. $W = p_1 x_1 + p_2 x_2$ and check all the second-order conditions.

8. Solve the utility maximization problem and check the second order conditions.: $U(x_1, x_2) = x^2 + y^2$ subject to $W = p_x x + p y y$.

9. Solve the utility maximization $U(x_1, x_2) = x_1 + \alpha \ln(x_2)$ s.t. $W = p_1 x_1 + p_2 x_2$ and check all the second-order conditions.

10. Take the derivative with respect to x of the following general functions:

(a) $F(x, y)$

(b) $F(x, y(x))$

(c) $F(x, x^2)$

(d) $F(x, y, z(x))$

(e) $F(x, ln(x))$

6.8 Math Appendix

Here, we list some key math definitions, theorems, and formulae in formal math for your reference.

Cobb–Douglas production function: The general form of a Cobb–Douglas productions function for a firm (or economy) producing Y using N inputs $x_1, x_2, ...x_N$ is:

$$Y \quad = \quad f(x_1, x_2, ..., x_i, ...x_N) = x_1^{\alpha_1} x_2^{\alpha_2} \cdots x_N^{\alpha_N} = \Pi_{i=1}^{N} x_i^{\alpha_i}.$$

If the α exponent is less than one, then the input exhibits diminishing marginal returns. If the α_i exponents sum to one, the function overall exhibits contant returns to scale and is homogenous of degree 1. In addition, the elasticity of substitution between inputs in a Cobb–Douglas production function is exactly 1, and it is constant. This functional form is also common for utility functions because its mathematical properties make it easy to work with and understand.

Cross–derivative: The second derivative of a multivariate function, $y = f(x_1, x_2, \ldots, x_n)$, that results from taking the partial derivative with respect to one right-hand side variable, some x_i, first, and then the derivative of the first partial derivative with respect to a different right-hand side variable, some $x_j \neq x_i$. The result is the cross-derivative written as: $\frac{\partial^2 y}{\partial x_i \partial x_j}$ or $f_{x_i, x_j}(x_1, x_2, \ldots, x_n)$.

Indifference curves: A set of combinations or "bundles" of goods that yield the same utility value to the consumer. Given a choice between two bundles of goods on the same indifference curve, the consumer is said to be "indifferent" between the choices. Formally, an indifference curve is a level set of all $x = (x_1, x_2, \ldots, x_N)$ such that $U(x) = \bar{U}$, where $\bar{U}$ is a fixed utility level.

Level sets: Combinations of the right-hand side variables that yield the exact same value of the function. Indifference curves are one applied example of the mathematical concept of level sets. For a production function, the level sets are often referred to as isoquants. Level sets are used to help us represent three or higher dimensional functions in two dimensions. For example with a topological map that shows mountain and valley highs and lows with lines representing specific heights above sea level, those lines are level sets. Weather maps that show temperatures across an area often have lines representing specific temperatures and those are also level sets.

Marginal product of labor (MPL): The additional amount of output generated by a marginal increase in the amount of labor used in production. For a general production function with labor, L, as an input, $Y = F(L, X)$, and X represents the set of other inputs (e.g., physical capital, raw materials, human capital, energy, etc.) and the marginal product of labor is $MPL = \partial Y/\partial L = F(L, X)/\partial L$.

Marginal rate of substitution (MRS): The rate at which a consumer is willing to give away one good in exchange for another good such that they are just as happy before and after the trade, (i.e. utility does not change). For a general utility function over N goods, the MRS between goods i and j would be given by

$$U \quad = \quad U(x_1, x_2, ..., x_i, ...x_N), \text{ for } i \in [1, N]$$
$$MRS_{ij} \quad = \quad -\frac{\partial U/\partial x_j}{\partial U \partial x_i}.$$

Multivariate function: A function whose input variables (or right-hand side variables or independent variables) consists of more than one variable. $y = f(x_1, x_2, ...x_N)$ is a multivariate function provided $N > 1$.

Partial derivative: The partial derivative of a multivariate function is the derivative of the function with respect to a single right-hand side variable holding all other right-hand side variables constant. For the function z=f(x,y), the partial derivative of z with respect to x is defined as

$$\frac{\partial z}{\partial x} \quad = \quad f_x = \frac{\partial f}{\partial x} = \lim_{h \to 0} \frac{f(x+h, y) - f(x, y)}{h}.$$

Single–variable function: A function whose input variables (or right–hand side variables or independent variables) consists of only one variable. $y = f(x)$ is a single–variable function provided x represents only one variable.

Total derivative or **total differential**: The total differential of a function is the sum of the partial derivatives of all the independent variables times the differential of the respective variable. The total differential of $y = f(x_1, x_2, ..., x_N)$ is

$$dy \quad = \quad \frac{\partial f}{\partial x_1} dx_1 + \frac{\partial f}{\partial x_2} dx_2 + \cdots + \frac{\partial f}{\partial x_N} dx_N.$$

The total derivative of a function is the sum of the partial derivatives of all the independent variables times the derivatives of the independent variables with respect to some other variable. The total derivative of $y = f(x_1, x_2, ..., x_N)$ is

$$\frac{dy}{dt} = \frac{\partial f}{\partial x_1}\frac{dx_1}{dt} + \frac{\partial f}{\partial x_2}\frac{dx_2}{dt} + \cdots + \frac{\partial f}{\partial x_N}\frac{dx_N}{dt}.$$

Young's theorem: For a twice–differentiable function, $y = f(x_1, x_2, ..., x_n)$, the order in which partial derivatives are taken does not affect the resulting cross-derivatives. That is, $\frac{\partial^2 y}{\partial x_i \partial x_j} = \frac{\partial^2 y}{\partial x_j \partial x_i}$.

7 Lagrangians: Optimization With Constraints. Again.

As mentioned repeatedly before, our models frequently involve maximizing or minimizing a function subject to a constraint. There are two methods for dealing with a constraint when optimizing a function: (1) Substitute the constraint into the objective function (the function you are maximizing or minimizing), the method we explored in Chapter 4, or (2) use a *Lagrangian*. We now turn to the latter method, which is a little trickier, but reveals more information tand economic intuition in our models.[8]

The last chapter was long on introducing new math terms and tools for multivariate calculus. In contrast, this chapter is focused on one tool alone, the Lagrangian method for solving constrained optimization problems.

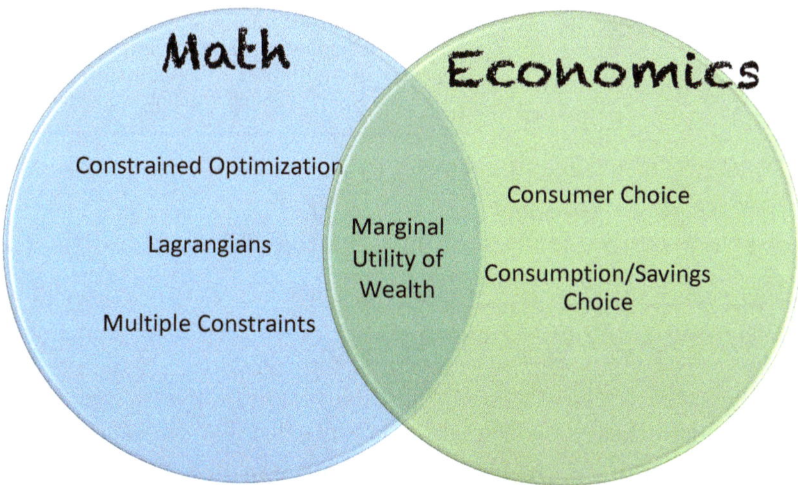

---

[8]Named for Italian mathematician Joseph–Louis Lagrange (1736-1813), who analyzed oscillations of the moon and contributed immensely to the mathematics used in dynamic analysis.

7.1 Introduction to the Lagrangian Method

The second method for solving a constrained optimization problem is to set up a **Lagrangian**. In a Lagrangian, we add the constraint to the objective function, but multiply the constraint by a new variable called the Lagrangian multiplier, commonly represented by the lower-case Greek letter lambda, λ. The workings of a Lagrangian are probably easiest to understand through an example, and we will use the same example from Chapter 4 when we did the problem by substituting the constraint in the objective function.

You may or may not recall the example problem was a two–good utility maximization problem, with Jiji choosing donuts, x, and fish, y. Here is the basic problem again:

$$\max_{x,y} U = x^{1/2} + y^{1/2}$$

$$s.t. \quad 4x + 2y = 32,$$

Definition

Lagrangian method: The Lagrangian method is a calculus technique for solving constrained optimization problems. It was developed by French-Italian mathematician and physicist Joseph-Louis Lagrange in the 18th century to describe the motion of mechanical systems, including oscillations of the moon. A Lagrangian consists of the objective function to be minimized or maximized, added to a series of products of multipliers times constraints, where there is a unique multiplier for each constraint.

where $p_x = 4$ and $p_y = 2$ are the prices of donuts and fish, respectively. In Chapter 4, we solved the budget constraint for y and substituted it into the objective function to get

$$\max_{x} U = x^{1/2} + (16 - 2x)^{1/2}.$$

This substitution did several things. First, it accounts for the constraint by reflecting y as a function of x as determined by the budget constraint. Second, it reduces the number of choice variables in the problem from two (x and y) to one (x). Third, since we only have one choice variable that means we could solve it using single– variable calculus methods.

With a Lagrangian, we combine the objective function and constraint in a different manner, *increase* the number of variables, and therefore need to use multivariate calculus methods.

So, rewriting this problem as a Lagrangian, we have the following:

$$\mathcal{L} = \overbrace{x^{1/2} + y^{1/2}}^{\text{Objective Function}} + \overbrace{\lambda}^{\text{Multiplier}} \overbrace{[32 - 4x - 2y]}^{\text{Constraint}}.$$

The script letter, $\mathcal{L}$, indicates that this representation is a Lagrangian. The Lagrangian is equal to the objective function (the function we are trying to maximize or minimize) plus λ times the constraint. The Greek letter lambda, λ, is the traditional variable for a Lagrangian multiplier (though others are used in other contexts). The multiplier, λ, has a particularly useful interpretation and will be discussed in more detail later. The addition of the Lagrangian multiplier here is the reason for the increase in the number of variables mentioned.

Notice that the constraint has been rearranged such that the constraint is equal to zero. You must always rearrange the constraint to be equal to zero. Doing so means that when we form the Lagrangian and add it to the objective function, we are effectively adding zero to the objective function and therefore not changing its value. In the next chapter, we explore this feature in more detail and show that either the Lagrangian multiplier or the constraint must be zero. Thus, because they are multiplied, the entire term added to the objective function always has a value of zero.

In the problem we wrote the constraint as: $32 - 4x - 2y = 0$. Alternatively, we could have written the constraint as $4x + 2y - 32 = 0$ to form the following Langrangian:

$$\mathcal{L} = \overbrace{x^{1/2} + y^{1/2}}^{\text{Objective Function}} + \overbrace{\lambda}^{\text{Multiplier}} \overbrace{[4x + 2y - 32]}^{\text{Constraint}}.$$

Both ways will ultimately yield the same result for the optimal choice for x^* and y^*. In that sense it does not matter how we enter the constraint, just so long as it is written equal to zero. Where the choice does make a difference is in the sign (positive or negative) of λ. In the first version, $32 - 4x - 2y$; λ will be positive, and if we use $4x + 2y - 32$, then λ will be negative. Both results are valid, but they will change the interpretation of λ. We will return to the interpretation of the Lagrangian multiplier and why it is positive or negative after we go through how to solve these problems for x^*, y^*, and λ^*.

Returning to the first version of the Lagrangian problem shown, notice that we are maximizing this utility function $x^{1/2} + y^{1/2}$ choosing two variables, x and y. Therefore we will need first-order conditions for both variables. Taking the partial derivatives of the Lagrangian with respect to x and y, respectively, we have:

$$\frac{\partial \mathcal{L}}{\partial x} = \overbrace{\frac{1}{2}x^{-1/2}}^{\text{Marginal benefit}} - \overbrace{\lambda(4)}^{\text{Marginal cost}} \quad ; \text{ and}$$

$$\frac{\partial \mathcal{L}}{\partial y} = \overbrace{\frac{1}{2}y^{-1/2}}^{\text{Marginal benefit}} - \overbrace{\lambda(2)}^{\text{Marginal cost}} \quad .$$

Both first-order conditions must be equal to zero for a maximum. Technically, here we have two equations and three unknowns: x, y, and λ. Therefore, we will need a third equation. The third equation comes from taking the derivative of $\mathcal{L}$ with respect to λ. Doing so reproduces the constraint:

$$\frac{\partial \mathcal{L}}{\partial \lambda} = 32 - 4x - 2y.$$

Since it reproduces the constraint, which we had from before, this step is often omitted, but we do need the constraint to solve the problem. Our standard first-order conditions for a maximum point hold here. The two FOCs on x and y must be equal to zero for a maximum. Furthermore, the budget constraint must also be equal to zero by assumption. We can write our three equations as

$$\frac{\partial \mathcal{L}}{\partial x} = \overbrace{\frac{1}{2}x^{-1/2}}^{\text{Marginal Benefit}} - \overbrace{\lambda\,(4)}^{\text{Marginal Cost}} = 0$$

$$\frac{\partial \mathcal{L}}{\partial y} = \overbrace{\frac{1}{2}y^{-1/2}}^{\text{Marginal Benefit}} - \overbrace{\lambda\,(2)}^{\text{Marginal Cost}} = 0$$

$$\frac{\partial \mathcal{L}}{\partial \lambda} = 32 - 4x - 2y = 0.$$

Look at the derivatives, $\frac{\partial \mathcal{L}}{\partial x}$ and $\frac{\partial \mathcal{L}}{\partial y}$. They, too, have the marginal benefit and marginal cost components. In the first, the marginal benefit of consuming a little bit more of x is $\frac{1}{2}x^{-1/2}$, that is, the marginal utility of x (donuts). Similarly, in the second equation, $\frac{1}{2}y^{-1/2}$ is the marginal utility of fish. The marginal cost components, however, are expressed in terms of the multiplier and the price of the good. If the price of one of the goods increases, it means the marginal cost will increase. Before we explore the meaning of the λ, let's solve this problem using these equations. Rewrite the two first-order conditions such that we have marginal benefits equals marginal costs, as follows:

$$\frac{1}{2}x^{-1/2} - \lambda\,(4) = 0$$
$$\frac{1}{2}y^{-1/2} - \lambda\,(2) = 0.$$

That becomes

$$\frac{1}{2}x^{-1/2} = 4\lambda$$
$$\frac{1}{2}y^{-1/2} = 2\lambda.$$

Divide the first condition by the second condition:

$$\frac{\frac{1}{2}x^{-1/2}}{\frac{1}{2}y^{-1/2}} = \frac{4\lambda}{2\lambda}$$

$$\frac{x^{-1/2}}{y^{-1/2}} = 2.$$

The lambda divides out, and then solving for y in terms of x we get

$$x^{-1/2} = 2y^{-1/2}$$
$$2x^{1/2} = \frac{1}{2}y^{1/2}$$
$$4x = y.$$

To solve, we need to use the third condition, the budget constraint. Replace y in the budget constraint with $4x$ and we have

$$32 - 4x - 2y = 0$$
$$32 - 4x - 2(4x) = 0$$
$$32 - 4x - 8x = 0$$
$$32 - 12x = 0$$
$$x = \frac{32}{12}$$
$$x^* = \frac{8}{3}.$$

This result is the same as what we got in Chapter 4. Now that we have x^* we can find y^*. We found that $y = 4x$, so $y^* = 4(8/3) = 32/3$. It's the same answer again.

What about lambda? We could use either of the first-order conditions to get a value for it. Let us use the condition on x:

$$\frac{1}{2}x^{-1/2} - \lambda(4) = 0$$
$$\lambda = \frac{1}{8}x^{-1/2} = \frac{1}{8x^{1/2}}.$$

We now know that $x^* = 8/3$, so

$$\lambda^* = \frac{1}{8(8/3)^{1/2}} = \frac{1}{16(2/3)^{1/2}}.$$

That does not look very intuitive. However, the economic interpretation is the *marginal utility of wealth*, which means the marginal increase in utility our agent would get from an infinitessimally small increase in his budget. Right now he has $32 dollars to spend. The multiplier tells us how fast utility will rise if we increase his wealth by a very small amount.

The more general and mathematical way of saying the same thing would be: *how much does the objective function change when we relax the constraint just slightly?* That is, rephrasing and putting it in bold/italitcs font so you know this point is important, ***the Lagrangian multiplier represents the marginal change in the value of the objective function with a marginal relaxation of the constraint.***

In our context, recall that the constraint is $32 = 4x + 2y$, which we could enter into the Lagrangian problem in one of two ways, either as $32 - 4x - 2y$ or $4x + 2y - 32$. What does a "relaxation of the constraint" mean here? A relaxation would mean increasing the wealth level, the \$32 Jiji has to spend, just slightly (marginally). If we gave Jiji just a little bit more cash, how much would utility rise? That is what the Lagrangian multiplier represents, the marginal utility of wealth in this context. If we use $32 - 4x - 2y$ for the constraint, recall that λ^* was positive and equal to $\frac{1}{16(2/3)^{1/2}}$. That means if we increased the \$32 to, say, \$32.01, Jiji's utility would increase at the rate of $\frac{1}{16(2/3)^{1/2}}$.

Marginal utility of wealth: The amount by which a consumer's utility increases because of a marginal increase in the consumer's level of wealth.

If, on the other hand, we entered the constraint as $4x + 2y - 32$, notice that the exogenous wealth level enters negatively here. A very small increase from $-\$32$ would be to $-\$31.99$, which is actually a *decrease* in wealth. Now when we solve for λ^* we would get $-\frac{1}{16(2/3)^{1/2}}$, indicating that a marginal increase in the exogenous variable would *decrease* utility at a rate of $-\frac{1}{16(2/3)^{1/2}}$, which makes sense. If we take money away, utility goes down.

Typically, we put the constraint into the objective function with the key exogenous variable (wealth=\$32 in this example) entering positively such that we can interpret the result as the marginal change in the value of the objective function that results from a marginal increase in the variable that relaxes the constraint. As you work through problems of these types in microeconomics, macroeco-

Given the following objective function and constraint, write out the Lagrangian optimization equation and find the first-order conditions. You do not need to solve:
a) $\max_{x,y} U = xy$ s.t. $W = 4x + 5y$
b) $\min_{K,L} rK + wL$ s.t. $(KL)^{0.5} = 100$.

nomics, and other advanced courses, eventually you will begin to see the patterns and learn to set these problems up this way. However, even if you set the Lagrangians up the other way, you will still get the correct answers for the choice variables, but remember the sign of the Lagrangian multiplier will change.

7.2   Another Example

We will do more examples that will further illustrate the meaning of λ. Consider a slightly more general form of the consumer problem of choosing between two goods, x and y, where the utility function is

$$U = \ln x + \ln y$$

and the budget constraint is

$$W = p_x x + p_y y,$$

where W is wealth (or income), p_x is the price of good x, and p_y is the price of good y. Thus, we have

$$\begin{aligned}
\max_{x,y} U &= \ln x + \ln y \\
s.t. \ W &= p_x x + p_y y.
\end{aligned}$$

In Lagrangian form, the problem becomes

$$\mathcal{L} = \overbrace{\ln x + \ln y}^{\text{Objective function}} + \lambda \overbrace{[W - p_x x - p_y y]}^{\text{Constraint}}.$$

The first-order conditions are

$$\frac{\partial \mathcal{L}}{\partial x} = \overbrace{\frac{1}{x}}^{\text{Marginal benefit}} - \overbrace{\lambda p_x}^{\text{Marginal cost}} = 0,$$

$$\frac{\partial \mathcal{L}}{\partial y} = \overbrace{\frac{1}{y}}^{\text{Marginal benefit}} - \overbrace{\lambda p_y}^{\text{Marginal cost}} = 0,$$

$$\frac{\partial \mathcal{L}}{\partial \lambda} = W - p_x x - p_y y = 0.$$

Combining the first two equations and solving for y in terms of x we have

$$\frac{1}{x} = \lambda p_x$$

$$\frac{1}{y} = \lambda p_y$$

$$\frac{1/x}{1/y} = \frac{p_x}{p_y}$$

$$\frac{y}{x} = \frac{p_x}{p_y}$$

$$y = \frac{p_x}{p_y}x.$$

Substituting into the budget constraint

$$W = p_x x + p_y y$$

$$W = p_x x + p_y \left(\frac{p_x}{p_y}x\right)$$

$$W = p_x x + p_x x$$

$$W = 2p_x x$$

$$x^* = \frac{W}{2p_x}.$$

Then, y^* is

$$W = p_x x + p_y y$$

$$W = p_x \left(\frac{W}{2p_x}\right) + p_y y$$

$$W = \frac{W}{2} + p_y y$$

$$p_y y = \frac{W}{2}$$

$$y^* = \frac{W}{2p_y}.$$

We can use either of the first-order conditions to find a value for λ:

$$\frac{1}{x} = \lambda p_x$$

$$\frac{1}{(W/2p_x)} = \lambda p_x$$

$$\lambda^* = \frac{2}{W}.$$

The marginal utility of wealth is $2/W$. To see this idea more clearly, let's go back to the original utility function,

$$U = \ln x + \ln y,$$

and replace the x and y with our solutions $x^* = \frac{W}{2p_x}$ and $y^* = \frac{W}{2p_y}$. Putting the optimized values into the objective function forms what we call a **value function**, which tells us the value of the objective function at the optimum based entirely on the exogenous variables, p_x, p_y, and W, in this case. In microeconomics we refer to this particular step as forming the **indirect utility function**.[9] The left-hand side is still utility, but it is determined by the optimal choices the consumer made given the wealth level and the prices, as follows:

$$U^* = V(p_x, p_y, W) = \ln\left(\frac{W}{2p_x}\right) + \ln\left(\frac{W}{2p_y}\right).$$

This equation is the indirect utility function. It is equal to the *optimized* value for utility given the prices of the two goods and the wealth level. Stated differently, given the exogenous variables (p_x, p_y, and W), the indirect utility function tells us the maximimum level utility obtainable. We call it *indirect* because the idea is the consumer derives utility (happiness) from consuming the actual goods, x and y, but not from the wealth level or the prices of the goods. Think back to the example of the previous section, with Jiji choosing donuts and fish. Eating fish and donuts is what makes Jiji happy. Jiji does not get utility directly from his wealth or from the prices, *but* Jiji's wealth level and the prices of donuts and fish affect Jiji's consumption levels. That is, the wealth and price levels *indirectly* determine Jiji's utility. We will see more of the indirect utility function in Chapter 9 on duality.

The marginal utility of wealth, W, can be found directly from this expression by taking the derivative of U with respect to W:[10]

$$\frac{dU}{dW} = \frac{1}{W/2p_x}\left(\frac{1}{2p_x}\right) + \frac{1}{W/2p_y}\left(\frac{1}{2p_y}\right)$$

$$= \frac{2p_x}{W}\left(\frac{1}{2p_x}\right) + \frac{2p_y}{W}\left(\frac{1}{2p_y}\right)$$

$$= \frac{1}{W} + \frac{1}{W} = \frac{2}{W}.$$

---

[9]We will see this function in more detail in Chapter 9.

[10]Note the use of the chain rule and how we take a derivative of the ln of a function.

Thus

$$\frac{dU}{dW} = \frac{2}{W} = \lambda^*,$$

which means that the added utility our consumer would get from an infinitessimally small increase in W (a relaxing of the budget constraint) is $\frac{2}{W}$, which was our solution for λ^*, the Lagrangian multiplier.

7.3   3 Good Example

In all of our consumer choice examples so far, the consumer has chosen between two goods with one constraint, the budget constraint. Before we introduced the Lagrangian, we discussed substituting the constraint into the objective function in Section 4.3. One of the consequences of doing so was reducing the number of choice variables by one. For example, using the same example from Section 4.3, Jiji's consumer choice problem, when he has utility over donuts, x, and fish, y, given by $U(x, y) = x^{1/2} + y^{1/2}$ subject to $W = 4x + 2y$ as the budget constraint. To solve that back in Chapter 4, we substituted the constraint into the objective function by solving for y in terms of x to get

$$\max_x U(x, y) = x^{1/2} + \left(\frac{W - 4x}{2}\right)^{1/2}.$$

Substituting the constraint into the objective function accomplished two things. First, it accounted for the constraint. Second, it reduced the number of choice variables from two, x and y, to just one, x. That meant we could use single–variable calculus techniques. That enabled us to avoid worrying about multivariate calculus and the notation that we learned in Chapter 5.

Now suppose we have a consumer with three choices. Even if we subtitute the constraint into the objective function by solving the budget constraint in terms of one of the goods, we would still be left with two goods in our objective function. That means we would have to use multivariate calculus to solve it. We could also use a Lagrangian and not substitute the constraint into the objective function at all. This method, too, relies on multivariate calculus to solve.

Let's do an example with three goods, and we will do it both ways: (1) substituting the constraint into the objective function and (2) using a Lagrangian.

Our consumer's name is Korra, and Korra likes nuts: earth nuts, (E), fire nuts (F), and water nuts (W). Korra has 100 copper pieces and the nuts are sold at her local grocery store at the prices of p_E, p_F, and p_W, respectively. Korra's utility maximization problem is

$$\max_{E,F,W} U = \ln E + \ln F + \ln W$$
$$s.t. \ \ 100 = p_E E + p_F F + p_W W.$$

7.3.1 Substitution Method

First, solve the budget constraint for one good. We could solve for any of the three goods to make the substitution, and here we will solve for W to get

$$W = \frac{100 - p_E E - p_F F}{p_W}.$$

Substituting that into Korra's utility function, we have

$$\max_{E,F} U = \ln E + \ln F + \ln\left(\frac{100 - p_E E - p_F F}{p_W}\right).$$

Observe that the number of choice variables reduced from three to two. Because there is more than one choice variable, we are still in the multivariate world. We need a first-order condition for each choice variable. Each first-order condition will be the partial derivative with respect to one of the choice variables, as follows:

$$\frac{\partial U}{\partial E} = \frac{1}{E} - \left(\frac{p_W}{100 - p_E E - p_F F}\right)\frac{p_E}{p_W} = 0$$
$$\frac{\partial U}{\partial F} = \frac{1}{F} - \left(\frac{p_W}{100 - p_E E - p_F F}\right)\frac{p_F}{p_W} = 0.$$

These first-order conditions must be equal to zero for a maximum. The first parts of the expressions, $1/E$ and $1/F$, are the marginal utilities of goods, E and F, respectively. The more complex right-hand portions are the costs in marginal utility of choosing more of E or F, which reduces the amount of W Korra can consume and therefore lowers her utility. Notice that, from our solution for W, the expressions in parentheses are both $1/W$. If we substitute that back into the first-order conditions, we can quickly solve for both E and F as expressions of W:

$$\frac{\partial U}{\partial E} = \frac{1}{E} - \left(\frac{1}{W}\right)\frac{p_E}{p_W} = 0$$

$$\frac{\partial U}{\partial F} = \frac{1}{F} - \left(\frac{1}{W}\right)\frac{p_F}{p_W} = 0$$

$$E = \frac{p_W}{p_E}W, \qquad F = \frac{p_W}{p_F}W.$$

Substituting these back into the budget constraint, we can solve for W^* and then for E^* and F^*:

$$E^* = \frac{100}{3p_E}, \qquad F^* = \frac{100}{3p_F}, \qquad W^* = \frac{100}{3p_W}.$$

The partial derivatives from the calculus problem produced two first-order conditions, one for each choice variable. But since we have three unknowns, E, F, and W, we need to use a third condition to solve, and that was the budget constraint we used in that last step.

7.3.2 Lagrangian Method

Now, let's do that exact same problem again using the Lagrange method. In Lagrangian form, the problem becomes:

$$\mathcal{L} = \overbrace{\ln E + \ln F + \ln W}^{\text{Objective function}} + \overbrace{\lambda\big[100 - p_E E - p_F F - p_W W\big]}^{\text{Constraint}}.$$

The first-order conditions are

$$\frac{\partial \mathcal{L}}{\partial E} = \overbrace{\frac{1}{E}}^{\text{Marginal benefit}} - \overbrace{\lambda p_E}^{\text{Marginal cost}} = 0,$$

$$\frac{\partial \mathcal{L}}{\partial F} = \overbrace{\frac{1}{F}}^{\text{Marginal benefit}} - \overbrace{\lambda p_F}^{\text{Marginal cost}} = 0,$$

$$\frac{\partial \mathcal{L}}{\partial W} = \overbrace{\frac{1}{W}}^{\text{Marginal benefit}} - \overbrace{\lambda p_W}^{\text{Marginal cost}} = 0,$$

$$\frac{\partial \mathcal{L}}{\partial \lambda} = 100 - p_E E - p_F F - p_W W = 0.$$

Notice again we have one first-order condition for each choice variable. Those derivatives are partial derivatives because we hold the other variables constant. The final condition is again the partial derivative of the Lagrangian with respect to the Lagrangian multiplier, and reproduces the constraint. We now have four equations to solve for four unknowns,: E, F, W, and λ.

The marginal benefit and marginal costs are labeled here to make it clear. The partial derivative of the utility function with respect to one of the choice variables yields its marginal utility (e.g., $1/E$.) The marginal cost of choosing each good is its price, the amount paid for each good, times the Lagrangian multiplier. The multiplier here, too, represents the marginal utility of wealth. When multipied by the price, it converts the price, which is in dollar or currency terms, into utility terms. *For each dollar spent, how much utility does that cost*, is what those negative terms in the first–order conditions represent. You can read the first order condition on E as the marginal benefit is the marginal utility to Korra of consuming more earth nuts while the price paid for earth nuts times the marginal utility of wealth gives the marginal cost in utility terms from giving up the other goods. When the marginal benefit is equal to the marginal cost, the choice is optimal and maximized utility.

Combining the first two equations and solving for F in terms of E, we get $F = (p_E/p_F)E$. Now combine the first and third equations to solve for W in terms of E to get $W = (p_E/p_W)E$. Solving, we of course get the same result as before:

$$E^* \;=\; \frac{100}{3p_E}, \qquad F^* = \frac{100}{3p_F}, \qquad W^* = \frac{100}{3p_W}.$$

We also get a solution for λ^*. Using the solution for E^* in the first of the first-order conditions, we solve to get $\lambda^* = 100/3$. That means the marginal utility of wealth is $100/3$, or, if we increased Korra's wealth by just a little bit, her utility would rise at the rate of $100/3$.

7.4   Two-Period Consumer Model

Quick Check

e) Suppose Korra also likes a fourth good, air nuts, A, and her utility is $U = \ln A + \ln E + \ln F + \ln G$, subject to the budget constraint $100 = p_A A p_E E + p_F F + p_W W$. Set up the Langrangian and derive all the first-order conditions. Provide an economic interpretation of each first-order condtion.

For this chapter, we have one long, last example associated with macroeconomics addressing consumer choice between consumption and savings. This model features multiple constraints, so we can look at approaches for handling multiple constraints using the Lagrangian method.

Consider a single consumer, Asami, who lives for two periods and is choosing how much to spend on consumption in the first period, c_1, and how much to spend on consumption in the second period, c_2, and how much to save, b. Asami has the following utility function over consumption today and in the future:

$$U(c_1, c_2) = \ln c_1 + \beta \ln c_2.$$

β is a parameter and takes a value between zero and 1, $0 < \beta < 1$. β is called the subjective discount factor and represents the empirical observation that people place more weight on consumption today than on consumption in the future (due to risk, uncertainty, and impatience among other things). Asami receives income y_1 in the first period and income y_2 in the second period. Note that she just gets the income; no choices are involved such as how long to work or which job to take, so y_1 and y_2 are exogenous variables. To rephrase, those two variables, y_1 and y_2, are not choices.

Asami can also borrow or save at the real interest rate, r. She has a budget constraint in each period. The first period budget constraint is

$$y_1 = c_1 + b,$$

which simply says Asami uses first–period income to consume and to save. Note that b can be negative, which would mean Asami is borrowing against her future income. In the second period, the budget constraint is

$$y_2 + (1 + r) b = c_2.$$

Second–period consumption comes from second–period income plus savings and the interest earned on savings. r is the real interest rate, so it corrects the nominal interest rate for inflation. Thus, the quantity rb represents the amount of interest Asami earns on what she saves. If Asami borrowed, such that b is negative, then she needs to pay that back with interest, rb, using future income. What's left over after she pays the debt can be used for consumption in period 2.

We have two constraints. In this model, we have three options for accounting for the constraints: (1) we can combine the two constraints into one constraint using b, then solve for c_2 and substitute the result into the objective function to form an unconstrained problem in one choice variable, c_1; (2) we can combine the two constraints into one constraint using b, and then form the Lagrangian with the one constraint; or (3) we can form a Lagrangian by adding each constraint to the objective function separately using different Lagrangian multipliers for each constraint, λ_1 and λ_2. Since we are studying Lagrangians, we will skip option 1 (left as an exercise), but we will go through options 2 and 3 here.

7.4.1   One Constraint

Let's start by combining the constraints and setting up the problem with only one constraint, as we did in the previous examples. Notice that both constraints contain b, so if we solve for b in one equation we can substitute the result into the other, as follows:

$$
\begin{aligned}
b &= y_1 - c_1 \\
y_2 + (1+r)\,b &= c_2 \\
y_2 + (1+r)\,(y_1 - c_1) &= c_2.
\end{aligned}
$$

Rearranging such that the constraint is equal to zero, we have

$$
\begin{aligned}
y_2 + (1+r)\,y_1 - (1+r)c_1 - c_2 &= 0 \\
y_1 + \frac{y_2}{1+r} - c_1 - \frac{c_2}{1+r} &= 0.
\end{aligned}
$$

Notice that $y_1 + \frac{y_2}{1+r}$ is the present value of lifetime income or her lifetime wealth. Let that be equal to W such that

$$
W = y_1 + \frac{y_2}{1+r},
$$

which will become useful later.

Setting up the Lagrangian, remembering to add the constraint such that it is equal to zero, we have

$$
\mathcal{L} = \ln c_1 + \beta \ln c_2 + \lambda \left[y_1 + \frac{y_2}{1+r} - c_1 - \frac{c_2}{1+r} \right].
$$

In this instance, we chose to have the exogenous variables that govern the constraint (y_1 and y_2) enter positively, which will make interpreting the Lagrangian multiplier a little bit easier later. In this problem the consumer chooses c_1 and c_2. Asami also chooses b, but since we replaced it using the budget constraints, it does not appear in the problem now. So, we have reduced the number of choice variables from three to two. Once we have the optimal levels of c_1^* and c_2^*, we can use either single–period budget constraint to solve for b^*.

Since we have two choice variables, we need two first-order conditions, one for each choice variable, plus the partial derivative with respect to λ:

$$
\begin{aligned}
\frac{\partial \mathcal{L}}{\partial c_1} &= \frac{1}{c_1} - \lambda = 0 \\
\frac{\partial \mathcal{L}}{\partial c_2} &= \frac{\beta}{c_2} - \lambda \left(\frac{1}{1+r} \right) = 0 \\
\frac{\partial \mathcal{L}}{\partial \lambda} &= y_1 + \frac{y_2}{1+r} - c_1 - \frac{c_2}{1+r} = 0
\end{aligned}
$$

As always, there is a marginal benefit / marginal cost interpretation to the first-order conditions. The first says the marginal utility of first–period consumption must be equal to the marginal cost (which will turn out to be the marginal utility of wealth). The second constraint says the marginal utility of second–period consumption must be equal to the present value of the marginal utility of wealth. There is a present value adjustment: the division by $(1+r)$ takes the future values in period 2 and shows the first period equivalent.

Solving these two equations for c_2 in terms of c_1, we have

$$
\begin{aligned}
\frac{1}{c_1} &= \lambda \\
\frac{\beta}{c_2} &= \lambda \left(\frac{1}{1+r} \right) \\
\frac{1/c_1}{\beta/c_2} &= \frac{1}{1/(1+r)} \\
\frac{c_2}{\beta c_1} &= 1+r \\
c_2 &= (1+r)\,\beta c_1.
\end{aligned}
$$

A quick side comment about that last equation, which relates the optimal level of c_2 to the optimal level of c_1. That equation is an example of what is referred to as an **Euler equation**. An Euler equation characterizes the *optimal path* of a variable through time. These types of equations appear frequently in macroeconomic models, where much of the focus is on how and why variables change over time.

Substituting that Euler equation back into the budget constraint we can solve for c_1^*:

$$
\begin{aligned}
y_1 + \frac{y_2}{1+r} - c_1 - \frac{c_2}{1+r} &= 0 \\
y_1 + \frac{y_2}{1+r} - c_1 - \frac{((1+r)\,\beta c_1)}{1+r} &= 0 \\
y_1 + \frac{y_2}{1+r} &= c_1 + \frac{((1+r)\,\beta c_1)}{1+r} \\
y_1 + \frac{y_2}{1+r} &= c_1 + \beta c_1 \\
(1+\beta)\,c_1 &= y_1 + \frac{y_2}{1+r} \\
c_1^* &= \frac{1}{1+\beta} \left[y_1 + \frac{y_2}{1+r} \right].
\end{aligned}
$$

In terms of W, Korra's present value of lifetime wealth, we have

$$c_1^* = \frac{1}{1+\beta}W,$$

which means that first–period consumption is the fraction $\frac{1}{1+\beta}$ of lifetime income. Then, c_2 is

$$\begin{aligned}
c_2 &= (1+r)\,\beta c_1 \\
c_2 &= (1+r)\,\beta \left(\frac{1}{1+\beta}\left[y_1 + \frac{y_2}{1+r}\right]\right) \\
c_2^* &= (1+r)\frac{\beta}{1+\beta}\left[y_1 + \frac{y_2}{1+r}\right].
\end{aligned}$$

In terms of W, we have

$$c_2^* = (1+r)\frac{\beta}{1+\beta}W,$$

or, in present value terms,

$$\frac{c_2^*}{1+r} = \frac{\beta}{1+\beta}W.$$

Thus, second–period consumption is the fraction $\frac{\beta}{1+\beta}$ of wealth adjusted for the real interest rate. Finally, using our solution for c_1^* or c_2^* and either of the one–period budget constraints, we can solve for the optimal choice of b:

$$b^* = \frac{\beta}{(1+\beta)}y_1 - \frac{1}{(1+\beta)(1+r)}y_2.$$

The optimal amount of wealth for Asami to carry over from period 1 to period 2 is a positive fraction of current income minus a fraction of her future income. Intuitively, if her first–period income is very high relative to her future income, she will want to save a positive amount. On the other hand, if future income is very high relative to current income, she will want to borrow against her future income.

What about λ? Using the first–order condition on c_1 we have:

$$\begin{aligned}
\frac{1}{c_1} &= \lambda \\
\lambda &= \frac{(1+\beta)}{W}.
\end{aligned}$$

Notice that the first–order condition itself, $\frac{1}{c_1} = \lambda$, tells us that the marginal utility of wealth is equal to the marginal utility of first–period consumption. Putting the optimal solutions for c_1^* and c_2^* into the original utility function, let's take the derivative with respect to wealth to find the marginal utility of wealth that way:

$$U = \ln c_1 + \beta \ln c_2$$
$$U = \ln\left(\frac{W}{1+\beta}\right) + \beta \ln\left(\frac{(1+r)\beta W}{1+\beta}\right)$$
$$\frac{dU}{dW} = \frac{1+\beta}{W}\left(\frac{1}{1+\beta}\right) + \beta\left(\frac{1+\beta}{(1+r)\beta W}\right)\left(\frac{(1+r)\beta}{1+\beta}\right)$$
$$\frac{dU}{dW} = \frac{1}{W} + \frac{\beta}{W} = \frac{1+\beta}{W}.$$

Thus,

$$\frac{dU}{dW} = \frac{1+\beta}{W} = \lambda.$$

Again, the Lagrangian multiplier is the marginal utility of wealth (i.e., the marginal value of relaxing the constraint).

7.4.2 Two Constraints

That was fun. Let's do it again but without combining the constraints.

Now when we set up the Lagrangian, we will include the two one-period budget constraints separately, and each will have its own Lagrangian multiplier. Our two constraints again are

$$y_1 = c_1 + b$$
$$y_2 + (1+r)\,b = c_2.$$

and we will assign λ_1 and λ_2 to the constraints, respectively, as the multipliers. Now setting up the Lagrangian we have

$$\mathcal{L} = \ln c_1 + \beta \ln c_2 + \lambda_1 \left[y_1 - c_1 - b\right] + \lambda_2 \left[y_2 + (1+r)b - c_2\right].$$

Again, see how the constraints are written. They are equal to zero, and the exogenous variables in each, y_1 and y_2, are written positively to ease interpretation of the Langrangian multipliers.

It is important to notice here that we have all three choice variables in play: c_1, c_2, and b. That means we must take a first-order condition with respect to each of them, so we will have three first–order conditions and two constraints for a five equation system with five unknowns: the three choice variables and the two Lagrangian multipliers.

Setting our partial derivatives equal to zero to form the first-order conditions, we have

$$\frac{\partial \mathcal{L}}{\partial c_1} = \frac{1}{c_1} - \lambda_1 = 0$$

$$\frac{\partial \mathcal{L}}{\partial c_2} = \frac{\beta}{c_2} - \lambda_2 = 0$$

$$\frac{\partial \mathcal{L}}{\partial b} = -\lambda_1 + \lambda_2(1+r) = 0$$

$$\frac{\partial \mathcal{L}}{\partial \lambda_1} = y_1 - c_1 - b = 0$$

$$\frac{\partial \mathcal{L}}{\partial \lambda_2} = y_2 + (1+r)b - c_2 = 0.$$

The interpretations of the first two FOCs are identical to what we had before, except in the second one we have λ_2 instead of $\lambda(1+r)$, when we did it with only one constraint. λ_2 is the marginal value of relaxing the second constraint, which would mean marginally increasing future (second–period) income. That is the same as the marginal utility of wealth, but it is multiplied by $(1+r)$ again to put it in present value terms because the utilty value corresponds to the first period. Indeed the third first-order condition directly relates the values of the two Lagrangian multipliers such that:

$$\lambda_1 = \lambda_2(1+r).$$

The final two first-order conditions, the partial derivatives with respect to λ_1 and λ_2, reproduce the first– and second–period budget constraints. Thus, we have five equations for five unknowns. To solve for the optimal values we start working with the three FOCs on the choice variables. Use the third equation to replace λ_1 in the first equation with λ_2, then we will divide by the second FOC as we did before:

$$\frac{1}{c_1} - \lambda_2(1+r) = 0$$

$$\frac{\beta}{c_2} - \lambda_2 = 0$$

$$\frac{1}{c_1} = \lambda_2(1+r)$$

$$\frac{\beta}{c_2} = \lambda_2$$

$$\frac{c_2}{\beta c_1} = 1+r$$

$$c_2 = \beta(1+r)c_1.$$

Once again, we have found the Euler equation describing the optimal path of consumption over time. Now we can substitute that result into the either budget constraint, and with the other budget constraint we can solve for the optimal values of c_1^*, c_2^*, and b^*. We leave it to you, the reader, to complete that algebra to get the following:

$$c_1^* = \frac{1}{1+\beta}\left[y_1 + \frac{y_2}{1+r}\right]$$

$$c_2^* = (1+r)\frac{\beta}{1+\beta}\left[y_1 + \frac{y_2}{1+r}\right]$$

$$b^* = \frac{\beta}{(1+\beta)}y_1 - \frac{1}{(1+\beta)(1+r)}y_2.$$

All that remains is to solve for the values of λ_1^* and λ_2^*. Going back to our first-order conditions and using our optimal solutions for consumption, we have

$$\lambda_1 = \frac{1}{c_1}$$

$$\lambda_2 = \frac{\beta}{c_2}$$

$$\lambda_1^* = \frac{1+\beta}{\left[y_1 + \frac{y_2}{1+r}\right]}$$

$$\lambda_2^* = \frac{1}{(1+r)}\frac{1+\beta}{\left[y_1 + \frac{y_2}{1+r}\right]}.$$

The Lagrangian multipliers are the marginal utility value of relaxing each constraint, respectively. Here, again, we see that the second multiplier, λ_2, because it is the marginal value of future period wealth, is divided by $1+r$ to put it in current period terms (i.e., present value).

7.5 Exercises

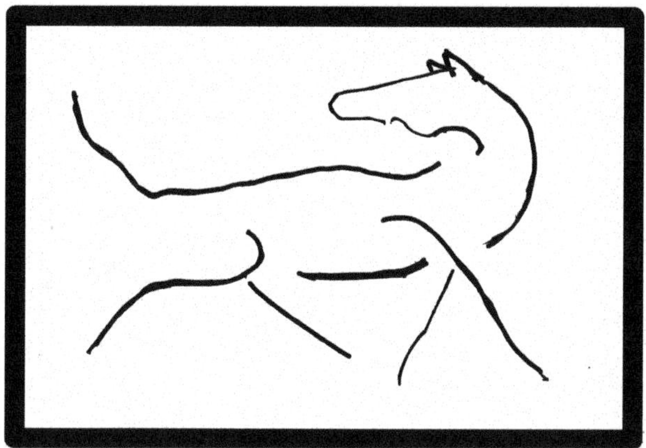

7.5.1 Quick Check Answers

a) & c) $\max_{x,y} U = xy$ s.t. $W = 4x + 5y$. Lagrangian: $\mathcal{L} = xy + \lambda[W - 4x - 5y]$, FOCs: $\frac{\partial \mathcal{L}}{\partial x} = y - 4\lambda = 0$, $\frac{\partial \mathcal{L}}{\partial y} = x - 5\lambda$ and $\frac{\partial \mathcal{L}}{\partial \lambda} = W - 4x - 5y = 0$. Solution: $x^* = W/8$, $y^* = W/10$, and $\lambda^* = W/40$

b) & d) $\min_{K,L} rK + wL$ s.t. $(KL)^{0.5} = 100$. Lagrangian: $\mathcal{L} = rK + wL + \lambda[100 - (KL)^{0.5}]$. FOCs: $\frac{\partial \mathcal{L}}{\partial K} = r - \lambda(0.5)K^{-0.5}L^{1/2} = 0$, $\frac{\partial \mathcal{L}}{\partial L} = w - \lambda(0.5)K^{0.5}L^{-0.5} = 0$, and $\frac{\partial \mathcal{L}}{\partial \lambda} = 100 - (KL)^{0.5} = 0$. Solution: $K^* = 100(w/r)^{0.5}$, $L^* = 100(r/w)^{0.5}$, $\lambda^* = 2(wr)^{0.5}$

e) $\mathcal{L} = \ln A + \ln E + \ln F + \ln W + \lambda[100 - p_A A - p_E E - p_F F - p_W W]$. FOCs: $\frac{\partial \mathcal{L}}{\partial A} = \frac{1}{A} - \lambda p_A = 0$, $\frac{\partial \mathcal{L}}{\partial E} = \frac{1}{E} - \lambda p_E = 0$, $\frac{\partial \mathcal{L}}{\partial F} = \frac{1}{F} - \lambda p_F = 0$, and $\frac{\partial \mathcal{L}}{\partial W} = \frac{1}{W} - \lambda p_W = 0$. Each FOC has the marginal benefit as the marginal utility of the good (A, E, F, or W), and the marginal cost is the price of the respective good times the Lagrangian multiplier.

f) Lagrangian: $\mathcal{L} = M^{1/2} + S^2 + \lambda_1[W - p_M M - p_S S] + \lambda_2[M - 5]$. FOCs: $\frac{\partial \mathcal{L}}{\partial M} = (1/2)M^{-1/2} - \lambda_1 p_M + \lambda_2 = 0$, $\frac{\partial \mathcal{L}}{\partial S} = 2S - \lambda_1 p_S = 0$, $\frac{\partial \mathcal{L}}{\partial \lambda_1} = W - p_M M - p_S S = 0$, and $\frac{\partial \mathcal{L}}{\partial \lambda_2} = M - 5 \geq 0$. That last condition stays a weak inequality and, again, we will adress these types of constraints in more detail in Chapter 8.

7.5.2 Practice Problems

Solve the following problems for their optimal solutions using a Lagrangian. Be sure to also solve for λ^*. For all the first-order conditions, identify the part that is the marginal benefit and the part that is the marginal cost. Find values for the Lagrangian multiplier and provide an interpretation.

1.

$$\max_{x,y} U = x^{1/2} + y^{1/2}$$

$$s.t. \quad x + y = 100$$

2.

$$\max_{x,y} U = 3x^{1/2} + y^{1/2}$$

$$s.t. \quad x + y = 100$$

3.

$$\max_{x,y} U = \ln x + 2\ln y$$

$$s.t. \quad x + y = 100$$

4.

$$\max_{x,y} U = x^{1/2}y^{1/2}$$

$$s.t. \quad x + y = 100$$

5. Consider the following two–period consumer problem:

$$\max_{c_1,c_2} = \ln c_1 + \beta \ln c_2$$
$$s.t. \quad c_1 + c_2 = W,$$

where c_1 is consumption today, c_2 is future consumption, and W is lifetime income. Note that the one period constraints are the same as in Section 7.4, but here $r = 0$.

(a) Set up the problem using a Lagrangian on the budget constraint.

(b) Get the first-order condition and interpret it in marginal benefit / marginal cost terms.

(c) Solve for c_1^*, c_2^*, b^*, and λ^*.

(d) Interpret λ^* in economics terms.

6. Reconsider Korra's three good problem from Section 7.3. But now suppose that her local grocery store has way more than three kinds of nuts. In fact, they have K kinds of different nuts, and Korra likes them all. Let the number of kinds of nuts be K, and let i be the counter from 1 to K, such that $i \in ([1, K]$. Then, the quantity of type i nuts that Korra eats is given by N_i. She also now has C copper pieces instead of 100. Her problem becomes the following:

$$\max_{N_i, i \in [1,K]} U = \sum_{i=1}^{K} \ln N_i$$

$$s.t. \quad C = \sum_{i=1}^{K} p_i N_i.$$

where p_i is the corresponding price of nut type i. In Lagrangian form the problem becomes

$$\mathcal{L} = \overbrace{\sum_{i=1}^{K} \ln N_i}^{\text{Objective function}} + \lambda \overbrace{\left[C - \sum_{i=1}^{K} p_i N_i \right]}^{\text{Constraint}}.$$

(a) Derive the first-order condition for N_1.

(b) Derive the first-order condition for a general kind of nut, N_i.

(c) Derive the first-order condition for a different general kind of nut, N_j, where $i \neq j$.

(d) Use the first order–conditions you derived in (a), (b), and (c) to find the solution to Korra's optimization problem for a general type of nut, i. Your answer for N_i^* should be in terms of C, K, and the full set of prices, p_i, for $i \in [1, K]$ only.

7. Redo the two-period model from Section 7.4 but by combining the constraints and substituting the constraint into the objective function for c_2 to make it a single–variable optimization problem. Interpret your first-order condition and solve it fully. Show that you get the same optimal solution.

8. Zhu Li is stranded on an island. She catches fish and spends time admiring the ocean view (leisure). Her preferences and fish catching production function are:

$$U(c, h) = \ln c + \alpha \ln h; \quad y = \ell^{1/3},$$

where c is consumption, h is leisure, ℓ is her labor effort, y is fish caught, and α is a parameter such that $0 < \alpha < 1$. Her time endowment is 24, and her time constraint is

$$\ell + h = 24.$$

(a) Substitute the production function into the objective function by assuming that $c = y$, that is, she will consume what she catches. Set up this problem as a Lagrangian, where the Lagrangian multiplier is on the time constraint.

(b) Find Zhu Li's optimal choice of labor, leisure, and consumption.

(c) How does Zhu Li's labor/leisure choice vary with α? How would you interpret α in economic terms?

(d) Solve for and interpret the Lagrangian multiplier.

9. Consider the following two–period consumer problem:

$$\max_{c_1, c_2} U(c_1, c_2)$$

$$\text{s.t.} \quad c_1 + \frac{c_2}{1+r} = a + y_1 + \frac{y_2}{1+r},$$

where the variables are all the same as in Section 7.3, but the utility function is general. Assume that $U_c > 0$ and $U_{cc} < 0$ for both c_1 and c_2.

(a) Set up the problem using a Lagrangian on the budget constraint.

(b) Get the first-order conditions and interpret them in marginal benefit / marginal cost terms.

(c) Solve one first-order condition for λ and substitute the result into the other first-order condition such that you have combined the two first-order conditions and replaced λ. Interpret this condition.

7.6   Math Appendix

Here, we list some key math definitions, theorems, and formulae in formal math for your reference.

Lagrangian method or Lagrange method: For a maximization problem with N choice variables and M constraints such that we have:

$$\max_{x_1, x_2, \ldots, x_N} F(x_1, x_2, \ldots, x_N) \text{ s.t. } g^1(x_1, x_2, \ldots, x_N) = 0, g^2(x_1, x_2, \ldots, x_N) = 0, \ldots, g^M(x_1, x_2, \ldots, x_N) = 0.$$

We derive the optimal set of x values by constructing the Lagrange function and taking all the first-order condtions with respect to the choice variables and the Lagrangian multipliers, λ_m, where $m \in [1, M]$:

$$\mathcal{L} = F(x_1, x_2, \ldots, x_N) + \lambda_1 g^1(x_1, x_2, \ldots, x_N), + \lambda_2 g^2(x_1, x_2, \ldots, x_N) + \cdots + \lambda_M g^M(x_1, x_2, \ldots, x_N)$$

$$\frac{\partial \mathcal{L}}{\partial x_1} = F_1(x_1, x_2, \ldots, x_N) + \lambda_1 g_1^1(x_1, x_2, \ldots, x_N) + + \lambda_2 g_1^2(x_1, x_2, \ldots, x_N) + \cdots + \lambda_M g_1^M(x_1, x_2, \ldots, x_N)$$

$$\frac{\partial \mathcal{L}}{\partial x_2} = F_2(x_1, x_2, \ldots, x_N) + \lambda_1 g_2^1(x_1, x_2, \ldots, x_N) + + \lambda_2 g_2^2(x_1, x_2, \ldots, x_N) + \cdots + \lambda_M g_2^M(x_1, x_2, \ldots, x_N)$$

$$\vdots$$

$$\frac{\partial \mathcal{L}}{\partial x_N} = F_N(x_1, x_2, \ldots, x_N) + \lambda_1 g_N^1(x_1, x_2, \ldots, x_N) + + \lambda_2 g_N^2(x_1, x_2, \ldots, x_N) + \cdots + \lambda_M g_N^M(x_1, x_2, \ldots, x_N).$$

These conditions characterize the optimal solution: $(x_1^*, x_2^*, \ldots, x_N^*; \lambda_1^*, \lambda_2^*, \ldots, \lambda_M^*)$.

Present value: The basic present value formula presented in the text was $PV = \frac{V}{1+r}^t$. That gives the value in today's terms of a payment or asset in the future with value V if the interest rate is r. The same concept can be applied to a stream of payments using $PV = \sum_{t=0}^{T} \frac{V_t}{1+r}^t$, where the stream lasts for T periods. The amount in each period is given by V_t and has the subscript t, indicating the amount could change from period to period. r in that formula is held constant, but were it to also vary with time, it would be $PV = \sum_{t=0}^{T} \frac{V_t}{1+r_t}^t$.

Net present value (NPV): NPV incorporates the idea that in order to obtain the future payment(s) some costs, C, need to be paid. So, the *net* present value subtracts off that cost as follows: $NPV = -C + \sum_{t=0}^{T} \frac{V_t}{1+r_t}^t$.

8   Corner Solutions

Thus far in this exploration, we have only looked at problems with what we call *interior* solutions. Loosely, the math all worked out, and we got nice, neat answers for everything. However, in some cases and, in fact, most of the really interesting cases in economics, the math does not work out so cleanly, and we have *corner* solutions, or other breakdowns, instead of interior solutions. These types of issues come in a number of forms. We will go through each type and explain what they mean, how to identify them, and how to solve for them.

Before we go into the details, it may be helpful to list the types we will explore and provide a brief description:

1. **Boundary constraints**: The choice variable may only have a limited range of permissible values. For example, a firm with a capacity constraint may not be able to exceed a certain level of output even though the profit–maximizing level is higher.

2. **Corner solutions**: In a constrained optimization problem, it is optimal to choose zero for at least one of the choice variables. With a linear utility function (e.g., $U = 2x + y$), for example, the consumer will often not want to buy one good because the other good always yields a higher utility level or one good is more expensive.

3. **Nonbinding constraints**: The constraint itself does not hold with equality. The capacity constraint example is also a good one here. If a firm is constrained to a maximum amount of output, but the profit maximization choice is less than that upper bound, the constraint is considered nonbinding.

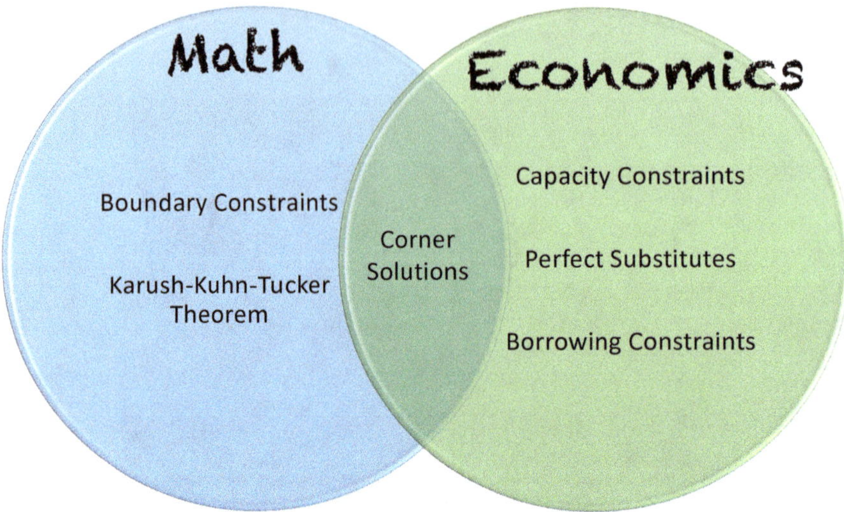

8.1 Boundary Constraints

Consider the following optimization problem:

$$\min_{x} y = 6x^2.$$

Without even doing any calculus, you can probably see that the answer will be $x^* = 0$, but let's take the derivative and form the first-order condition anyway:

$$\frac{dy}{dx} = 12x = 0$$
$$\Rightarrow x^* = 0.$$

> **Definition**
>
> **Boundary constraints** apply when the domain of the right-hand side variable(s) has either a lower limit, an upper limit, or both.

$y = 6x^2$ forms a parabola with its lowest point right at the origin, so the solution is pretty obvious if you graph it. Now suppose the problem is the following, where x has a limited range (i.e., a *boundary condi-*

tion):

$$\min_{x} y = 6x^2$$
$$s.t. -2 \leq x \leq 2.$$

We know from the first version of the problem that $x^* = 0$ yields the global minimum of $y^* = 0$. x has to lie within a particular range between -2 and 2, and 0 is in that range. Thus, that added boundary condition on x has no effect. The solution is still $x^* = 0$, which minimizes the function.

But now suppose the problem reads as follows:

$$\min_{x} y = 6x^2$$
$$s.t. 2 \leq x \leq 4.$$

Now clearly $x^* = 0$ cannot be the solution, because zero is not a valid value for x. It is obvious here that the first-order condition alone does not yield the correct answer because of the impact of the boundary condition. What do we do? Check the endpoints of the boundary to see where the minimum lies. That is, compare the value of function at $x = 2$ and $x = 4$, which yield 24 and 96, respectively. Thus, $x^* = 2$ is the solution to the minimization problem.

Although the first-order condition did not yield the solution, the first derivative does give us a strong hint as to what the *constrained optimum* will be. Recall from Chapter 2 how the first derivative tells us whether the function is increasing or decreasing depending on whether it is positive or negative. In our example problem, notice that for all values of x that are permissible, and the endpoints in particular, the derivative $12x$ is always positive. That means the function is *increasing* in x. Since we are trying to minimize y, and y is increasing in x, it follows that the smallest value of x allowed will be the answer.

Now suppose the problem was slightly different and the boundary condition only permitted negative values of x, as follows:

$$\min_{x} y = 6x^2$$
$$s.t. -4 \leq x \leq -2.$$

The first-order condition, again, fails because it still yields $x^* = 0$, which is not a permissible value of x. But now the first derivative, $12x$, is always negative, which means that the function is *decreasing* in x. Therefore, since y is decreasing in x, the largest permissible value of x will be the solution at $x^* = -2$.

Notice that our solutions to all the variations were one of two types: (1) an **interior solution**, where the first-order condition was equal to zero and the optimizing value of x lied inside, in the interior, of the boundary conditions, that is, within the range of permissible x, or (2) a **corner solution**, where the first derivative is not equal to zero at the optimal solution but the solution is at the *boundary*, an endpoint, of the permissible range of x, in other words, at a *corner*.

One last exercise with this example: Suppose the problem is now a maximization problem instead of a minimization problem, as follows:

$$\max_{x} y = 6x^2.$$

Definition

Corner solution: A corner solution occurs when the optimal value of the right-hand side variable(s) falls on either the lower and upper bound of the domain. More formally, if $y = f(x)$ and x has lower and upper bounds, $\underline{x}$ and $\bar{x}$, respectively, then, if x^* is an extreme point of y, then x^* is a corner solution if $x^* = \underline{x}$ or $x^* = \bar{x}$.

With no constraints or boundary conditions governing the range of x what happens? First, if we blindly use the first-order condition,

$$\frac{dy}{dx} = 12x = 0$$
$$\Rightarrow x^* = 0,$$

we would again get $x^* = 0$. However, checking our second-order condition

$$\frac{d^2y}{dx^2} = 12 > 0,$$

which tells us the function is convex and therefore the extreme point we found from the first-order condition is a minimum. How do we find the maximum? Again, look at the endpoints for the range of x, which in the unbound case are $-\infty$ and ∞. Those are the solutions to the problem.

Let's see how things change with boundaries on x. Let's look at one of the easier cases first:

$$\max_{x} y = 6x^2$$
$$s.t.\, 2 \leq x \leq 4.$$

We know that the first-order condition will not help us from the work we did previously. So, we look at the endpoints on x and note that the first-order condition tells us that the function is increasing at both endpoints, $12(2) = 24 > 0$ and $12(4) = 48 > 0$. If the function is increasing over this range, then the larger x, the larger y is. Thus, the upper bound at $x = 4$ would be the solution.

Next, look at the case of the range over negative values in our example:

$$\min_{x} y = 6x^2$$
$$s.t.\, -4 \leq x \leq -2.$$

Now the first-order condition tells us the function is decreasing over that range. Therefore, the largest possible value of x at $x = -2$ would be the solution for a minimum.

To be clear, here we write out the formal statement (Theorem 8.1) for boundary conditions and a maximization or a minimization.

Theorem 8.1: *If x^* is a solution to the problem $\max_x f(x)$ s.t. $a \le x \le b$, then it satisfies one or both of the following conditions. If the solution to the* max *or* min *problem lies strictly within the boundaries, $a < x^* < b$, then both conditions will hold:*

$$f'(x^*) \le 0 \text{ and } (x^* - a)f'(x^*) = 0$$
$$f'(x^*) \ge 0 \text{ and } (b - x^*)f'(x^*) = 0.$$

If x^* is a solution to the problem $\min_x f(x)$ s.t. $a \le x \le b$, then it satisfies one or both of the following conditions. If the solution to the max *or* min *problem lies strictly within the boundaries, $a < x^* < b$, then both conditions will hold:*

$$f'(x^*) \ge 0 \text{ and } (x^* - a)f'(x^*) = 0$$
$$f'(x^*) \le 0 \text{ and } (b - x^*)f'(x^*) = 0.$$

Roughly, that means the solution is either where the first-order condition is zero or at one of the endpoints. If it is an interior solution, the first-order condition is zero. If it is a corner solution at one of the endpoints, which of the two endpoints is the solution depends on whether the function is increasing or decreasing.

8.1.1 Applied Example: Capacity Constraint

Consider the following profit maximization problem Tanjiro's charcoal firm faces:

$$max_Q \Pi = AQ - BQ^2 - CQ$$
$$\text{s.t.}$$
$$0 \le Q \le \bar{Q}.$$

A, B, and C are all positive exogenous parameters. $\bar{Q}$ is also an exogenous parameter and represents a capacity constraint, so it is the upper bound for Q. Implicitly, there is also a lower bound for Q, and that lower bound is 0. So, we have that $0 \le Q \le \bar{Q}$. Taking the derivative and forming the first-order condition, we have

$$\frac{d\Pi}{dQ} = A - 2BQ - C = 0$$

$$\frac{d^2\Pi}{dQ^2} = -2B < 0.$$

The second–order condition is strictly negative since we know that $B > 0$ by assumption. So, any solution we derive from the first-order condition must be a maximum. Now, returning to the first-order condition and attempting to solve the problem, the boundaries may or may not come into play. It depends on the parameters. The solution is to provide an answer that shows *how* it depends on the parameters. To do so, we show the different optimal solutions depending on the values the parameters take.

First, if we look for the interior solution, that means we are looking for a solution when the first-order condition will hold with equality. We can solve the equation above to get

$$Q^* = \frac{A - C}{2B}.$$

What if $A < C$? That would imply that $Q^* < 0$, which is out of the range. If $A < C$, then $Q^* = 0$, and we are at a boundary. Note also if $A < C$, it implies the first-order condition is also always negative.[11] In economics terms, it means that the marginal cost *always* exceeds the marginal beneift. What about the upper bound capacity constraint? If $\frac{A-C}{2B} > \bar{Q}$, then $Q^* = \bar{Q}$.

Therefore, if we formally write the solution to Tanjiro's profit maximization problem, we would write

$$Q^* = \begin{cases} 0 & \text{if } A - C \leq 0 \\[2ex] \frac{A-C}{2B} & \text{if } 0 < \frac{A-C}{2B} < \bar{Q} \\[2ex] \bar{Q} & \text{if } \frac{A-C}{2B} \geq \bar{Q}. \end{cases}$$

This solution characterizes Tanjiro's optimal choice, which depends on the parameters. We refer to the middle solution as the *interior* solution because it falls inside the boundary constraints. In contrast, the other two solutions are *corner* solutions because they exist at one of the boundaries.

---

[11]Setting the first-order conditon equal to zero is not actually correct. Keep that in mind; we will explore that case more fully in the next subsection.

8.2 Corner Solutions

The case considered here is when there may or may not be an explicit range for a choice variable, but it is implicitly understood that a lower bound of zero exists. We saw an example of this case in the previous subsection with Tanjiro's charcoal firm. There were conditions under which the optimal choice was to produce nothing and choose $Q^* = 0$ to maximize profits. Lots of variables we consider make no sense as negative values, for example, the amount of a good consumed or the amount of labor hired.

8.2.1 Specific Example

Let's look at a consumer choice problem to illustrate corner solutions. Consider Nezuko, who gets utility from two things: reading manga, x, and eating pocky, y. Her utility and budget constraint are given by the following:

$$\max_{x,y} U = x + y$$
$$\text{s.t. } W = p_x x + p_y y.$$

Initially, let's work with a very specific example in which $W = 12$, $p_x = 3$, and $p_y = 2$. Her problem is now

$$\max_{x,y} U = x + y$$
$$\text{s.t. } 12 = 3x + 2y.$$

Before jumping into the calculus, notice a key feature of the objective function: *It is linear.* That is, if you graph U against either x or y, you will get a straight line. If we graph the indifference "curves," they are also straight lines. See Figure 8.1.

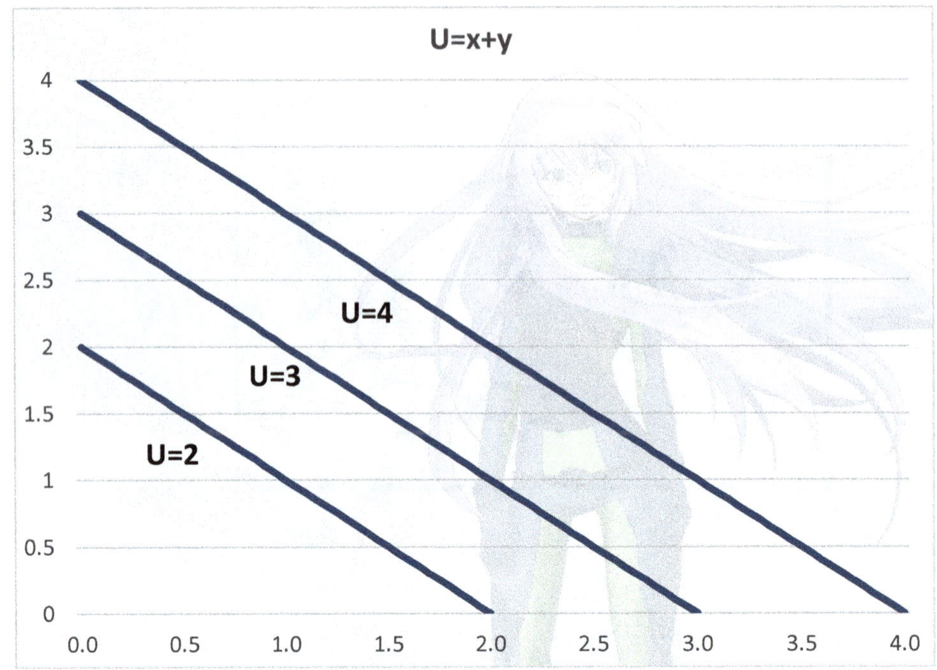

Figure 8.1: Indifference curves for $U = x + y$.

Unlike the indifference curves from Figure 6.1 there are no "curves," just straight lines. The interpretation is the same. All points along each line represent the same level of utility, so the consumer is indifferent (i.e., gets equal utility) from all points on that same line. Lines further from the origin represent higher utility, so the consumer tryies to get to the highest indifference curve possible. Unlike the truly curved indifference curves of Figure 6.1, the slopes of the indifference curves here are constant.

Now, think about what would happen if you drew in a budget constraint. Budget constraints are also linear. Thus, in order for the budget constraint and indifference curve to be tangent (and show you the optimal point), they have to have the same slope. But if they have the same slope, they will lie on top of each other. More-over, there is no guarantee or reason to believe that the indifference curves and budget constraint will have the same slope. The budget constraint could have a steeper or flatter slope. In either case, that means there will be no points that are tangent.

> ### Definition
>
> **Perfect substitutes**: Perfect substitutes are goods a consumer finds equally acceptable. For example, Zenitsu does his homework using either a blue or a black pen. He does not care which color; both are equally acceptable. They are perfect substitutes. The linear utlity function expresses the idea of perfect substitutes: $U = x_1 + x_2$.

Before trying to optimize anything, consider what having a linear utility function means. Basic calculus can help us understand the utility function. Taking first and second partial derivatives gives us the following:

$$U = x + y$$

$$\frac{\partial U}{\partial x} = 1$$

$$\frac{\partial U}{\partial y} = 1$$

$$\frac{\partial^2 U}{\partial x^2} = \frac{\partial^2 U}{\partial y^2} = 0.$$

In words, that means that for Nezuko, reading one more manga always increases utility by 1 point, and eating one more Pocky increases utility by 1 point. That is, marginal utility is *constant*. That is to say, there is no diminishing marginal utility here. Nezuko would enjoy the billionth manga just as much as the first one. It is the same for eating Pocky; the additional happiness is always the same. That feature will be critical as we look for the solution.

Let's set up a Lagrangian and see what happens:

$$\mathcal{L} = x + y + \lambda\left[12 - 3x - 2y\right]$$

$$\frac{\partial \mathcal{L}}{\partial x} = 1 - 3\lambda \leq 0$$

$$\frac{\partial \mathcal{L}}{\partial y} = 1 - 2\lambda \leq 0$$

$$\frac{\partial \mathcal{L}}{\partial \lambda} = 12 - 3x - 2y = 0.$$

The first thing to notice is that instead of setting the derivatives with respect to x and y equal to zero, they are written as *less than or equal* to zero. Now, the marginal benefit – marginal cost interpretation will help us immensely in understanding what the math is trying to say.

Let's focus on the first-order condition on x: $1 - 3\lambda \leq 0$. The 1 is still the marginal benefit. As discussed, that is Nezuko's marginal utility of manga. It is a constant 1. -3λ is the marginal cost, which is the price of manga, $p_y = 3$, times the Lagrangian multiplier, which is the marginal utility of wealth as we had in examples before. Similarly, the second of the first-order conditions states that the constant marginal utility that Nezuko gets from Pocky, again 1, is the marginal benefit. The marginal cost here is the price of Pocky, $p_x = 2$ times λ.

Now notice that those two conditions cannot possibly be equal to zero simultaneously. If they were, from the first one on x we would get that $\lambda = 1/3$ and from the second one on y we would get that $\lambda = 1/2$. The Lagrangian multiplier is a single variable and cannot be two different numbers simultaneously. Therefore, if both first-order conditions are equal to zero, we have a contradiction and can rule out this possibility.

Next, let's think about why the *less than or equals* is there and how to interpret it. We know that when the equation holds with equality, the marginal benefit is equal to the marginal cost. But what if it is strictly less than, as in $1 - 3\lambda < 0$? What does that mean? That means that the marginal cost is always larger than the marginal benefit, $1 < 3\lambda$. It is always more costly to have that good than the good's worth in terms of utility, so the optimal choice will be the lower bound, which is zero. If we were looking at the second equation and read that as not holding with equality such that $1 < 2\lambda$, we would interpret that as the marginal cost of Pocky is always larger than the marginal benefit.

The discussion thus far has only covered interpretation, but not how to figure out whether those equations are either equal to or less than zero, and ultimately how to solve a problem of this type. To solve such problems, we need to consider all the possible combinations and see which ones are consistent, that is which ones do not produce any logical inconsistencies and identify which ones can be ruled out because of a contradiction. The three cases we need to consider here are: (1) Both first-order conditions hold with strict equality; (2) the first-order condition on x holds with strict equality but the first-order condition on y does not; and (3) vice versa, the first-order condition on y holds with strict equality but the first-order condition on x does not.

Case 1:

$$\frac{\partial \mathcal{L}}{\partial x} = 1 - 3\lambda = 0$$

$$\frac{\partial \mathcal{L}}{\partial y} = 1 - 2\lambda = 0$$

$$\frac{\partial \mathcal{L}}{\partial \lambda} = 12 - 3x - 2y = 0$$

We started this case. If you solve for λ in both equations, you get $\lambda = 1/3$ *and* $\lambda = 1/2$. That is a clear contradiction, and thus it cannot be the case that both first-order conditions hold with strict equality. We have a contradiction; therefore, we rule out case 1 as characterizing the optimal solution.

Case 2:

$$\frac{\partial \mathcal{L}}{\partial x} = 1 - 3\lambda = 0$$

$$\frac{\partial \mathcal{L}}{\partial y} = 1 - 2\lambda < 0$$

$$\frac{\partial \mathcal{L}}{\partial \lambda} = 12 - 3x - 2y = 0$$

Here, we can solve the first-order condition on x for $\lambda = 1/3$. If a first-order condition has a strict inequality, as does the one on y, the marginal cost should always exceed the marginal benefit, and the agent will choose zero of that good (i.e., Nezuko chooses no Pocky.) But, when we put the solution for λ into the condition on y, we get $1 - 2(1/3) <$

0, which reduces to $1/3 < 0$, which is also a contradiction. It is not true. Therefore, we can rule out case 2 as being an optimal solution.

Case 3:

$$\frac{\partial \mathcal{L}}{\partial x} = 1 - 3\lambda < 0$$

$$\frac{\partial \mathcal{L}}{\partial y} = 1 - 2\lambda = 0$$

$$\frac{\partial \mathcal{L}}{\partial \lambda} = 12 - 3x - 2y = 0$$

Here, Nezuko chooses to buy Pocky but zero manga. So, solving for λ from the first-order condition on y, which holds with strict equality in this case, we have $\lambda = 1/2$. Putting that value into the other first-order condition, we get $1 - 3(1/2) < 0$, or $-1/2 < 0$, and there is no contradiction. So, in the optimal solution, Nezuko purchases Pocky but no manga. To finish the problem, go to the budget constraint now. Since she chooses $x^* = 0$, the budget constraint then becomes $12 - 3(0) - 2y = 0$, which is easily solved to get $y^* = 6$. The final answer to Nezuko's problem is $x^* = 0$, $y^* = 6$, and $\lambda^* = 1/2$.

What does $\lambda = 1/2$ mean? It is the same as before, the marginal utility of wealth. Suppose we increase Nezuko's wealth by \$1. What would she do? She would also spend that on Pocky. For \$1 she can buy $1/2$ a Pocky, which would give her $1/2$ points more of utility: $\Delta U / \Delta W = 1/2$.

Now let's go through this without calculus to build our intuition and see how the calculus and math reflects the logic of the model. Think again about Nezuko's utility function, $U = x + y$. That says she gets equal marginal utility from both manga and Pocky. Six manga yield the same utility as six Pocky. However, the price of manga is three while the price of Pocky is two. For every manga she buys, she could have bought 1.5 Pockies. In pure economics terms, the opportunity cost of a manga is 1.5 Pockies, and 1.5 Pockies makes her happier than just 1 manga. That intuition is what you see in the set of equations , Nezuko makes the obvious choice of spending all her money on Pocky and none on manga.

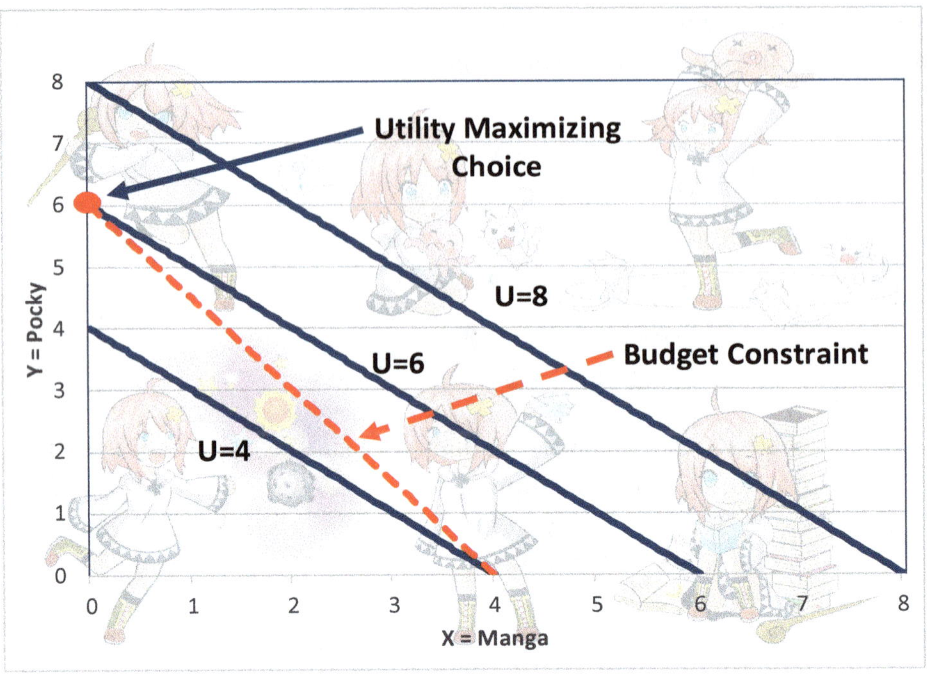

Figure 8.2: Corner solution with linear indifference curves.

Look at this problem graphically in Figure 8.2. Indifference curves have been drawn in for three levels of utility at $U = (4, 6, 8)$. The budget constraint has also been drawn after rearranging such that $y = 6 - 1.5x$. The optimal choice at $x^* = 0$, $y^* = 6$ is marked with a large red circle. First notice that the slope of the indifference curve differs from the slope of the budget constraint. Using what we learned about total derivatives, let's find the marginal rate of substitution for Nezuko:

$$
\begin{aligned}
dU &= 1dx + 1dy \\
\text{Set} \quad dU &= 0 \\
0 &= 1dx + 1dy \\
MRS &= -\frac{dy}{dx} = -1.
\end{aligned}
$$

The slope of the indifference curves is a constant negative 1, meaning that Nezuko will always be willing to trade a manga for a Pocky or vice versa, and that trade will leave her equally well-off. The slope of the budget constraint is the (negative of the) ratio of the prices 3:2, which is where the 1.5 comes from in the point slope version of the budget constraint, $y = 6 - 1.5x$. That means that for every 2 x (manga), Nezuko can buy 3 y, Pocky. If we were looking at an interior solution, the two slopes would be equal, and the budget constraint and the indifference curve would be tangent at the optimum. Here, the two slopes are not equal, even at the optimum, where they meet to form a "corner" solution.

Visually you can see that the highest indifference curve that Nezuko can reach given the budget constraint is the one that intersects with the budget constraint right on the y-axis where $y = 6$ and $x = 0$. Any higher indifference curve would require a larger budget, as it would be above the

budget constraint and therefore not feasible. Note how the $U = 4$ indifference curve also touches the budget constraint but at the opposite end, where $x = 4$ and $y = 0$. That would actually be the minimum value of utility Nezuko could obtain provided the budget constraint holds with equality.

8.2.2 General Example

Let's examine Nezuko's problem again, but use a more general form for her utility function, $U = U(x, y)$, and let the prices be variables, p_x and p_y, rather than specific values. We will assume that $U_x > 0$ and $U_y > 0$, meaning that utility is strictly increasing in both x (manga) and y (Pocky). Furthermore, we will assume that $U_{xx} \leq 0$ and $U_{yy} \leq 0$. That means that her utility may exhibit diminishing marginal utility *or* that marginal utility may be constant, linear. Moreover, the utility function could exhibit diminishing returns for some values of x and y and constant returns for other values. As a result, a corner solution may be the optimal solution, but it could also be an interior solution. Now, her utility and budget constraint are given by the following:

$$U = U(x, y)$$
$$W = p_x x + p_y y.$$

Setting up the Lagrangian, taking the appropriate partial derivatives, and forming the first-order conditions, we have

$$
\begin{aligned}
\mathcal{L} &= U(x, y) + \lambda \left[W - p_x x - p_y y \right] \\
\frac{\partial \mathcal{L}}{\partial x} &= U_x(x, y) - p_x \lambda \leq 0 \\
\frac{\partial \mathcal{L}}{\partial y} &= U_y(x, y) - p_y \lambda \leq 0 \\
\frac{\partial \mathcal{L}}{\partial \lambda} &= W - p_x x - p_y y = 0.
\end{aligned}
$$

Let's interpret the two first-order conditions on x and y in economics terms. Starting with $U_x(x, y) - p_x \lambda \leq 0$, $U_x(x, y)$ is the marginal utility of x (manga). That is the marginal benefit to Nezuko from choosing a tiny bit more manga. $p_x \lambda$ is the marginal cost of manga. It is comprised of the price of manga times the marginal utility of wealth, as represented by the Lagrangian multiplier, λ. That should make sense. The marginal utility of wealth shows how much utility Nezuko gains from an additional unit of wealth. If she spends p_x, she loses that in utility from consuming less y. So, we have the usual marginal benefit and marginal cost in the first-order condition. But now, that equation could be either equal to zero or less than zero. If equal, it means that the marginal benefit equals the marginal cost and there is an *interior* solution such that $x^* > 0$. If it does not hold and is less than zero, then the marginal cost always exceeds the marginal benefit, and therefore the optimal choice is $x^* = 0$.

The first-order condition on y has the parallel interpretation. In $U_y(x, y) - p_y \lambda \leq 0$, $U_y(x, y)$ is the marginal utilty of y and therefore the marginal benefit of consuming more Pocky. The marginal cost of more Pocky is $p_y \lambda$, the marginal utility of wealth times the cost of Pocky, which represents the foregone utility from consuming more manga if she chooses more Pocky. Again, if the equation holds with equality, then we have an interior solution and $y^* > 0$. If it does not hold with equality, then we have a corner solution and $y^* = 0$.

What is the solution? The solution is the x^*, y^* and λ^* that satisfies the two first-order conditions and the budget constraint. We cannot take the problem any further and derive a more specific answer, but that is fine. Note that in this general problem we could have an interior solution with $x^* > 0$ and $y^* > 0$, or we could have two different types of corner solutions: one where $x^* = 0$ and the other where $y^* = 0$. We can rule out the possibility of *both* x^* and y^* being equal to zero. Why? Because our initial assumption ($U_x > 0$ and $U_y > 0$) that marginal utility is always positive for both x and y means that utility increases with x and y. She would never choose zero for both because by spending on at least one good she can increase her utility from zero if she consumes nothing.

We will fully summarize everything in the final subsection, but to recap this sub-section so far, when we take first-order conditions, we cannot automatically set them equal to zero to find the solution because of the potential for corner solutions. If the first-order condition is equal to zero, the optimal value of the associated choice variable will be strictly positive, an interior solution. If the first-order condition is less than zero, then the optimal value of the associated choice variable is zero. Writing that compactly in math terms we can write

$$\left(\frac{\partial \mathcal{L}}{\partial x_n} \right) x_n^* = 0,$$

where we suppose there are N choice variables. x_n represents any choice variable among the N possibilities. $\left(\frac{\partial \mathcal{L}}{\partial x_n} \right)$ is the first-order condition from the Lagrangian with respect to x_n. That times the optimal choice must be zero. Or, in other words, either the first-order condition is equal to zero *or* the optimal choice is equal to zero.

If in building our model we *assume* an interior solution (and this assumption is sometimes made), then we can set the first-order conditions equal to zero. However, we do often use utility functions (or objective functions more broadly) that are strictly concave such that the marginal value is always increasing but diminishing and that is usually sufficient to rule out corner solutions.

Consider the utility function we used before for the two-period consumer model in

Section 7.4:

$$U(c_1, c_2) = \ln c_1 + \beta \ln c_2$$

That utility function will not yield a corner solution. Why?

Let's set up the Lagrangian using one combined constraint and get the first-order conditions again to see why:

$$\mathcal{L} = \ln c_1 + \beta \ln c_2 + \lambda \left[y_1 + \frac{y_2}{1+r} - c_1 - \frac{c_2}{1+r} \right]$$

$$\frac{\partial \mathcal{L}}{\partial c_1} = \frac{1}{c_1} - \lambda \leq 0$$

$$\frac{\partial \mathcal{L}}{\partial c_2} = \frac{\beta}{c_2} - \lambda \left(\frac{1}{1+r} \right) \leq 0.$$

$$\frac{\partial \mathcal{L}}{\partial \lambda} = y_1 + \frac{y_2}{1+r} - c_1 - \frac{c_2}{1+r} = 0.$$

We have allowed for possible corner solutions by using the less than or equals signs in the two first-order conditions.

Now, consider the first term in the condition on c_1, which is $\frac{1}{c_1}$. That is the marginal utility of consumption in period 1. If there were a corner solution here, that would mean $c_1^* = 0$. But then the marginal benefit of consumption in period 1 becomes $\frac{1}{c_1} = \frac{1}{0}$, which is undefined. However, $\lim_{c_1 \to 0} \frac{1}{c_1} = +\infty$! The marginal benefit goes to infinity, while the marginal cost is finite. The marginal cost is λ, which is a finite value. Therefore, we do not have a situation in which the marginal cost always exceeds the marginal benefit and we do not have a corner solution. Some $c_1^* > 0$ will always be optimal, and the first-order condition will then always hold with equality. On your own, go through the same logic for c_2 and convince yourself that a corner solution is not possible.

Quick Check

e) Suppose that we have the following production maximization problem subject to a budget constraint: $\max_{x,y} Z = x^{1/2} + y$ s.t. $20 = 1x + 10y$. The firm has a budget of $B = 20$ and has to pay for inputs x and y at prices $p_x = 1$ and $p_y = 10$. The Lagrangian is $\mathcal{L} = x^{1/2} + y + \lambda[20 - 1x - 10y]$. Find the first-order conditions and determine which FOCs do and do not hold.

8.3   Nonbinding Constraints

Thus far in this section, we have looked at cases where the first-order conditions on the choice variables do not hold with equality. Either the bounds on the choice variable create a corner solution or the first-order condition fails to hold because the marginal cost always exceeds the marginal benefit and the agent chooses the lower bound of zero. There remains one more important case to cover, and that is when the first-order condition on the Lagrangian multipler does not hold, that is, when the **constraint** does not hold.

In almost everything we have done so far, when there was some kind of constraint, we implicitly assumed it would hold. In many cases, it will hold and sometimes has to hold. For example, the very first constraint we encountered was Margo's time constraint in Chapter 1, when we had $T = H + N$. The economic logic of that statement means it must hold with equality. Margo cannot escape time.

Definition

Binding/nonbinding: A constraint that can logically hold with equality or be an inequality $\leq$ or $\geq$ is considered "binding" when it holds equality and its presence affects the optimal choices made. A constraint is "nonbinding" if its presence does not affect the optimal choices made and the same solution would be obtained if the constraint were not there at all.

However, when we got to Chapter 4 and introduced Jiji's utility maximization problem, we had the following budget constraint: $4x + 2y = 32$, where x was donuts and y was fish. By using the equal sign here, we are implicitly assuming that Jiji uses all of his money. In fact, when we use the equal sign we are forcing Jiji to spend all his money on fish and donuts.

But why shouldn't a consumer have the option to not spend everything? We can allow for that option by rewriting the budget constraint a little more generally as $4x + 2y \leq 32$. Now, Jiji could spend everything or spend less than \$32 but cannot spend more than

\$32. That seems like a reasonable starting point and might matter considerably for some analyses where the constraint might "bind" (be equal to zero) in some cases or not bind (be not equal to zero) in other cases.

How do we handle that? First, recall when we introduced the Lagrangian approach that we made a big deal out of rearranging the constraint such that it is equal to zero when we add it to the objective function. Here is Jiji's problem again as we set it up in Chapter 7:

$$\mathcal{L} = \overbrace{x^{1/2} + y^{1/2}}^{\text{Objective function}} + \overbrace{\lambda}^{\text{Multiplier}} \overbrace{[32 - 4x - 2y]}^{\text{Constraint}}.$$

The set up does not change even if we allow the constraint to be something other than zero. It turns out that if the constraint does not bind, then the Lagrangian multiplier will be equal to zero, $\lambda = 0$. So, we are always adding zero to the objective function when we set up a Lagrangian because either the constraint is equal to zero or the multiplier is equal to zero. Intuitively, the idea that the multiplier goes to zero should make sense if you recall what the multiplier represents; *how much the objective function increases when we relax the constraint just slightly.* If the constraint is not binding, a small relaxation of the constraint would make no difference and therefore have no value.

To put that logic in the context of our example, suppose Jiji has some different utility function. He still has \$32 but chooses optimally to spend on \$20 on donuts and fish. That choice gives him the maximum utility he can achieve. If we "relax the constraint" and give Jiji \$1 more, and he now has \$33, Jiji would not change his choice. The extra dollar has no effect on his decision. Thus, the marginal value to Jiji's utility of one extra dollar is zero (i.e., the marginal value of relaxing the constraint is zero.) Morever, if we set up and solved the full problem, we would find that $\lambda^* = 0$.

8.3.1 Single–Variable Example

To illustrate, let's consider an example in single–variable calculus. Suppose that Tamayo enjoys train rides, x. Each train ride cost ¥200, and she has ¥600. Her constraint, allowing for the possibility that it is not binding, is $600 \geq 200x$. Her decision over how many train rides to take is represented by the following problem:

$$\max_{x} U = 10x - x^2$$
$$s.t.$$
$$200x \leq 600.$$

Setting up the Lagrangian we have:

$$\mathcal{L} = 10x - x^2 + \lambda[600 - 200x].$$

The first-order conditions are

$$\frac{\partial \mathcal{L}}{\partial x} = 10 - 2x - 200\lambda \leq 0$$

$$\frac{\partial \mathcal{L}}{\partial \lambda} = 600 - 200x \geq 0.$$

We have three possibilities here: (1) the interior solution, where $x > 0$ and the budget constraint binds, $600 - 200x = 0$; (2) another interior solution, where $x > 0$ but the constraint does not bind, $600 - 200x > 0$; or (3) a corner solution where $x = 0$ and the constraint does not bind, $600 - 200x > 0$.

Let's work our way backward through those options. We can rule out the third possibility quite quickly. If the constraint does not bind, then $\lambda = 0$. If $\lambda = 0$, then the first-order condition becomes $10 - 2x - 200\lambda \leq 0 \Rightarrow 10 - 2x - 200(0) \leq 0 \Rightarrow 10 - 2x \leq 0$. But if that's true then we cannot have $x = 0$ or we would have a contradiction ($10 \leq 0$). In economics terms, there is positive marginal utility at low levels, so choosing $x = 0$ is not optimal.

Now consider case 2, where the constraint does not bind. Again, that would mean $\lambda = 0$, but since $x > 0$, the first-order condition does bind: $10 - 2x - 200\lambda = 0 \Rightarrow x^* = 5$. But if $x^* = 5$, then it would violate the budget constraint since five train tickets would cost ¥1,000 but Tamayo only has ¥600.

That leaves us with option 1: an interior solution with a binding constraint. If the constraint binds, then $x^* = 3$, which we get directly from the budget constraint. Then the first-order condition becomes $10 - 2(3) - 200\lambda \Rightarrow \lambda^* = 1/50$. Everything there is consistent with the first-order conditions, so we have an interior solution, and the constraint is binding. We have nothing out of the ordinary thus far.

Quick Check

f) Suppose that a monopolist over Wisteria flowers has the following profit function, $\Pi = (21 - Q)Q - Q$, and faces a capacity constraint of $Q \leq 15$. Does the constraint bind? What if the constraint is $Q \leq 5$?

To illustrate the nonbinding constraint case, suppose Tamayo's budget is ¥2,000. Now the Lagrangian and the first-order condition become

$$\mathcal{L} = 10x - x^2 + \lambda\,[2,000 - 200x]$$

$$\frac{\partial \mathcal{L}}{\partial x} = 10 - 2x - 200\lambda \leq 0$$

$$\frac{\partial \mathcal{L}}{\partial \lambda} = 2,000 - 200x \geq 0.$$

The same three possibilities exist. The third possibility of a nonbinding constraint and a corner solution can be ruled out for the same reason. But now consider case 2 again. If the constraint does not bind, then $\lambda = 0$, and the first-order condition yields $x^* = 5$. That is consistent with the budget constraint because five train tickets only costs ¥1,000 and Tamayo has ¥2,000 now. That solution generates a utility level of $U = 10x - x^2 = 10(5) - 5^2 = 25$.

However, option 1 will now generate a contradiction. Under option 1 the budget constraint holds, which means that Tamayo buys 10 train tickets and spends her entire budget. But if that is entered into the first-order condition, which must hold with equality for an interior solution, it becomes $10 - 2x - 200\lambda = 0 \Rightarrow 10 - 2(10) - 200\lambda \Rightarrow -20 - 200\lambda = 0 \Rightarrow \lambda^* = -0.1$. That means that increasing Tamayo's budget *decreases* her utility. In other words, the marginal utility of wealth is negative. That is easy to see if you keep in mind the marginal benefit / marginal cost interpretation of the first-order condition, Marginal utility is given by the derivative which is $10 - 2x$. If x is 10, then marginal utility is -10, meaning Tamayo's utility decreases with more train rides. Indeed, compare the utility level to what we got in the previous case. Now, $U(10) = 10x - x^2 = 10(10) - 10^2 = 0$. By spending all her wealth, Tamayo is not maximizing her utility, and therefore $x = 10$ cannot be a solution to the maximization problem. As a result, case 2 is now the solution, where $x^* = 5$, the budget constraint does not hold since $2,000 - 200(5) = 1,000 > 0$, and thus $\lambda^* = 0$.

8.3.2   Two–Variable Example

Suppose our consumer, Inosuke, is trying to book the perfect vacation from his point of view. The perfect vacation is the right number of days on vacation and the distance traveled. His utility over the vacation is

$$U(x, y) = 12x - x^2 + 8y - y^2,$$

where x is the number of days on vacation and y is the driving distance to the vacation spot in hundreds of kilometers. Inosuke has a budget of $550, each day of vacation costs $100, and every 100 kilometers driven is an additional $50, so his budget constraint is:

$$550 \geq 100x + 50y.$$

Setting up the Lagrangian, we have

$$\mathcal{L} = 12x - x^2 + 8y - y^2 + \lambda[550 - 100x - 50y].$$

The first-order conditions are

$$\frac{\partial \mathcal{L}}{\partial x} = 12 - 2x - 100\lambda \leq 0$$

$$\frac{\partial \mathcal{L}}{\partial y} = 8 - 2y - 50\lambda \leq 0$$

$$\frac{\partial \mathcal{L}}{\partial \lambda} = 550 - 100x - 50y \geq 0.$$

The numerical example in this initial problem has been chosen such that there is an interior solution. However, the reader is strongly encouraged to go through all the possible cases and show that all three conditions must hold at the optimum. The solution one gets after some algebra is $x^* = 4$ and $y^* = 3$. You might wish to take the time to verify that for yourself right now, as it may be helpful in understanding what comes next.

But now consider the following variation, where Inosuke's budget increases to \$1,050 to spend. The set–up and first–order conditions are

$$\begin{aligned}
\mathcal{L} &= 12x - x^2 + 8y - y^2 + \lambda[1050 - 100x - 50y] \\
\frac{\partial \mathcal{L}}{\partial x} &= 12 - 2x - 100\lambda \leq 0 \\
\frac{\partial \mathcal{L}}{\partial y} &= 8 - 2y - 50\lambda \leq 0 \\
\frac{\partial \mathcal{L}}{\partial \lambda} &= 1050 - 100x - 50y \geq 0.
\end{aligned}$$

Now, if one goes through the same algebra, assuming all three conditions hold, the results are $x^* = 8$ and $y^* = 5$. Without any further investigation, that would seem to be a reasonable answer. However, it would be incorrect unless one assumed that Inosuke had to spend his entire budget.

Consider the case where the budget constraint does not bind and $\frac{\partial \mathcal{L}}{\partial \lambda} > 0$. Inosuke chooses to spend less than his budget constraint. That means λ goes to zero and the first-order conditions reduce to

$$\begin{aligned}
\frac{\partial \mathcal{L}}{\partial x} &= 12 - 2x \leq 0 \\
\frac{\partial \mathcal{L}}{\partial y} &= 8 - 2y \leq 0 \\
\frac{\partial \mathcal{L}}{\partial \lambda} &= 1050 - 100x - 50y > 0.
\end{aligned}$$

The Lagrange multiplier goes to zero and disappears completely from the first-order conditions. Solving the resulting first-order conditions for an interior solution we get $x^* = 6$ and $y^* = 4$. Notice these values are both less than the values we found when we assumed the budget constraint held with equality. A quick comparison of the utility levels quickly reveals that $x^* = 6$ and $y^* = 4$ yields the higher utility, $U^* = 52$, while if $x^* = 8$ and $y^* = 5$, then $U^* = 47$.

What is happening? Figure 8.3 illustrates this problem. The point in the center is what we call a "bliss" point, and it represents a maximum point for utility. The agent cannot do better than that specific point. Recall our discussion of indifference curves in Section 6.3 and that one can think of the indifference curve graphs as similiar to topological maps in that the lines represent equal heights. The point in the center here is akin to the top of the mountain. When the budget constraint is \$550, the closest that Inosuke can get to the peak is $x^* = 4$ and $y^* = 3$. However, when his budget expands to \$1,100, he can reach the peak; that now becomes a feasible choice. If we assume the budget constraint holds, that means we are forcing Inosuke to choose a point along the upper budget constraint, and he picks a point as close as he can get to the peak, but clearly suboptimal. In fact, he gets the same level of utilty at $(x^*, y^*) = (8, 5)$ as he did at $(x^*, y^*) = (4, 3)$ when he had half the budget.

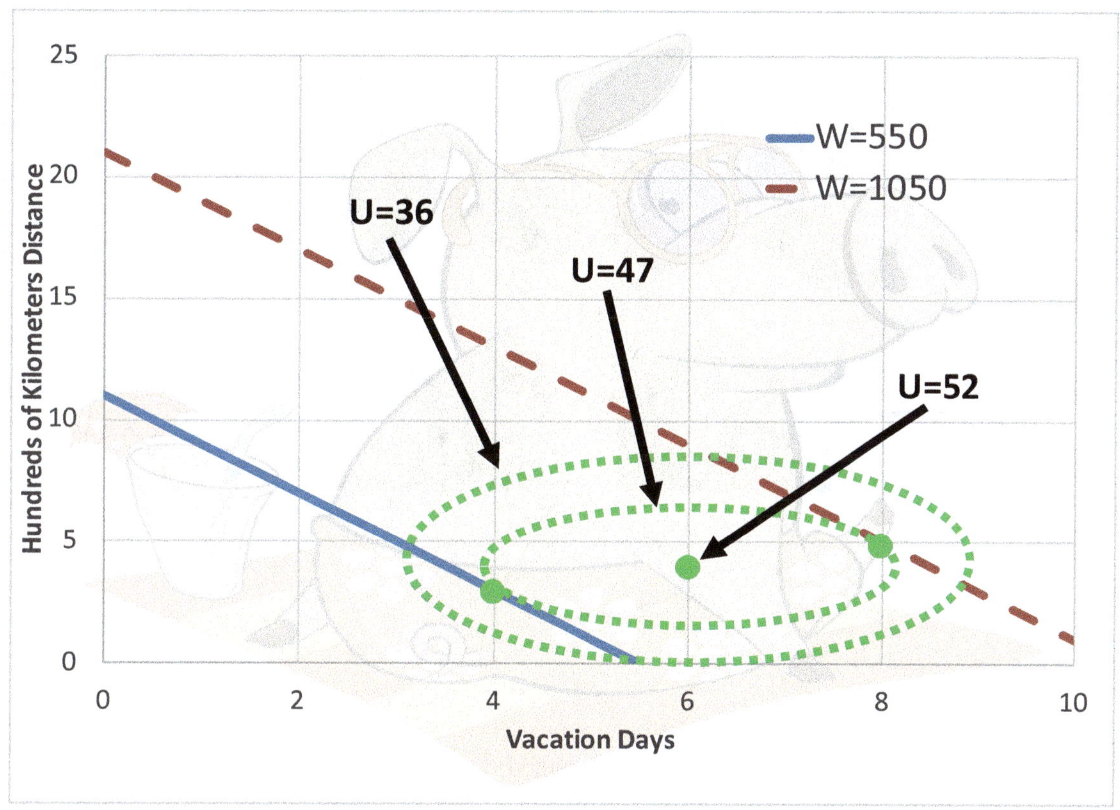

Figure 8.3: Circular indifference curves and peak utility.

How do you know when utility is higher if the constraint does not bind? The most direct way is to solve for both the constrained and unconstrained solutions, then compare the utility levels, of course, checking that the solution without the constraint is still feasible (less than the whole budget). But there is another more direct clue to look at for whether the constraint will bind or not, the sign of the Lagrangian multiplier.

In the example when Inosuke's budget was only \$550, using the first-order condition, you quickly see that $\lambda^* = 1/25$, a positive value. The positive value matters because, recalling how we interpret the multiplier as the marginal value of relaxing the constraint, the positive value means utility would increase with a relaxation of the constraint. That is, Inosuke's utility would rise with a larger budget.

However, in the second example, when we forced the budget constraint to hold, λ flips signs and becomes $\lambda^* = -1/25$, a negative value. That, in effect, says Inosuke's utility would *decrease* with a larger budget. Why? Because by forcing him to remain on the constraint we would be moving him further from the bliss point. So, the true optimal value of λ is zero, which means the constraint is not binding in this case.

8.4 Karush-Kuhn-Tucker (KKT)

All of our cases can be summarized by what is known at the Karush-Kuhn-Tucker theorem (KKT). The formal statement is presented here, with more explanation to follow. The reader is encouraged to take this definition and go back through the examples in this section to see how it applies in each case.

Theorem 8.2: Karush-Kuhn-Tucker theorem (KKT). *Consider the following general constrained optimization problem with $n = (1, ..., N)$ choice variables and $m = (1, ..., M)$ constraints:*

$$\max_{x_1, x_2, ..., x_N} f(x_1, x_2, ..., x_N)$$

$$\text{s.t.}$$

$$g^1(x_1, x_2, ..., x_N) \geq 0$$
$$g^2(x_1, x_2, ..., x_N) \geq 0$$
$$\vdots$$
$$g^M(x_1, x_2, ..., x_N) \geq 0.$$

If all the functions, f and g^m, $m = 1, ..., M$, are concave and differentiable, then for the following Lagrangian problem

$$\mathcal{L} = f(x_1, x_2, ..., x_N) + \sum_{1}^{M} \lambda_m g^m(x_1, x_2, ..., x_N)$$

there exists m Lagrangian multipliers λ_m^* such that the following conditions are necessary and sufficient for the point $(x_1^*, x_2^*, ..., x_N^*)$ to be a solution to the problem:

$$\overbrace{x_n^*}^{\text{Choice variable}} \quad \overbrace{\left[f_n(x_1^*, x_2^*, ..., x_N^*) - \sum \lambda_m^* g_n^m(x_1^*, x_2^*, ..., x_N^*) \right]}^{\text{First-order condition on } x_n^*} \;=\; 0 \quad \forall n$$

$$\overbrace{\lambda_m^*}^{\text{Lagrange multiplier } m} \quad \overbrace{g^m(x_1^*, x_2^*, ..., x_N^*)}^{\text{Constraint } m} \;=\; 0 \quad \forall m.$$

Writing those two conditions more compactly without showing all the arguments in the objective function, $f(\cdot)$, or the constraints, $g_m(\cdot)$, for clarity, we have

$$\overbrace{x_n^*}^{\text{Choice variable}} \quad \underbrace{\left[f_n - \sum \lambda_m^* g_n^m \right]}_{\text{First-order condition on } x_n^*} \;=\; 0 \quad \forall n$$

$$\overbrace{\lambda_m^*}^{\text{Lagrange multiplier } m} \quad \underbrace{g^m}_{\text{Constraint } m} \;=\; 0 \quad \forall m.$$

A quick summary in words of the two conditions is: (1) $x_n^* \left[f_n - \sum \lambda_m^* g_n^m \right]$: Either the choice variable or the first-order condition must be zero for all choice variables; and (2) $\lambda_m^* g^m$: either the Lagrange multiplier or the constraint must be zero for all constraints.

To elaborate and pull apart the notation, in the first condition we have the nth choice variable at the optimal level, x_n^*, multiplying the first-order condition that results from taking the derivative of the Langrangian with respect to x_n. Since they are multiplying each other and equal to zero, at least one of the two has to be zero: Either the choice variable is zero, or the first-order condition is zero. The second condition has the optimal value of the mth Lagrange multiplier, λ_m^*, and that is multiplying its associated constraint, g^m. The product is again zero, which means at least one of the two must be zero: either the Lagrange multiplier or the constraint. These are sometimes referred to as *complementary slackness conditions*.

8.5 Exercises

8.5.1 Quick Check Answers

a) FOC: $10x^{-1/2} - 2 = 0 \Rightarrow x^* = 25$. $x^* = 25$ lies within the boundaries so it is an interior solution.

b) FOC: $10x^{-1/2} - 2 = 0 \Rightarrow x^* = 25$. $x^* = 25$ exceeds the upper bound; therefore, the solution is $x^* = 10$, the upper bound.

c) $\mathcal{L} = 2G + 3P + \lambda[42 - 5G - 2P]$, FOCs: $\frac{\partial \mathcal{L}}{\partial G} = 2 - \lambda 5 \leq 0$, $\frac{\partial \mathcal{L}}{\partial P} = 3 - \lambda 2 \leq 0$. They cannot hold simultaneously because that would imply $\lambda^* = 2/5$ and $\lambda^* = 3/2$, which is a contradiction. If the first FOC holds, then $2 - \lambda 5 = 0$ and $\lambda^* = 2/5$. But using the second FOC that implies $3 - (2/5)2 = 11/5 \leq 0$, which is a contradiction. If the second FOC holds, then $3 - \lambda 2 = 0$ and $\lambda^* = 3/2$. That is consistent with the first condition, $2 - (3/2)5 = -11/2 \leq 0$. Therefore, $G^* = 0$, $P^* = 21$, and $\lambda^* = 3/2$.

d) $\mathcal{L} = 2G + 3P + \lambda[42 - 1G - 6P]$, FOCs: $\frac{\partial \mathcal{L}}{\partial G} = 2 - \lambda 1 \leq 0$, $\frac{\partial \mathcal{L}}{\partial P} = 3 - \lambda 6 \leq 0$. They cannot hold simultaneously because that would imply $\lambda^* = 2$ and $\lambda^* = 1/2$, which is a contradiction. If the first FOC holds, then $2 - \lambda 1 = 0$ and $\lambda^* = 2$. That is consistent with the second condition, $3 - (2)6 = -9 \leq 0$. If the second FOC holds, then $3 - \lambda 6 = 0$ and $\lambda^* = 1/2$. But using the first FOC implies $2 - (1/2)1 = 3/2 \leq 0$, which is a contradiction. Therefore, $G^* = 42$, $P^* = 0$, and $\lambda^* = 2/5$.

e) $\mathcal{L} = x^{1/2} + y + \lambda[20 - x - 10y]$. FOCs: $\frac{\partial \mathcal{L}}{\partial x} = (1/2)x^{-1/2} - \lambda \leq 0$, $\frac{\partial \mathcal{L}}{\partial y} = 1 - 10\lambda \leq 0$, $\frac{\partial \mathcal{L}}{\partial \lambda} = 20 - x - 10y = 0$. The first one, on x, must hold because the marginal benefit portion has $\frac{1}{2x^{1/2}}$, and if $x* \to 0$ that will approach infinity. If the second condition also holds, we have $\lambda^* = 1/10$. Using that value in the first condition, we get that $x^* = 25$. But putting that into the budget constraint gives us $20 = 25 + 10y$, which cannot hold regardless of the value of y since y cannot be negative. Therefore, the second condition does not hold. The result is $x^* = 20$, $y^* = 0$, and $\lambda^* = 1/4\sqrt{5}$.

f) $\max_Q \Pi = (21 - Q)Q - Q$ s.t. $Q \leq 15$. FOC: $\frac{d\Pi}{dQ} = 21 - 2Q - 1 \leq 0 \Rightarrow Q^* = 10$. $Q^* = 10$ lies within the boundaries, so that solution is the profit-maximizing solution. If, however, the constraint is $Q \leq 5$, then $Q^* = 10$ cannot be the solution as it lies outside the boundaries. The solution is then at the upper boundary: $Q^* = 5$.

g) FOCs: $\frac{\partial \mathcal{L}}{\partial x} = 12 - 2x - 100\lambda \leq 0$, $\frac{\partial \mathcal{L}}{\partial y} = 8 - 2y - 50\lambda \leq 0$, and $\frac{\partial \mathcal{L}}{\partial \lambda} = 550 - 100x - 50y \geq 0$.

Case 1: $\frac{\partial \mathcal{L}}{\partial x} < 0$; $\frac{\partial \mathcal{L}}{\partial y} = 0$; and $\frac{\partial \mathcal{L}}{\partial \lambda} = 0$. That implies $x^* = 0$, $y^* = 11$, and $\lambda^* = -7/25$. The negative value indicates a problem. Note that in the case in which all the conditions hold, we got $x^* = 4$ and $y^* = 3$, so his utility would be $U = 12(4) - 4^2 + 8(3) - 3^2 = 47$. However, if $x^* = 0$, $y^* = 11$, then his utility is $U = 12(0) + 0^2 + 8(11) - 11^2 = -33$, and clearly not the maximum. So case 1 is ruled out.

Case 2: $\frac{\partial \mathcal{L}}{\partial x} = 0$; $\frac{\partial \mathcal{L}}{\partial y} < 0$; and $\frac{\partial \mathcal{L}}{\partial \lambda} = 0$. Following the same logic in case 1, we now get $x^* = 5.5$, $y^* = 0$, and $\lambda^* = 1/100$. $\lambda^* > 0$, but the utility generated is $U = 12(5.5) - (5.5)^2 = 30.25 > 0$, which is less than the case where all the first-order conditions hold. So case 2 is ruled out.

Case 3: $\frac{\partial \mathcal{L}}{\partial x} = 0$; $\frac{\partial \mathcal{L}}{\partial y} = 0$; and $\frac{\partial \mathcal{L}}{\partial \lambda} > 0$. Now since the constraint does not bind, $\lambda = 0$ and we get $x^* = 6$ and $y^* = 4$. But putting those values back into the budget constraint results in: $100(6) + 50(4) = 800 \leq 500$, which is obviously a contradiction, as the spending exceeded the budget constraint. Case 3 is ruled out. That leaves the case of all the constraints binding, as shown in the text.

h) FOCs: $\frac{\partial \mathcal{L}}{\partial x} = 2(x - 1) - \lambda \leq 0$; $\frac{\partial \mathcal{L}}{\partial y} = 2(y - 1) - \lambda \leq 0$; and $\frac{\partial \mathcal{L}}{\partial \lambda} = 10 - x - y \geq 0$. If the constraint holds, then the solution is $x^* = 5$, $y^* = 5$, and $\lambda^* = 8$. That implies that $z^* = 32$. Recall, we are *minimizing* here, so if the constraint does not hold, then $\lambda^* = 0$, $x^* = 1$, and $y^* = 1$. That yields $z^* = 0$, which is clearly lower and therefore the optimal solution.

8.5.2 Practice Problems

1. Solve the following problems with boundary constraints on the choice variable. For each solution, explain how your answer satisfies Theorem 8.1.

   (a) $\max y = -2x^2$ subject to $x \geq -5$

   (b) $\max y = -2x^2$ subject to $x \geq 1$

   (c) $\min y = (x - 4)^2 + 5$ subject to $x \leq 10$

   (d) $\min y = (x - 4)^2 + 5$ subject to $x \leq 2$

   (e) Cost minimization: $\min_Q TC = 8Q^2 - 4Q + 10$ subject to $Q \geq 0$

   (f) Cost minimization: $\min_Q TC = 8Q^2 - 4Q + 10$ subject to $Q \geq 10$

2. Redo Nezuko's problem from Section 8.2.1 with the linear utility function, but use the following prices: $p_x = 2$ and $p_y = 3$. Solve the problem and be sure to interpret the first-order conditions. Create a graph like the one in Figure 8.2 for this problem.

3. Solve the following problems under the two different constraints given and compare your answers. Be sure to use a Lagrangian and compare the values of λ.

   (a) $\min Z = x^2 + y^2$ subject to $x + y = 10$ or $x + y \leq 10$

   (b) $\max \Pi = 100K^{1/2} + 100L^{1/2} - 10K - 5L$ subject to $K + L = 100$ or $K + L = 200$

4. Kyojuro's water ice shop has the following production function:

$$y = 3L^{1/2} \text{ for } L \leq 6$$
$$y = 6 \text{ for } L > 6,$$

where L is the amount of labor he hires.

(a) If the price of the water ice is ¥6 and the wage rate for labor is ¥4, set up Kyojuro's profit–maximization problem (revenues minus costs) and solve for L^*.

(b) The weather is really hot, demand for Kyojuro's water ice skyrockets, and he raises his price to ¥10; solve for L^*.

5. Redo Kyojuro's water ice shop problem in the preceding problem but in general terms, as follows:

$$y = AL^{1/2} \text{ for } L \leq \bar{L}$$
$$y = \bar{Y} \text{ for } L > \bar{L},$$

where L is the amount of labor he hires.

(a) If the price of the water ice is P and the wage rate for labor is w, set up Kyojuro's profit maximization problem and solve for L^*. Clearly, define the various cases and how they depend on the exogenous variables and parameters A, P, $\bar{Y}$, and w. (Hint: You should find two cases.)

(b) The weather is really hot, demand for Kyojuro's water ice skyrockets, and he raises his price to $\bar{P}$, where $\bar{P} > P$; solve for L^*. Discuss what happens to his production of water ice in light of the two cases you found in part (a).

6. Consider the following two–period consumer problem that Muzan faces:

$$\max_{c_1, c_2} = \ln(c_1) + \beta \ln(c_2)$$
$$\text{s.t.} \quad c_1 + \frac{c_2}{1+r} = a + y_1 + \frac{y_2}{1+r},$$

where the variables and parameters are all the same, as in Section 7.4. However, assume that $y_1 = 100$, $y_2 = 100$, $a = 0$, $\beta = 0.95$, and $r = 0.05$. Note that b is still part of the problem but has been substituted out when the first– and second–period budget constraints were combined.

(a) Solve Muzan's problem using a Lagrangian with a single constraint on lifetime consumption and wealth. Solve for specific optimal values of c_1^*, c_2^*, and b^*. Interpret the first-order condition in economic terms.

(b) Now, suppose that Muzan's employer cannot pay him this period, but can pay him the full wages in period 2, such that $y_1 = 0$, $y_2 = 120$, $a = 0$, and $r = 0.05$. Resolve Muzan's problem and compare with your answers in part (a).

(c) Now, suppose that Muzan's employer still cannot pay until the second period. But, in addition, the Bank of Edo has doubts about Muzan's creditworthiness and limits his borrowing to $b \geq -10$. Before attempting to solve this problem, draw the budget constraint implied by this borrowing limit.

(d) Once you have the budget constraint drawn, solve for Muzan's new optimal consumption levels and borrowing under this new constraint. Be sure to use a Lagrangian and add another Lagrangian multiplier to account for the new borrowing constraint as $c_1 - y_1 + 10 \geq 0$. Interpret all the first-order conditions.

7. **Quasi-linear utility.** One useful variant on our objective functions, particularly for utility, is a functional form known as quasi-linear. For example, $U(x,y) = x + y^{1/2}$ is a quasi-linear utility function over two goods, x and y. One good (or many goods) enter linearly, x in this case, but at least one other good enters nonlinearly, y.

(a) Before getting into the problem, create an indifference curve graph for this objective function, $U(x,y) = x + y^{1/2}$, for utility levels of $U = 2$, $U = 3$, and $U = 4$. It is suggested that you use Excel or some other graphing program. What is special about this set of indifference curves?

(b) Now suppose that Mitsuri consumes two things: hearts, x, and diamonds, y, and her preferences are given by the utility function discussed: $U(x,y) = x + y^{1/2}$. Her budget constraint is: $W = p_x x + p_y y$. Set up this problem as a Lagrangian.

(c) Get the first-order conditions and interpret them in economic terms.

(d) Suppose that $p_x = 2$, and $p_y = 2$. What are the optimal levels of consumption of hearts and diamonds for Mitsuri when $W = 1$, $W = 1/4$, and $W = 10$? Explain these answers and how the quasi-utility function affects the results.

8.6   Math Appendix

Here, we list some key math definitions, theorems, and formulae in formal math for your reference.

Second–order conditions for an *unconstrained* two–variable problem using matrix algebra.
One approach to checking the second-order conditions of a multivariate optimization problem is to use matrix algebra (Chapter 12). To do so, we form a **Hessian** matrix of all the 2nd derivatives. Consider a general multivariate function we seek to either maximize of minimize

$$\max_{x,y} \left(\text{ or } \min_{x,y} \right) z = f(x,y).$$

Suppose that (x^*, y^*) satisfies the first-order conditions such that

$$\frac{\partial z}{\partial x} = f_x(x^*, y^*) = 0 \text{ and } \frac{\partial z}{\partial y} = f_y(x^*, y^*) = 0.$$

Take the full set of second derivatives, including the cross-derivative,

$$\frac{\partial^2 z}{\partial x^2} = f_{xx}(x,y), \quad \frac{\partial^2 z}{\partial y^2} = f_{yy}(x,y), \text{ and } \frac{\partial^2 z}{\partial x \partial y} = f_{xy}(x,y),$$

and arrange them into the Hessian matrix, H, as follows:

$$H = \begin{bmatrix} f_{xx}(x,y) & f_{xy}(x,y) \\ f_{xy}(x,y) & f_{yy}(x,y) \end{bmatrix},$$

with the own second derivatives along the diagonal and the cross–derivative in the off–diagonal positions. The solution (x^*, y^*) is a maximum if $f_{xx}(x^*, y^*) < 0$, $f_{yy}(x^*, y^*) < 0$, and the determinant of the Hessian matrix at (x^*, y^*) is positive:

$$|H| = \begin{vmatrix} f_{xx}(x,y) & f_{xy}(x,y) \\ f_{xy}(x,y) & f_{yy}(x,y) \end{vmatrix} = f_{xx}(x^*, y^*)f_{yy}(x^*, y^*) - (f_{xy}(x^*, y^*))^2 > 0.$$

The solution (x^*, y^*) is a minimum if $f_{xx}(x^*, y^*) > 0$, $f_{yy}(x^*, y^*) > 0$, and the determinant of the Hessian matrix at (x^*, y^*) is positive:

$$|H| = \begin{vmatrix} f_{xx}(x,y) & f_{xy}(x,y) \\ f_{xy}(x,y) & f_{yy}(x,y) \end{vmatrix} = f_{xx}(x^*, y^*)f_{yy}(x^*, y^*) - (f_{xy}(x^*, y^*))^2 > 0.$$

Notice that the matrix determinants produce the saddle point condition discussed in Chapter 6.

Second–order conditions for a *constrained* two–variable, one–constraint problem using matrix algebra. Compare the previous definition to the representation for a constrained optimization problem. Now, suppose we have

$$\max_{x,y}\left(\text{ or }\min_{x,y}\right) z = f(x,y) \text{ subject to } g(x,y) = 0.$$

Suppose that (x^*, y^*) satisfies the first-order conditions of the Lagrangian such that

$$\frac{\partial \mathcal{L}}{\partial x} = f_x(x^*, y^*) + \lambda g_x(x^*, y^*) = 0 \text{ and } \frac{\partial \mathcal{L}}{\partial y} = f_y(x^*, y^*) + \lambda g_y(x^*, y^*) = 0.$$

Taking the full set of second derivatives including the cross-derivative *and* the lagrangian multiplier, λ,

$$\frac{\partial^2 \mathcal{L}}{\partial x^2} = f_{xx}(x,y) + \lambda g_{xx}(x,y), \quad \frac{\partial^2 \mathcal{L}}{\partial y^2} = f_{yy}(x,y) + \lambda g_{yy}(x,y), \text{ and } \frac{\partial^2 \mathcal{L}}{\partial x \partial y} = f_{xy}(x,y) + \lambda g_{xy}(x,y)$$

$$\frac{\partial^2 \mathcal{L}}{\partial x \partial \mathcal{L}} = g_x(x,y), \quad \frac{\partial^2 \mathcal{L}}{\partial y \partial \mathcal{L}} = g_y(x,y), \text{ and } \frac{\partial^2 \lambda}{\partial \mathcal{L}^2} = 0.$$

Arrange these into a *bordered* Hessian matrix, $\bar{H}$, such that the derivatives with respect to λ form the last column and the last row, as follows:

$$|\bar{H}| = \begin{bmatrix} f_{xx}(x,y) + \lambda g_{xx}(x,y) & f_{xy}(x,y) + \lambda g_{xy}(x,y) & g_x(x,y) \\ f_{xy}(x,y) + \lambda g_{xy}(x,y) & f_{yy}(x,y) + \lambda g_{yy}(x,y) & g_y(x,y) \\ g_x(x,y) & g_y(x,y) & 0 \end{bmatrix}. \tag{1}$$

For a two–choice variable, one–constraint problem, the point (x^*, y^*) satisfying the first-order conditions will be a maximum if $|\bar{H}| > 0$ and a minimum if $|\bar{H}| < 0$. Note that those conditions only apply to a one–constraint, two–variable problem. The conditions change with more variables and/or more constraints. However, we will not go into those conditions here.

9 Duality

Duality is an approach to constrained optimization problems that makes use of the fact that we can look at our problems in a different way by reversing the roles of the objective function and the constraint. For example, in our consumer's utility maximization problem, the **UMP**, the objective function is the utility function, and the budget constraint is, well, the constraint. The consumer chooses quantities of goods to maximize utility but must fit those choices within the budget constraint.

But suppose we made the budget the objective function and utility the constraint. What does that mean? Consider the idea that our consumer needs to figure out the quantities to buy such that utility is at least $\bar{U}$. Suppose Paddi likes two goods, cake and candy, and he wants to get to a certain level of utility by consuming them. He could spend trillions of dollars or more on cake and candy and easily reach $\bar{U}$ level of utility. However, that would likely be overkill (and possibly make him sick if he ate trillions of dollars worth of cake and candy all at once). Instead, the goal now is to reach $\bar{U}$ and do so by spending the least amount. That is, what is the most efficient allocation of spending on goods that gets our consumer to $\bar{U}$? We call this problem, the **Expenditure Minimization Problem** or **EMP**. The EMP is the "dual" of the UMP (the utility maximization problem).

Analogously, consider our firm's problem with a constraint added. Suppose, that the firm has a limited budget for production. Given, that it can not exceed the budget, what is the maximimum level of output it can generate? We refer to that problem as the **Production Maximization Problem, PMP**, and it is the primal problem. Alternatively, suppose that the firm has a target production level, so the firms seeks to reach that level by minimizing the cost of inputs. We call that the **Cost Minimization Problem, (CMP)**. The CMP is the "dual" of the PMP.

The pair of dual problems, one for the consumer and the other for the firm, are extremely similar, so similar that it is quite common in textbooks to go through duality in one context but not the other, leaving it to the students to work out on their own how to do the other. That is a good exercise and approach to see how well you understand the material. Here, however, we are going to do both such that, if you get stuck in one, perhaps by working through the other you will be able to solve the issue.

The concept of duality is often presented with a series of proofs using convex sets. The proofs are elegant, but perhaps tedious and not terribly intuitive for many. If you are inclined to like rigorous proofs, apologies; you may find that approach fascinating and intuitive, and you are encouraged to seek out that exposition. For those who are not theorem-proof inclined, here is a reduced version of it that gets to the applications and economic logic.

Furthermore, it turns out that the solutions to the two dual problems have interesting properties that allow us better investigate and understand consumer and firm behavior. They provide useful relations that can be insightful when we have data on consumer spending or firm production, that we might not be able to directly observe. These properties can be exploited by econometricians to characterize real–world consumer and firm behavior in ways that are useful not only for economists, but for any business person or government regulatory agency. These topics are beyond what we do here, but serve as motivation for exploring the dual problems as a useful application of the math to real–world issues.

> **Definition**
>
> **Key Abbreviations:**
>
> UMP = Utility Maximization Problem (Consumer)
> EMP = Expenditure Minimization Problem (Consumer)
> PMP = Production Maximization Problem (Firm)
> CMP = Cost Minimization Problem (Firm)
> **We will use these abbreviations throughout this chapter.**

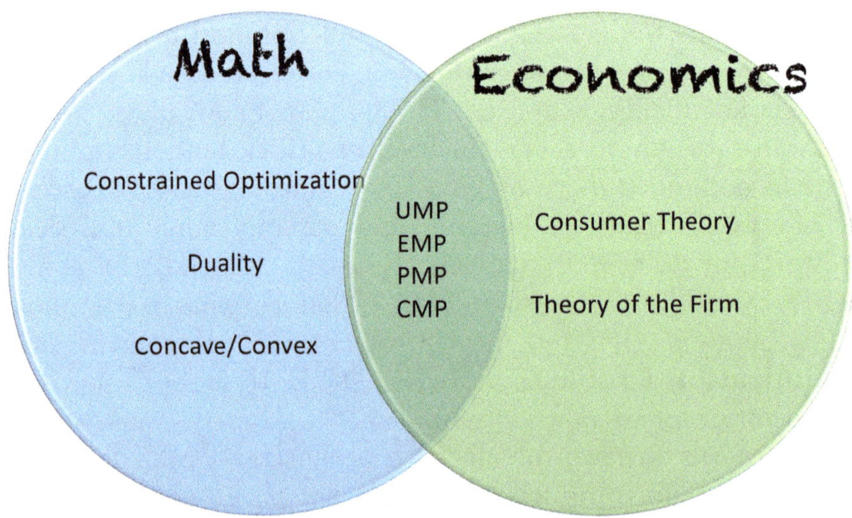

9.1 Consumers: The UMP and the EMP

Let's start with the most basic consumer UMP, a review, basically, and then we will look at the dual EMP. Once we are done, we will compare the solutions and then show the relationship mathematically and graphically.

9.1.1 Quick Review of the UMP

Xiyangyang wishes to maximize his utility. He likes to consume dumplings, D, and mooncakes, M. At his favorite restaurant, Jia Chang's Homestyle Restaurant, dumplings cost \$1 and mooncakes \$2 each. He has \$24, and his utility is given by $U = ln(D) + ln(M)$. Setting up the UMP we have:

$$\max_{D,M} \quad ln(D) + ln(M), \text{ subject to } 1D + 2M = 24$$

$$\mathcal{L} = \overbrace{ln(D) + ln(M)}^{\text{Objective function}} + \lambda \overbrace{\left[24 - 1D - 2M\right]}^{\text{Constraint}}.$$

Notice the labeling in the equation in that the utility function is the objective function (i.e., what we are trying to maximize,) while the budget constraint serves its familiar role as the constraint, which is accounted for using a Lagrangian multiplier. Forming the first-order conditions (assuming all hold with equality) and solving we have:

> **Definition**
>
> **Utility Maximization Problem (UMP)**: The primal decision problem of consumers who maximize their subjective utility while being constrained by a budget limitation.

$$\frac{\partial \mathcal{L}}{\partial D} = \frac{1}{D} - \lambda = 0$$

$$\frac{\partial \mathcal{L}}{\partial M} = \frac{1}{M} - 2\lambda = 0$$

$$\frac{\partial \mathcal{L}}{\partial \lambda} = 24 - 1D - 2M = 0$$

$$D^* = 12, \ M^* = 6, \ \lambda^* = 1/12.$$

With that solution, Xiyangyang gets a level of utility equal to $U = ln(12) + ln(6) = 2.48 + 1.79 = 4.27$. That is, $U = 4.27$ is the maximum level of utility Xiyangyang can achieve with \$24 at Jia Chang's restaurant.

9.1.2 Expenditure Minimization

Now, let's turn the problem around. Suppose that Xiyangyang goes to back to Jia Chang's restaurant the next day seeking be just as happy as he was the day before. Xiyangyang wants to figure out how much money he needs to bring to Jia Chang's (they only take cash), but he doesn't want

to overspend. Now his problem is *what is the minimum amount of money he must spend to achieve a utility level of 4.27?* Notice how the problem in words is inverted from the standard UMP: *"What is the maximum utility that can be achieved given the spending limit?"* Now we have the EMP that asks: *"What is the minimum expenditure needed to achieve the utility requirement?"*

Setting up the EMP we have:

$$\min_{D,M} \quad 1D + 2M, \text{ subject to: } 4.27 = ln(D) + ln(M)$$

$$\mathcal{L} = \overbrace{1D + 2M}^{\text{Objective Function}} + \lambda \overbrace{[4.27 - ln(D) - ln(M)]}^{\text{Constraint}}.$$

Compare that with the UMP from before The objective function and constraint flip roles. In the UMP we maximize utility subject to a budget constraint, where the exogenous variable is the wealth level, W, or 24 in the example. In the EMP, we *minimize* the expenditure subject to a utility constraint where the exogenous variable is the minimum utility level, $\bar{U}$, which is 4.27 in the example.

When we are being more general and allow for the possibility that the utility constraint does not bind, we write something like $\bar{U} \geq 4.27$ or $ln(D) + ln(M) \geq 4.27$ which states utility must be greater than or equal to 4.27. That is analogous to what we did in Chapter 8 when we allowed the budget constraint to be an inequality such as $W \geq p_x x + p_y y$. In that case, W is the spending limit and cannot be exceeded, but there may be cases when the individual optimally chooses to spend something less than everything they have. In the EMP, $\bar{U}$ is a *minimum* requirement, so when we allow for a nonbinding constraint, we write it such that actual utility (e.g., $ln(D) + ln(M)$) could be equal to or greater than the requirement $\bar{U}$.

Returning to the problem and forming the first-order conditions we get

$$\frac{\partial \mathcal{L}}{\partial D} = 1 - \lambda \frac{1}{D} = 0$$

$$\frac{\partial \mathcal{L}}{\partial M} = 2 - \lambda \frac{1}{M} = 0$$

$$\frac{\partial \mathcal{L}}{\partial \lambda} = ln(D) + ln(M) = 4.27.$$

Solving, we have the following, starting by dividing the first condition by the second condition:

$$\frac{1}{2} = \frac{M}{D}$$

$$D = 2M$$

$$ln(M) + ln(2M) = 4.27$$

$$ln(M) + ln(2) + ln(M) = 4.27$$

$$2ln(M) = 4.27 - ln(2)$$

$$ln(M) = \frac{1}{2}(3.58) = 1.79$$

$$e^{ln(M)} = e^{1.79}$$

$$M = 6$$

$$D^* = 12, \ M^* = 6, \ \lambda^* = 12.$$

Therefore, the minimum expenditure needed to reach $\bar{U} = 4.27$ is $\$1(12) + \$2((6) = \$24$. Note that it is no coincidence, the answer here from the EMP is the same as the exogenous constraint in the UMP. There is some rounding going on, but the quantities of dumplings (12) and mooncakes (6) also gets us back the answers from the UMP. However, λ does differ. In the UMP λ was the marginal utility of wealth. Here, λ is the marginal increase in expenditure with a marginal increase in the required utility, $\bar{U}$. That indicates how quickly expenditure will rise with increases in the required utility. That makes sense, first because utility is in natural logs, so generating increases in utility requires more and more dumplings and mooncakes because of the diminishing marginal utility.[12]

Quick Check

Minimize the expenditures on x and y subject to the utility constraint:

a) $\min_{x,y} 1x + 1y$ s.t. $U = x^{1/2} + y^{1/2} \geq 8$, where $\bar{U} = 8$ and the prices are $p_x = 1$ and $p_y = 1$.

b) $\min_{x,y} 1x + 4y$ s.t. $U = x^{1/2}y^{1/2} \geq 1$, where $\bar{U} = 1$ and the prices are $p_x = 1$ and $p_y = 4$.

---

[12] Keep in mind that all the same rules about finding a minimum or maximum and examining the second-order sufficient conditions apply to both the primal and dual problems. We are not checking our second-order conditions here to prevent cluttering the chapter, but the reader should not forget those rules.

9.2 General Representation

Now that we have completed an introductory example with numbers, let's do that again more generally. Starting with the UMP, our problem is

$$\max_{x_1,x_2} \quad U(x_1, x_2), \text{ subject to: } W \geq p_1 x_1 + p_2 x_2$$

$$\mathcal{L} = \overbrace{U(x_1, x_2)}^{\text{Objective Function}} + \lambda \overbrace{[W - p_1 x_1 - p_2 x_2]}^{\text{Constraint}}.$$

The first-order conditions are therefore

$$\frac{\partial \mathcal{L}}{\partial x_1} = U_{x1} - \lambda p_1 = 0$$

$$\frac{\partial \mathcal{L}}{\partial x_2} = U_{x2} - \lambda p_2 = 0$$

$$\frac{\partial \mathcal{L}}{\partial \lambda} = W - p_1 x_1 - p_2 x_2 = 0.$$

The solutions take the following form:

$$x_1^*(P, W), \ x_2^*(P, W), \text{ and } \lambda^*(P, W).$$

These solutions are called the Marshallian demand functions after the microeconomist, Alfred Marshall, who developed this approach. The Marshallian demand functions are the solutions to the UMP.

Important: Notice how the solutions are written. First, the asterisk, as before, indicates the optimal choice. Second the optimal choices are written as unspecified

> **Definition**
>
> **Marshallian demand:** The solutions to the UMP, where the optimal demand functions for goods are functions of wealth and prices.

functions of the exogenous variables P and W. Third, capital P is the set of all the prices. It includes both p_1 and p_2. Technically, we refer to it as a *vector* of prices. Vectors are typically introduced in basic matrix algebra, but not hard at all to understand. Here, think of capital P as a list of all the prices. (For more, see Chapter 12.)

Now, let's turn to the general representation of the EMP:

$$\min_{x_1, x_2} \quad p_1 x_1 + p_2 x_2, \text{ subject to: } U \geq \bar{U}$$

$$\mathcal{L} = \overbrace{p_1 x_1 + p_2 x_2}^{\text{Objective Function}} + \lambda \overbrace{\left[U - \bar{U} \right]}^{\text{Constraint}}$$

$$\frac{\partial \mathcal{L}}{\partial x_1} = p_1 - \lambda U_{x_1} = 0$$

$$\frac{\partial \mathcal{L}}{\partial M} = p_2 - \lambda U_{x_2} = 0$$

$$\frac{\partial \mathcal{L}}{\partial \lambda} = U - \bar{U} = 0.$$

Now, the solutions take the following form:

$$x_1^{*H}(P, \bar{U}), \; x_2^{*H}(P, \bar{U}), \text{ and } \lambda^{*H}(P, \bar{U}).$$

Important: Notice *again* how the solutions are written. First, the asterisk, as before, indicates the optimal choice. Second, the optimal choices are written as unspecified functions of the exogenous variables P and $\bar{U}$. In contrast to the UMP, where wealth, W, was the exogenous variable that created the constraint, that is not the case in the EMP. It is the minimum utility required, $\bar{U}$, that constrains the EMP. Third, capital P is again the set of all the prices.

Fourth, there is a new superscript, H, on the solutions. That indicates these are solutions to the EMP. The H stands for *Hicksian*, and these are called the **Hicksian Demand Functions** after economist John Hicks. The Hicksian demand functions are super useful. They al-

Definition

Hicksian demand functions or *Compensated demand functions*: The solutions to the EMP where the optimal demand functions for goods are functions of the utility level and prices.

low us to ask and answer questions such as, "If we raise the price of some good through taxes, how much would we have to give the consumer in cash to keep them just as happy as they were before taxes were raised?" That is, "How much would we have to 'compensate' the consumer for the change in prices?" As such, Hicksian demand functions are sometimes referred to as compensated demand functions.[13]

Figure 9.1 summarizes our exploration of the dual problems in consumer theory so

---

[13]Some textbooks will use the letter "C" instead of "H" to indicate compensated demand functions. Furthermore, when discussing duality some textbooks will superscript (or subscript) the Marshallian demand functions with an "M" to help you distinguish between them in the notation. In yet other cases, the Hicksian demand function notation will be $h(p, u)$, dispensing with the x altogether to make it clearer.

far.

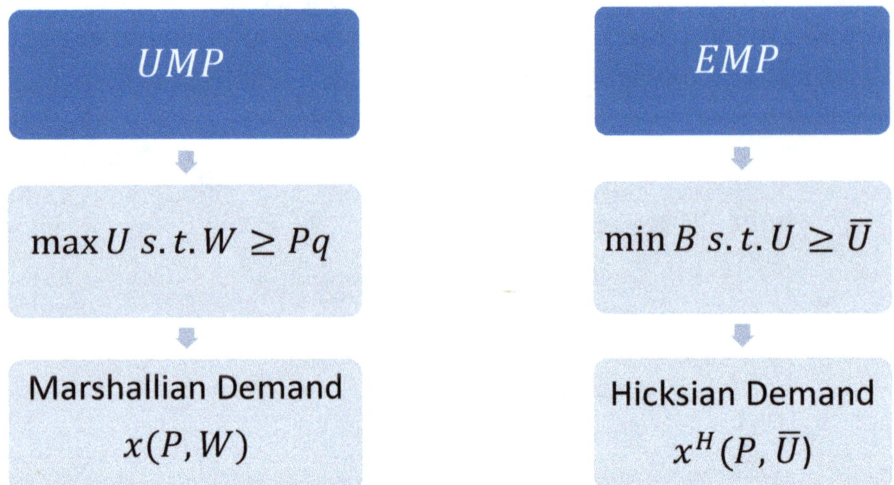

Figure 9.1: Basic consumer theory duality.

As Figure 9.1 indicates, we have two parallel forms of the problem. On the left-hand side is the Utility Maximization Problem (UMP) and on the right-hand side is the Expenditure Minimization Problem (EMP). The leftside, the UMP, is commonly referred to as the "primal" problem while the right side, the EMP, is called the "dual" problem. Under the UMP we maximize utility subject to the budget constraint. Under the EMP we minimize the expenditure subject to a minimum utility level constraint. We have two different exogenous sources of the constraints: W for utility maximization and $\bar{U}$ for expenditure minimization. The results under the UMP are the Marshallian demand functions, which give the optimal amount of the goods x to consume given the prices, P, and wealth level, W. Under the EMP are the Hicksian demand functions, which give the optimal levels of the goods x to consume given the prices, P, and the minimum utility level required, $\bar{U}$. However, we have only just begun to explore all the relationships we can utilize here. We will add more to this diagram as we discover how these two problems are further related.

Now that we have solved the two main dual problems in consumer theory, let's step back and think visually about what we just did. Consider Figures 9.2 and 9.3, which revisit Uncle Iroh's from Section 6.4.2. Uncle Iroh wants to maximize utility over ginseng tea and cake given his budget constraint.

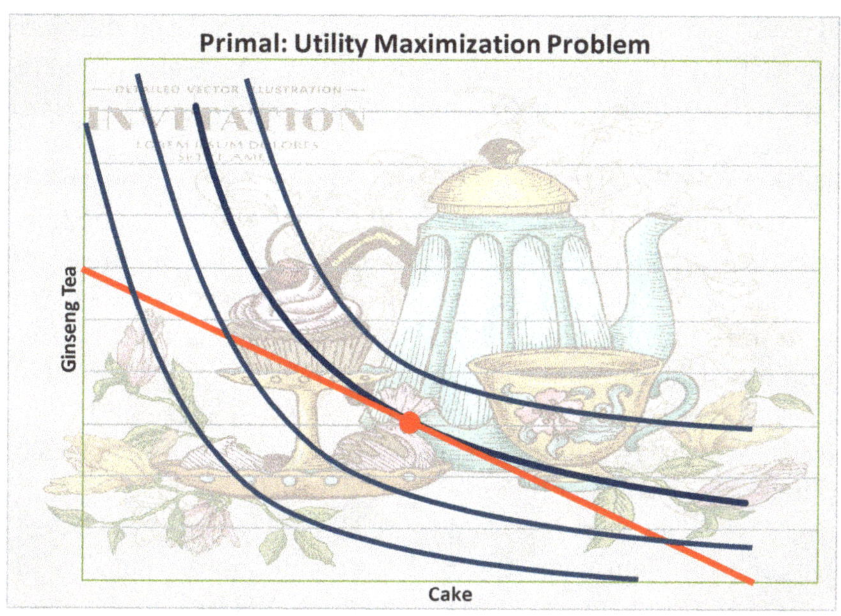

Figure 9.2: Primal problem.

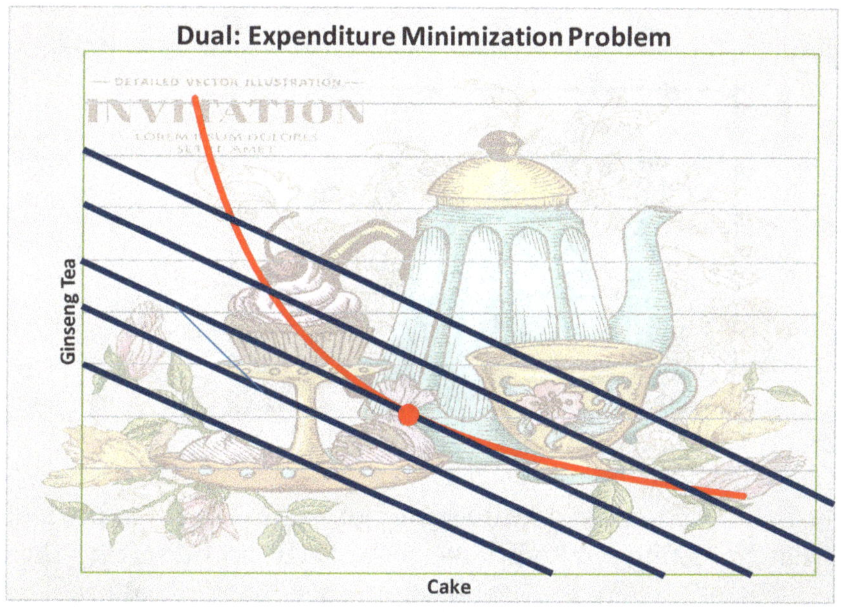

Figure 9.3: Dual problem.

The first diagram represents the UMP, the primal problem. What is our consumer doing here? The consumer faces a budget constraint, which is the downward sloping straight red line. The curved blue lines represent the indifference curves, with higher utility corresponding to the indifference curves further up and to the right, where each point has greater levels of tea and cakes. Uncle Iroh's goal is to find the combination of tea and cakes that get to the highest indifference curve. You can imagine the indifference curve can be moved – slid left and right – where the goal is to get the indifference curve as far right (up) as possible while still touching the budget constraint. Where they just touch reveals the optimal, or utility–maximizing, consumption levels of tea and cakes that fit within the budget constraint.

In the second diagram we represent the EMP, the dual problem. Now the indifference curve (red curve) is fixed. The blue straight-line budget constraints can be moved. Uncle Iroh's goal is to push the budget constraint down and to the left as far as possible, to minimize spending, while still touching the red indifference curve. That indifference curve represents the required utility level. At the point where the indifference curve and budget constraint just touch, we have the expenditure–minimizing levels of tea and cake that yield the required utility level.

In sum, and referring to the two figures, the two problems look very much the same graphically. The difference is which object is fixed: utility, $\bar{U}$, or wealth, W. Then, given the fixed object, what level of our choice variables (tea and cakes, x_T and x_C) most efficiently solves the problem? For the UMP, we call those solutions Marshallian demand functions, $x^*(P, W)$ and for the EMP, we call the solutions Hicksian demand functions, $x^{*H}(P, \bar{U})$.

9.3 The Indirect Utility Function and the Expenditure Function

Recall from Chapter 7 that we formed a "value function," also called an **indirect utility function**. We did that by taking our solutions to the UMP and inserting them directly into the utility function itself. There is a useful counterpart in the dual problem when we minimize expenditures. To refresh your memory, let's use the example with Xiyangyang, where his utility is still given by $U = ln(D) + ln(M)$, but we will leave his wealth, W, and the prices, p_D and p_M, general. The set–up, first-order conditions, and solutions are

$$\max_{D,M} \quad U(D, M) = ln(D) + ln(M), \text{ subject to: } W \geq p_D D + p_M M$$

$$\mathcal{L} = ln(D) + ln(M) + \lambda [W - p_D D - p_M M]$$

$$\frac{\partial \mathcal{L}}{\partial D} = \frac{1}{D} - \lambda p_D = 0$$

$$\frac{\partial \mathcal{L}}{\partial M} = \frac{1}{M} - \lambda p_M = 0$$

$$\frac{\partial \mathcal{L}}{\partial \lambda} = W - p_D D - p_M M = 0$$

$$D^*(P, W) = \frac{W}{2p_D}, \qquad M^*(P, W) = \frac{W}{2p_M}, \qquad \lambda^*(P, W) = \frac{W}{2}.$$

To form the indirect utility function we substitute our solutions, the Marshallian demand functions, into the original utility function. That yields

$$U(D, M) = ln(D) + ln(M), \qquad D^*(P, W) = \frac{W}{2p_D}, \qquad M^*(P, W) = \frac{W}{2p_M}$$

$$V(P, W) = ln(D^*) + ln(M^*)$$

$$= ln\left(\frac{W}{2p_D}\right) + ln\left(\frac{W}{2p_M}\right)$$

$$= ln\left(\frac{W^2}{4p_D p_M}\right)$$

$$= 2lnW - 2ln2 - ln(p_D) - ln(p_M).$$

where the last two lines come from the properties of the natural log (see Chapter 3). The interpretation of the indirect utility function, $V(P, W)$, is the consumer's maximized utility level given the prices, P, of all goods, and the consumer's wealth level, W. We call this *indirect* because the consumer does not derive utility or happiness from prices or wealth *directly* in our models. They get utility from consuming the goods (or services).

We can do the same exercise for the dual, expenditure minimization problem, where $B(D, M)$ is the budget expenditure as a function of the goods to be chosen:

$$\min_{D,M} B(D, M) \quad p_D D + p_M M, \text{ subject to: } \bar{U} = ln(D) + ln(M)$$

$$\mathcal{L} = p_D D + p_M M + \lambda[\bar{U} - ln(D) - ln(M)]$$

$$\frac{\partial \mathcal{L}}{\partial D} = p_D - \lambda \frac{1}{D} = 0$$

$$\frac{\partial \mathcal{L}}{\partial M} = p_M - \lambda \frac{1}{M} = 0$$

$$\frac{\partial \mathcal{L}}{\partial \lambda} = \bar{U} - ln(D) - ln(M) = 0$$

$$D^{*H} = \left(\frac{p_M}{p_D}e^{\bar{U}}\right)^{1/2}, \qquad M^{*H} = \left(\frac{p_D}{p_M}e^{\bar{U}}\right)^{1/2}.$$

Solving the system of first-order conditions is straightforward but does involve using the rules for natural logs and e (again see Chapter 3 if you need a refresher). In case you need help, we go through the steps here. Start, as usual, by combining the two first-order conditions on the choice variables:

$$\frac{p_D}{p_M} = \frac{M}{D}$$
$$p_D D = p_M M$$
$$M = \frac{p_D}{p_M}D.$$

Definition

Expenditure function: The indirect level of expenditures as a function of prices and the utility level at the optimal solution to the dual problem, the EMP.

Substitute that result into the constraint:

$$\bar{U} = ln(D) + ln(M)$$
$$\bar{U} = ln(D) + ln\left(\frac{p_D}{p_M}D\right)$$
$$\bar{U} = ln\left(\frac{p_D}{p_M}D^2\right)$$
$$e^{\bar{U}} = e^{ln\left(\frac{p_D}{p_M}D^2\right)}$$
$$e^{\bar{U}} = \frac{p_D}{p_M}D^2$$
$$D^2 = \frac{p_M}{p_D}e^{\bar{U}}$$
$$D^{*H} = \left(\frac{p_M}{p_D}e^{\bar{U}}\right)^{1/2}, \qquad M^{*H} = \left(\frac{p_D}{p_M}e^{\bar{U}}\right)^{1/2},$$

where we have skipped the steps to derive M^{*H}, but you are encouraged to follow these same steps to get that solution. You can also get the solution for M^{*H} by using the result for D^{*H} and substituting into the constraint.

These results, M^{*H} and D^{*H}, are the Hicksian demands, which are the expenditure–minimizing solutions. We can form a new function by substituting these solutions back into the original objective function, $B(D, M) = p_D D + p_M M$ as follows:

$$E(P,\bar{U}) = p_D\left(\frac{p_M}{p_D}e^{\bar{U}}\right)^{1/2} + p_M\left(\frac{p_D}{p_M}e^{\bar{U}}\right)^{1/2}$$
$$E(P,\bar{U}) = 2\left(p_D p_M e^{\bar{U}}\right)^{1/2}.$$

That function, $E(P, \bar{U})$, is referred to as the expenditure function. It shows the required minimum expenditure given the prices to achieve the level $\bar{U}$ of utility. It is indirect, as was the indirect utility function. Why? It generates expenditure as a function of utility and prices, but no one is spending *directly* on utility or prices. They are spending on goods. The expenditure function shows the minimum cost of getting to $\bar{U}$ given the prices on the goods.

The expanded in Figure 9.4 adds in the steps to form the indirect utility function and the expenditure function.

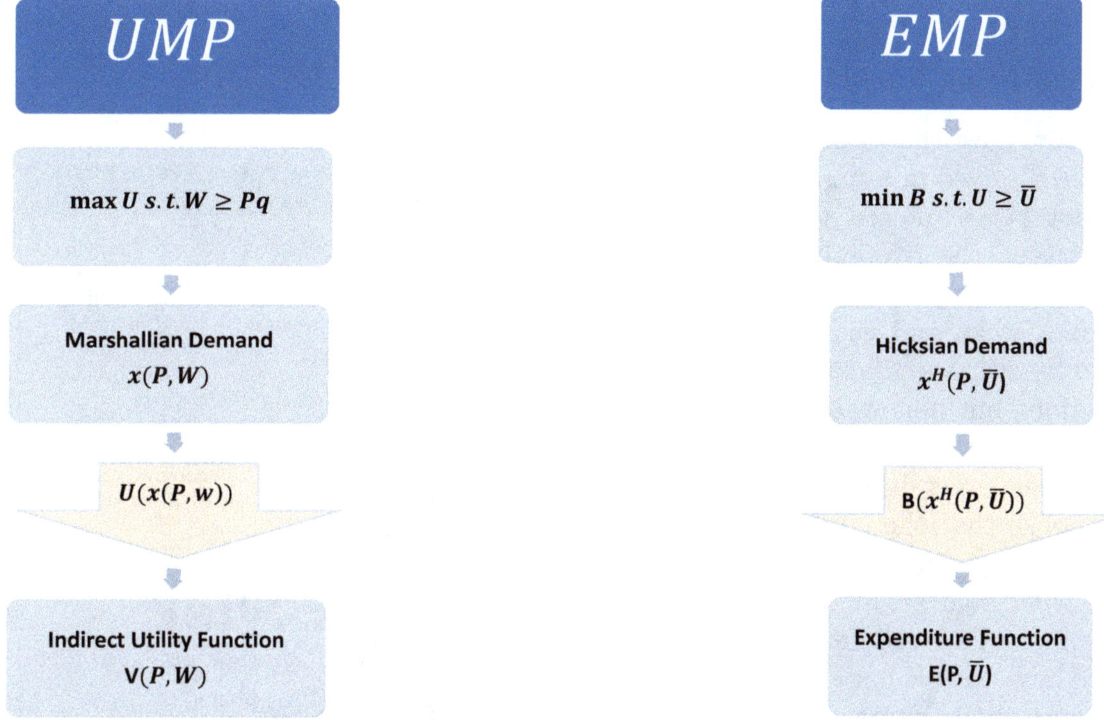

Figure 9.4: Consumer theory duality, extended.

Through the downward–pointing arrows, the diagram indicates that by substituting the Marshallian demand functions into the utility function, we form the indirect utility function. On the right side, by substituting the Hicksian demand functions into the original budget expression, we form the expenditure function.

9.4 Duality Relationships

9.4.1 Recovering Functions

While we now have the primal UMP and the dual EMP solved in a parallel fashion, we have not yet exploited the interconnection between the two solutions. The following relationships are highly useful if through data work and econometrics we are able to recover an estimate of some of these functions but not others. Then, using the duality relationships, we may be able to get the other functions. For example, suppose that the city government is considering imposing a sugar tax on all beverages with high sugar content. The city government wants to know how much they would have to give back to consumers to keep them equally happy, in effect what is the utility cost of imposing the tax. Using data, an econometrician may be able to recover the Marshallian demand functions for sugary drinks in the city, but that would not be the best tool for policy analysis, which would be the Hicksian demand functions.

On that note, first observe that the indirect utility function and the expenditure function can be used to recover the constraining exogenous variables, W and $\bar{U}$. We will use the previous example to illustrate. Given the indirect utility function

Soda tax. There is a substantial literature on the use of taxes on high–sugar content beverages. To learn more about this idea, or any topic in economics, the most comprehensive database to search is Econlit, operated by the American Economic Association (AEA). In addition, two journals published by the AEA, *The Journal of Economic Literature (JEL)* and *The Journal of Economic Perspectives (JEP)*, are often the best starting points for learning about existing research into a subject. You can find an excellent summary in the JEP on the soda tax policy by Allcott et al.

Most universities pay for the subscription to use Econlit, so it is advisable to get to the database through the website of your university's library. Moreover for any research into a topic, two journals published by the AEA specialize in summary/overview articles on specific topics written by the foremost experts in that subject.

from the UMP, if we substitute in the expenditure function from the EMP for W, we have the following:

$$V(P,W) \;=\; 2ln(W) - 2ln2 - ln(p_D) - ln(p_M), \qquad E(P,\bar{U}) = 2\left(p_D p_M e^{\bar{U}}\right)^{1/2}$$

$$V(P,W) \;=\; 2ln\left[2\left(p_D p_M e^{\bar{U}}\right)^{1/2}\right] - 2ln2 - ln(p_D) - ln(p_M)$$

$$= 2ln2 + 2ln\left(p_D p_M e^{\bar{U}}\right)^{1/2} - 2ln2 - ln(p_D) - ln(p_M)$$

$$= 2ln2 + ln\left(p_D p_M e^{\bar{U}}\right) - 2ln2 - ln(p_D) - ln(p_M)$$

$$= 2ln2 + ln(p_D) + ln(p_M) + lne^{\bar{U}} - 2ln2 - ln(p_D) - ln(p_M)$$

$$= lne^{\bar{U}}$$

$$V(P,W) \;=\; \bar{U}.$$

We can also substitute the other way to find W. This time, instead of substituting the expenditure function into the indirect utility function, substitute the indirect utility function into the expenditure function: $E(P,\bar{U}) = E(P,V(P,W)) = W$. Using our Xiyangyang example,

$$E(P,\bar{U}) \;=\; 2\left(p_D p_M e^{\bar{U}}\right)^{1/2}, \qquad V(P,W) = 2ln(W) - 2ln2 - ln(p_D) - ln(p_M)$$

$$= 2\left(p_D p_M e^{2ln(W) - 2ln2 - ln(p_D) - ln(p_M)}\right)^{1/2}.$$

We can make our algebra lives a little easier by recognizing that:

$$V(P,W) \;=\; 2ln(W) - 2ln2 - ln(p_D) - ln(p_M) = ln\left(\frac{W^2}{4p_D p_M}\right).$$

Now use that when substituting:

$$E(P,\bar{U}) \;=\; 2\left(p_D p_M e^{ln\left(\frac{W^2}{4p_D p_M}\right)}\right)^{1/2}.$$

But that font is so tiny within the exponent, so we will write the exponential function using this form: $e^x = exp(x)$

$$E(P,\bar{U}) \;=\; 2\left(p_D p_M exp\left[ln\left(\frac{W^2}{4p_D p_M}\right)\right]\right)^{1/2}$$

$$=\; 2\left(p_D p_M \left(\frac{W^2}{4p_D p_M}\right)\right)^{1/2}.$$

You can see why we used that form in these steps where the *exp* and *ln* function are the inverse of one another and thus cancel. Continuing to simplify,

$$E(P,\bar{U}) \;=\; 2\left(\frac{W^2}{4}\right)^{1/2} = 2\frac{W}{2} = W$$

$$E(P,\bar{U}) \;=\; W.$$

In addition, we can combine the Hicksian demand solutions with the indirect utility to recover the Marshallian deemands. *And.....* we can combine the Marshallian demand solutions with the expenditure function to get the Hicksian demand functions. First, let's add those relations to the diagram, then show these two new relations using our Xiyangyang example.

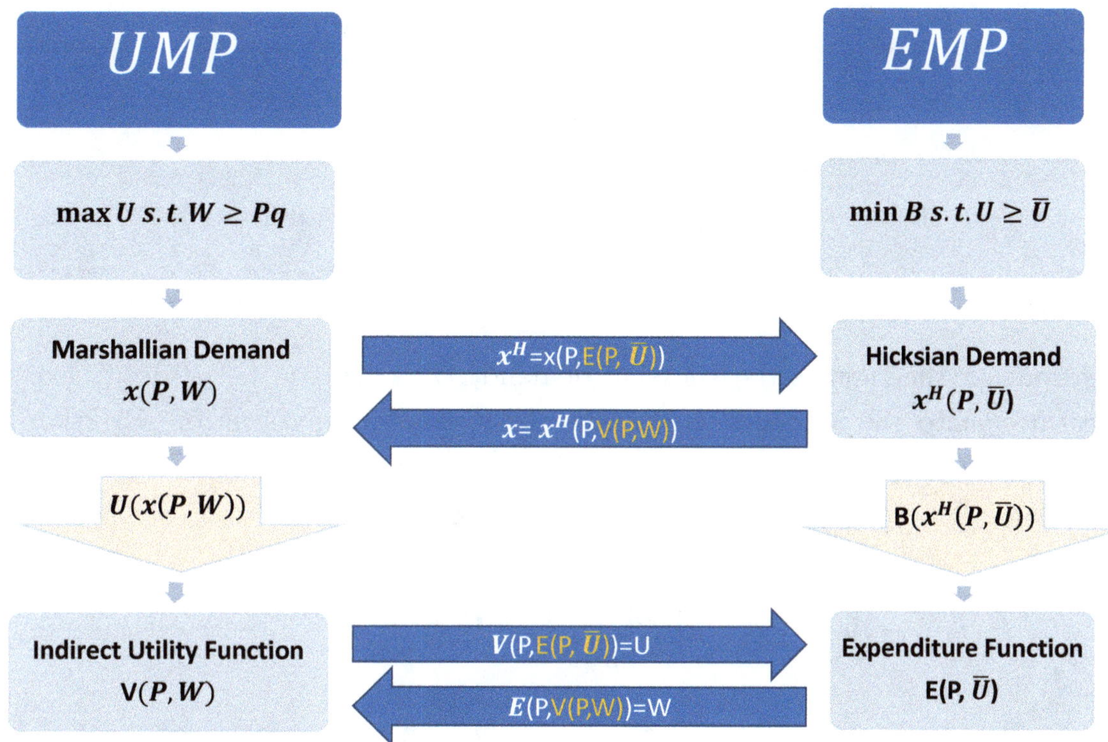

Figure 9.5: Consumer theory uality, version 3.

Now, let's explore those relations between the Marshallian demand functions, $x(P,W)$, and the Hicksian demand functions, $x^H(P,\bar{U})$. Like the connection between $V(P,W)$ and $E(P,\bar{U})$, we take advantage of the fact that $\bar{U}$ and $V(P,W)$ are in utility units, while W and $E(P,\bar{U})$ are in currency or dollar terms. So, let's start with the Marshallian demand functions and replace W with $E(P,\bar{U})$ to get the Hicksian demands. We show that next. Immediately following that, we'll substitute $V(P,W)$ into the Hicksian demand functions to get the Marshallian demand functions.

Our Marshallian demands from the example are $D^*(P,W) = \frac{W}{2p_D}$ and $M^*(P,W) = \frac{W}{2p_M}$, while the expenditure function is $E(P,\bar{U}) = 2\left(p_D p_M e^{\bar{U}}\right)^{1/2}$. Starting with the demand for dumplings, D,

$$D^*(P, W) \quad = \quad \frac{W}{2p_D}$$

$$= \quad \frac{2\left(p_D p_M e^{\bar{U}}\right)^{1/2}}{2p_D}$$

$$= \quad \left(\frac{p_M}{p_D} e^{\bar{U}}\right)^{1/2} = D^{H*}(P, \bar{U}),$$

where the result on the right is the Hicksian demand, $D^{*H}(P, \bar{U}) = \left(\frac{p_M}{p_D} e^{\bar{U}}\right)^{1/2}$. The same substitution into the Marshallian demand for mooncakes gives us the Hicksian demand for mooncakes:

$$M^*(P, W) \quad = \quad \frac{W}{2p_M}$$

$$= \quad \frac{2\left(p_D p_M e^{\bar{U}}\right)^{1/2}}{2p_M}$$

$$= \quad \left(\frac{p_D}{p_M} e^{\bar{U}}\right)^{1/2} = M^{H*}(P, \bar{U}),$$

where that is the Hicksian demand for mooncakes, $M^{*H}(P, \bar{U}) = \left(\frac{p_D}{p_M} e^{\bar{U}}\right)^{1/2}$.

The process for the other relationship looks quite similar. If we have the Hicksian demand functions and the indirect utility function, we can recover the Marshallian demand functions as follows. For the demand for dumplings, D, starting from:

$$D^{*H}(P, \bar{U}) \quad = \quad \left(\frac{p_M}{p_D} e^{\bar{U}}\right)^{1/2}, \qquad V(P, W) = ln\left(\frac{W^2}{4p_D p_M}\right).$$

Subsitute and simplify:

$$
\begin{aligned}
D^{*H}(P,\bar{U}) &= \left(\frac{p_M}{p_D} e^{\ln\left(\frac{W^2}{4p_D p_M}\right)}\right)^{1/2} \\
&= \left(\frac{p_M}{p_D}\left(\frac{W^2}{4p_D p_M}\right)\right)^{1/2} \\
&= \left(\frac{W^2}{4p_D^2}\right)^{1/2} = \frac{W}{2p_D}
\end{aligned}
$$

And for the mooncakes we have:

$$
\begin{aligned}
M^{*H}(P,\bar{U}) &= \left(\frac{p_D}{p_M} e^{\ln\left(\frac{W^2}{4p_D p_M}\right)}\right)^{1/2} \\
&= \left(\frac{p_D}{p_M}\left(\frac{W^2}{4p_D p_M}\right)\right)^{1/2} \\
&= \left(\frac{W^2}{4p_M^2}\right)^{1/2} = \frac{W}{2p_M}.
\end{aligned}
$$

The Marshallian demands have been recovered.

9.4.2 Roy's Identity and Shephard's Lemma

We will now look at two more relationships between our various results from the UMP and the EMP known as Roy's identity and Shephard's lemma. These properties allow us to recover the demand functions from either the indirect utility function or the expenditure function. Let's start by stating the properties up front in mathematical terms:

Roy's identity:

$$
x^{*}(P,W) = -\frac{\partial V(P,W)/\partial P}{\partial V(P,W)/\partial W}
$$

Shephard's lemma:

$$
x^{H}(P,U) = \frac{\partial E(P,\bar{U})}{\partial P}
$$

Roy's identity states that the Marshallian demand is the negative of the ratio of the derivative of the indirect utility function with respect to the price to the derivative of the indirect utility function with respect to wealth. So, if we have the indirect utility function, we can get the Marshallian demands from that. Shephard's lemma is more straightforward. It states that the Hicksian demand (the compensated demand) function is the derivative of the expenditure function with respect to the price. Analogously, if we have the expenditure function, we can recover the Hicksian demand functions.

In both of these expressions, a derivative is taken with respect to P, but recall that P is a vector, a list of prices. At the same time, x and x^H are also lists of the demand for each individual good. Do not be confused by the vector notation here. We can write these expressions for individual goods by labeling all our goods by an index variable. Let the index variable be i, where i goes from 1 to N, with N being the total number of goods. In the example above, as with most examples we have used, there are only two goods (e.g. dumplings and mooncakes). i could be 1 for dumplings and 2 for mooncakes. Or, we could let i be D for dumplings and M for mooncakes. In either case i takes on distinct values (or letters) so we know which good is being referred to at any time. Using that notation, we could then write Roy's identity and Shephard's lemma for individual goods as

$$x_i(P, W) = -\frac{\partial V(P, W)/\partial P_i}{\partial V(P, W)/\partial W}$$

$$x_i^H(P, W) = \frac{\partial E(P, \bar{U})}{\partial P_i}.$$

Notice that i appears as a subscript for both the good x and the price of x, which means we are taking the derivative with respect to the price of that same good, as opposed to the price of a different good.

Let's now use our example and apply Roy's identity and Shephard's lemma, where i takes on the values D and M for dumplings and mooncakes.

Roy's identity:

$$x(P, W) = -\frac{\partial V(P, W)/\partial P}{\partial V(P, W)/\partial W}$$

$$x_i(P, W) = -\frac{\partial V(P, W)/\partial P_i}{\partial V(P, W)/\partial W}$$

$$x_D(P, W) = -\frac{\partial V(P, W)/\partial P_D}{\partial V(P, W)/\partial W}$$

$$x_D(P, W) = -\frac{\partial\left[\ln\left(\frac{W^2}{4p_D p_M}\right)\right]/\partial P_D}{\partial\left[\ln\left(\frac{W^2}{4p_D p_M}\right)\right]/\partial W}.$$

Applying the rules for taking the derivative of the ln of a function, $ln(f(x))$, we have

$$= -\frac{\frac{4p_D p_M}{W^2}(-1)\frac{W^2}{4p_D^2 p_M}}{\frac{4p_D p_M}{W^2}(2)\frac{W}{4p_D p_M}}$$

$$= \frac{W}{2p_D}.$$

In the last step nearly everything cancels, leaving the Marshallian demand for D. To get the Marshallian demand for M, follow the same steps, but take the partial derivative with respect to p_M instead of p_D in the numerator.

For Shephard's lemma, we will derive the Hicksian demand for mooncakes and leave the derivation of the Hicksian demand for dumplings as a practice problem.

Shephard's lemma:

$$x^H(P,U) = \frac{\partial E(P,\bar{U})}{\partial P}$$

$$x_M^H(P,U) = \frac{\partial E(P,\bar{U})}{\partial P_M}$$

$$x_M^H(P,U) = \frac{\partial \left[2\left(p_D p_M e^{\bar{U}}\right)^{1/2}\right]}{\partial P_M}$$

Here, we have a chain rule application that yields

$$x_M^H(P,U) = \frac{1}{2}2\left(p_D p_M e^{\bar{U}}\right)^{-1/2}\left(p_D e^{\bar{U}}\right)$$

$$x_M^{*H} = \left(\frac{p_D}{p_M}e^{\bar{U}}\right)^{1/2}.$$

Figure 9.6 adds these two new relations, Roy's identity and Shephard's lemma to the visualization.

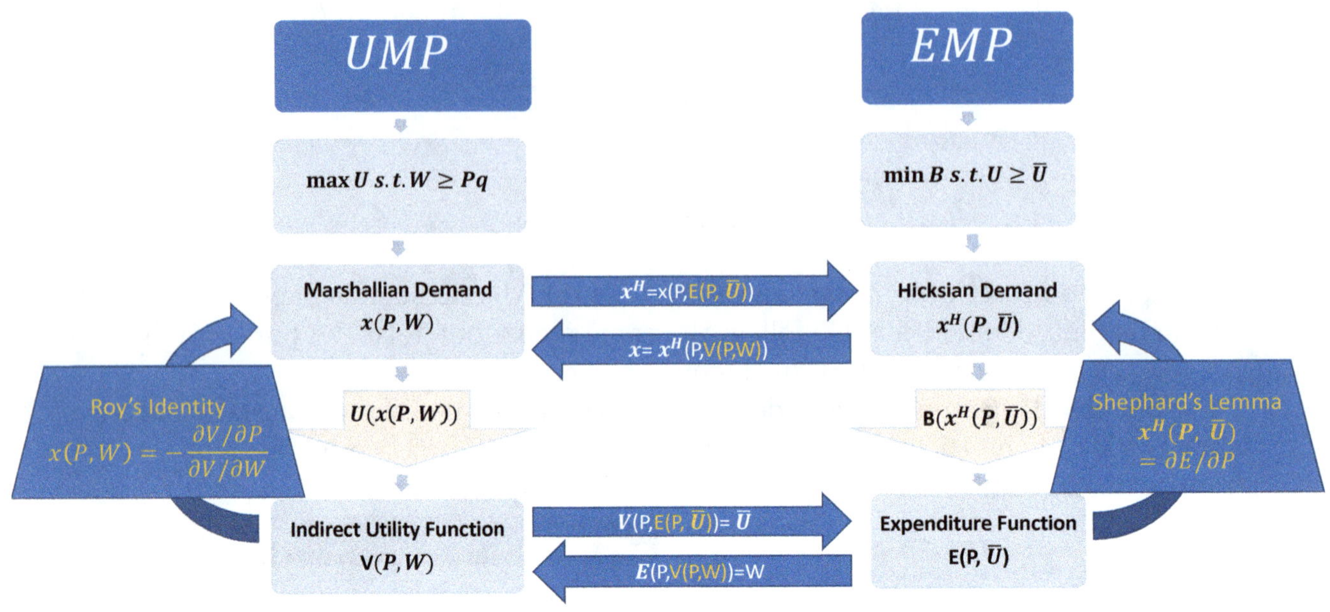

Figure 9.6: Consumer theory duality, final version.

Before moving on to dual problem for the firm, two additional notes are in order here. First, the treatment of duality here does not cover all the relations. There are more, but we have covered the ones most often used. Additional relations get more mathematically technical. Second, we do not cover the Slutsky substution matrix here. However, that is one of the most useful items that comes from microeconomic theory and is derived from examining the dual problem. It is what allows us to use data on consumers and spending to recover demand curves and understand the distinction between substitution effects (changes in buying behavior because of price changes) and wealth or income effects (changes in buying behavior derived from income changes). The Marshallian demand alone confounds the two effects. Should the price of, say, housing, rise significantly, consumers are faced with higher prices and will substitute toward the lower priced good. In the case of housing, they will shift demand toward likely cheaper, small homes, or renting instead of owning, and even then renting smaller lower quality places. At the same time, a rise in housing prices makes all the nonhome-owning consumers effectively poorer, which is a wealth effect. We can separate those effects out qualitatively and quantitatively using the dual relations and the Hicksian and Marshallian demand properties.

9.5 Producers: The PMP and the CMP

As with the basic consumer choice problem, there is a fully parallel set of dual analyses for producers. The primal problem maximizes production subject to a cost limit. The dual problem minimizes cost subject to an output requirement. Note, these are different,though related to the profit–maximization problem. In the profit maximization problem, the producer chooses the quantity inputs (typically capital, labor, and maybe raw materials) that are being sold at market prices in order to maximize profits given the producer's production technology represented by the production function.

To examine the producer's cost and production technology in more detail, we consider the **production maximization problem (PMP)** and the **cost minimization problem (CMP)**. In the PMP,

$$\max_{z_1, z_2} Q = F(z_1, z_2)$$
$$\text{s.t. } \bar{C} = w_1 z_1 + w_2 z_2.$$

Here, Q is the quantity produced and $F(z_1, z_2)$ is the production technology that uses z_1 and z_2, which represent the quantities of inputs used. In the most common representation, these inputs might be capital, K, and labor, L. However, they might also be raw materials, land, or energy, and there may be far more than just two inputs. w_1 and w_2 are the corresponding prices of those inputs and could be wages paid to labor, the rental rate of capital, the prices of materials in the commodities market, and so forth $F(z_1, z_2)$ we have seen plenty of times before in various forms as the production function with inputs of z_1 and z_2. As noted, this function is typically strictly concave, thus exhibiting diminishing marginal returns (more input leads to more output, but the increase in output gets smaller and smaller). It represents the technology of the firm: how well the firm combines inputs to

> ### Definition
>
> **Production maximization problem (PMP):** The primal decision problem of firms, where they maximize their output of goods and services subject to an input cost limit.

generate the output. If its technology or efficiency improves, the firm can produce more output using the same amount of inputs.

Before proceeding to an example, we redo the full duality diagram showing the relationships, but on the producer side.

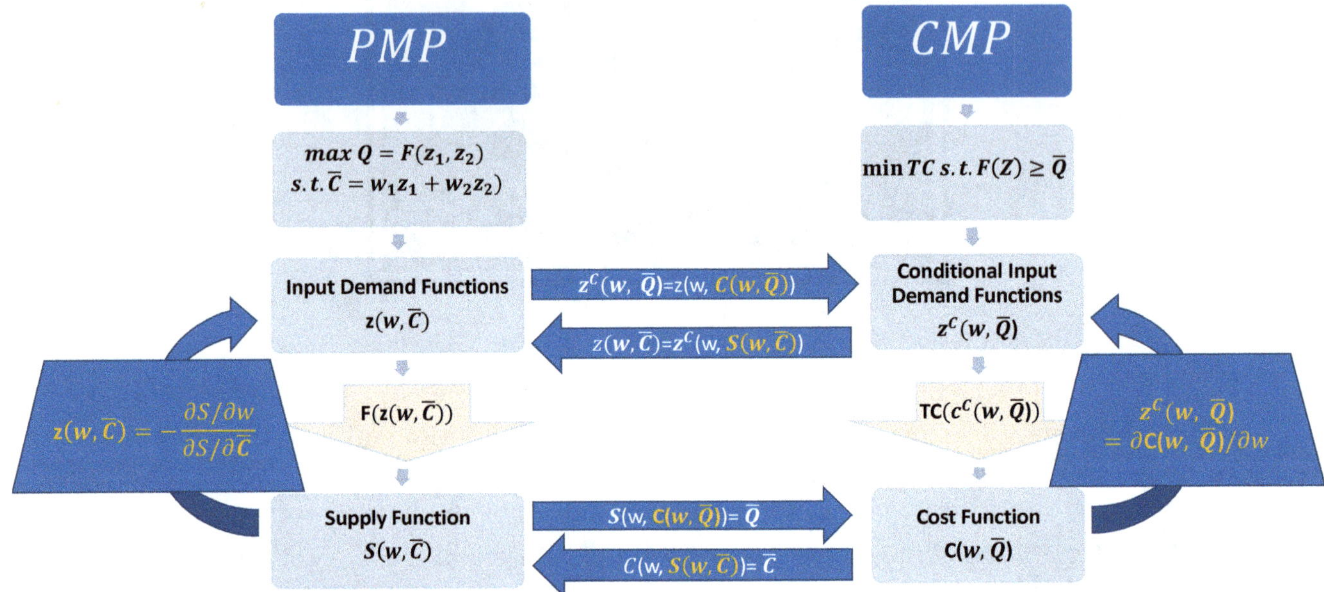

Figure 9.7: Producer theory duality

Note, we have the same relationship between the primal and dual problem. The only thing that changes are the names we give to the functions and the letters we assign to the variables. Now, on to an example.

Suppose Tibbie makes and sells tacos using two inputs, tortillas and vegetables. She wants to sell as many tacos as possible, but her budget for ingredients is limited to $\bar{C}$. We could set up Tibbie's problem as a Lagrangian as follows:

> **Definition**
>
> **Cost minimization problem (CMP)**: The dual decision problem of firms, where they minimize their expenditure on input costs subject to an output requirement.

$$\mathcal{L} \;=\; F(z_1, z_2) + \lambda[\bar{C} - w_1 z_1 - w_2 z_2].$$

The dual problem is the **cost minimization problem** (CMP). In the CMP, the producer is choosing the inputs to minimize the cost of attaining a specific level of production, $\bar{Q}$. This problem may be more intuitive for some. In our example, Tibbie knows she needs to produce 100 tacos, so she wants to find the cheapest way to make those 100 tacos by minimizing the costs on the inputs. Again, we use a Lagrangian:

$$\min_{z_1, z_2} TC \;=\; w_1 z_1 + w_2 z_2,$$

$$\text{s.t. } \bar{Q} \;=\; F(z_1, z_2).$$

Notice how the constraint and objective function have been rearranged to form the dual problem. In the PMP, output is the objective, while the costs are the constraint. In the CMP, the costs are the objective function, but the required level of output is the constraint.

9.5.1 Production Maximization

Now that we have the set–up and terminology down, let's proceed and solve an example of the dual producer problem, starting with the primary problem, the production maximization problem. We will use a somewhat general example to illustrate.

Paddi makes gingerbread. To make the gingerbread, he chooses the amount to use of two inputs: ginger, G, and bread, B. The quantity of gingerbread he produces is Q, and the gingerbread production function is given by $Q = AG^\alpha B^\beta$. Paddi buys ginger and bread at the market prices, w_G and w_B, respectively. When he goes to the market to buy his ingredients he only has $\bar{C}$ in cash with him, and Paddi does not use credit cards or payment apps, so $\bar{C}$ is all he has. Putting that information together we write the Lagrangian for the production maximization problem as

$$\max_{G,B} \mathcal{L} \;=\; AG^\alpha B^\beta + \lambda[\bar{C} - w_G G - w_B B].$$

We take partial derivatives on Paddi's two input choices and λ to form the first-order conditions for a maximum, as follows:

$$\frac{\partial \mathcal{L}}{\partial G} \;=\; \alpha AG^{\alpha-1} B^\beta - \lambda w_G = 0$$

$$\frac{\partial \mathcal{L}}{\partial B} \;=\; \beta AG^\alpha B^{\beta-1} - \lambda w_B = 0$$

$$\frac{\partial \mathcal{L}}{\partial \lambda} \;=\; \bar{C} - w_G G - w_B B = 0.$$

The algebra to solve these equations is pretty straightforward, and the resulting production maximizing levels of the inputs are:

$$G^*(W, \bar{C}) \;=\; \left(\frac{\alpha}{\alpha + \beta}\right)\left(\frac{1}{w_G}\right)\bar{C}$$

$$B^*(W, \bar{C}) \;=\; \left(\frac{\beta}{\alpha + \beta}\right)\left(\frac{1}{w_B}\right)\bar{C}.$$

These solutions are referred to as the "factor demands," and they represent the production maximizing input choices given the vector of input prices, $W = (w_G, w_B)$, and the cost constraint, $\bar{C}$. The analogous results with the consumer's utility maximization problem are the Marshallian demand functions.

We can form the supply function, the "indirect" production function, by substituting these factor demands into the original production function to get

$$
\begin{aligned}
AG^{\alpha}B^{\beta} &= A\left[\left(\frac{\alpha}{\alpha+\beta}\right)\left(\frac{1}{w_G}\right)\bar{C}\right]^{\alpha}\left[\left(\frac{\beta}{\alpha+\beta}\right)\left(\frac{1}{w_B}\right)\bar{C}\right]^{\beta} \\
&= A\left(\frac{\bar{C}}{\alpha+\beta}\right)^{\alpha+\beta}\left(\frac{\alpha}{w_G}\right)^{\alpha}\left(\frac{\beta}{w_B}\right)^{\beta} \\
S\left(\bar{C}, w_G, w_B\right) &= A\left(\frac{\bar{C}}{\alpha+\beta}\right)^{\alpha+\beta}\left(\frac{\alpha}{w_G}\right)^{\alpha}\left(\frac{\beta}{w_B}\right)^{\beta}.
\end{aligned}
$$

The general form of the supply function is written as $S(\bar{C}, W)$, where the capital "W" represents all the input prices, $w_1, w_2, w_3, \ldots$ in the problem. We call this function "indirect" because the firm does not use its cost limit to make anything, nor does it literally use the prices of the inputs to make anything. However, knowing the cost limit and the input prices, the supply function allows us to determine the efficient production level.

As with the consumer problem, we can recover the input factor demands from the indirect production function by taking derivatives. Specifically, to recover the input demand for input i, take the ratio of the derivative of $S\left(\bar{C}, w_G, w_B\right)$ with respect to $\bar{C}$ over the derivative of $S\left(\bar{C}, w_G, w_B\right)$ with respect to the input price w_i, or:

$$
z_i^* = -\left(\frac{\partial S\left(\bar{C}, w_G, w_B\right)/\partial w_i}{\partial S\left(\bar{C}, w_G, w_B\right)/\partial \bar{C}}\right).
$$

Using the supply function from Paddi's PMP, we could recover the input factor demands, B^* and G^*. Those steps are left as an exercise.

9.5.2 Dual: Cost Minimization

The dual of the firm's problem flips the objective function and the constraint found in the PMP. Here, minimizing the cost is the goal, but the constraint is to produce at least $\bar{Q}$ of the output. The general set up that we introduced earlier is

$$
\begin{aligned}
\min_{z_1, z_2} TC &= w_1 z_1 + w_2 z_2, \\
\text{s.t. } \bar{Q} &\leq F(z_1, z_2).
\end{aligned}
$$

Applying that to Paddi's gingerbread shop, we imagine Paddi needs to make at least $\bar{Q}$ loaves of gingerbread. He does that by minimizing his expenditure on the inputs G and B:

$$\min_{G,B} TC \;=\; p_G G + p_B B,$$

$$\text{s.t. } \bar{Q} \;\leq\; AG^\alpha B^\beta.$$

Setting up the Lagrangian and taking the first-order conditions, we have

$$\mathcal{L} \;=\; w_G G + w_B B + \lambda[\bar{Q} - AG^\alpha B^\beta]$$

$$\frac{\partial \mathcal{L}}{\partial G} \;=\; w_G - \lambda \alpha A G^{\alpha-1} B^\beta = 0$$

$$\frac{\partial \mathcal{L}}{\partial B} \;=\; w_B - \lambda \beta A G^\alpha B^{\beta-1} = 0$$

$$\frac{\partial \mathcal{L}}{\partial \lambda} \;=\; \bar{Q} - AG^\alpha B^\beta = 0.$$

Compare these first-order conditions for the CMP to the first-order conditions for the PMP. They are nearly identical except for the location of the Lagrangian multiplier. In the FOCs of the PMP the λ multiplies the prices and reflects the marginal increase in production that would occur with an increase in the cost limit. In the CMP, the Lagrangian multiplier reflects the marginal increase in cost that occurs with a marginal increase in the minimum required output. As always, the Lagrangian multiplier is the marginal value to the objective (production or costs) of increasing the constraint ($\bar{C}$ or $\bar{Q}$).

Solving the first-order conditions, yields the following solution,s called the *conditional factor demands*:

$$G^{C*} \;=\; \left(\frac{\alpha}{\beta}\frac{w_B}{w_G}\right)^{\alpha/(\alpha+\beta)} \left(\frac{\bar{Q}}{A}\right)^{1/(\alpha+\beta)}$$

$$B^{C*} \;=\; \left(\frac{\beta}{\alpha}\frac{w_G}{w_B}\right)^{\beta/(\alpha+\beta)} \left(\frac{\bar{Q}}{A}\right)^{1/(\alpha+\beta)}.$$

They are *conditional* factor demands in the sense that the optimal choices for those inputs are conditional on needing to produce at least $\bar{Q}$. Thus, we write the superscript C to distinguish them from the input factor demands that result in the PMP. These solutions are analogous to the Hicksian demand functions from the consumer's expenditure minimization problem. The Hicksian demands were the expenditure minimizing choices, *conditional* on needing to achieve $\bar{U}$ in utility.

Substituting the conditional factor demand functions into the original objective function $p_B B + p_G G$, we form the cost function, which tells us the cost of production given the output requirement and the input prices:

$$
\begin{aligned}
C(w,\bar{Q}) &= w_G G^{C*} + w_B B^{C*} \\
&= w_G \left(\frac{\alpha\, p_B}{\beta\, p_G}\right)^{\alpha/(\alpha+\beta)} \left(\frac{\bar{Q}}{A}\right)^{1/(\alpha+\beta)} + w_B \left(\frac{\beta\, p_G}{\alpha\, p_B}\right)^{\alpha/(\alpha+\beta)} \left(\frac{\bar{Q}}{A}\right)^{1/(\alpha+\beta)} \\
&= \left[\left(\frac{\alpha}{\beta}\right)^{\alpha/(\alpha+\beta)} + \left(\frac{\beta}{\alpha}\right)^{\beta/(\alpha+\beta)}\right] w_G^{\beta/(\alpha+\beta)} w_B^{\alpha/(\alpha+\beta)} \left(\frac{\bar{Q}}{A}\right)^{1/(\alpha+\beta)}.
\end{aligned}
$$

The last line comes from a bit of algebra to combine terms. As shown in Figure 9.7, the same relationships between the primal and dual are present in the producer's problem. Showing these relationships is left for the practice exercises.

The following table summarizes the basic problems. Notice that in this table, all the constraints were written as inequalities, as discussed in Chapter 8. Although we did not work out any problems in this chapter where a constraint might not bind or a first-order condition will not hold with equality, both cases are certainly possible, and the logic of Chapter 8 applies. Perhaps there is a problem like that in the exercises? Beware.

Table 9.1: Duality Summary

Problem	Agent	Primal/Dual	Objective Function	Constraint	Solution
UMP	Consumer	Primal	$\max U(x_1, x_2)$	$W \geq p_1 x_1 + p_2 x_2$	$x(P, W)$
EMP	Consumer	Dual	$\min p_1 x_1 + p_2 x_2$	$U(x_1, x_2) \geq \bar{U}$	$x^{*H}(P, \bar{U})$
PMP	Firm	Primal	$\max F(z_1, z_2)$	$\bar{C} \geq w_1 x_1 + w_2 x_2$	$z(w, \bar{C})$
CMP	Firm	Dual	$\min w_1 x_1 + w_2 x_2$	$F(z_1, z_2) \geq \bar{Q}$	$z(w, \bar{Q})$

A few concluding remarks on duality as used in economics textbooks are in order. The exposition can range from straightforward to really challenging and technical. There is a lot here, though much of it is quite subtle, and the treatment here is meant to provide a simple set of examples so that the reader can understand the basics. Duality and the relationships among the functions make conducting theory work easier because it helps impose restrictions on the functions, making them less general while at the same time providing various ways to look at the most important decisions consumers and firms make. We can use the properties and discipline imposed by these relations to see what consumers and firms are doing in the real world.

9.6 Exercises

9.6.1 Quick Check Answers

a) $\mathcal{L} = 1x + 1y + \lambda[8 - x^{1/2} - y^{1/2}] \Rightarrow \quad \frac{\partial L}{\partial x} = 1 - \frac{1}{2}\lambda x^{-1/2} = 0$ and $\frac{\partial L}{\partial y} = 1 - \frac{1}{2}\lambda y^{-1/2} = 0,$
$\Rightarrow x^{*H} = 16, y^{*H} = 16$

b) $\mathcal{L} = 1x + 4y + \lambda[1 - x^{1/2}y^{1/2}] \Rightarrow \quad \frac{\partial L}{\partial x} = 1 - \frac{1}{2}\lambda x^{-1/2}y^{1/2} = 0$ and $\frac{\partial L}{\partial y} = 4 - \frac{1}{2}\lambda x^{1/2}y^{-1/2} = 0, \Rightarrow x^{*H} = 2, y^{*H} = 1/2$

c) $\mathcal{L} = x^{1/2} + y^{1/2} + \lambda[W - p_x x - p_y y] \Rightarrow \quad \frac{\partial L}{\partial x} = \frac{1}{2}x^{-1/2} - \lambda p_x = 0,$ and $\frac{\partial L}{\partial y} = \frac{1}{2}y^{-1/2} - \lambda p_y = 0 \Rightarrow \quad x^* = \frac{p_x^2 W}{p_x^3 + p_y^3}, y^* = \frac{p_y^2 W}{p_x^3 + p_y^3} \quad V(P,W) = \frac{(p_x + p_y)W^{1/2}}{\left(p_x^3 + p_y^3\right)^{1/2}}$

d) $\mathcal{L} = p_x x + p_y y + \lambda[\bar{U} - \ln(x) - \ln(y)] \Rightarrow \quad \frac{\partial L}{\partial x} = p_x - \lambda\frac{1}{x} = 0, \frac{\partial L}{\partial y} = p_y - \lambda\frac{1}{y} = 0 \Rightarrow x^{H*} = \exp\left[\frac{1}{2}(\bar{U} - \ln p_x - \ln p_y)\right],$ and $y^{H*} = exp\left[\frac{1}{2}(\bar{U} - \ln p_x - \ln p_y)\right].$ $E(P,\bar{U}) = p_x x^{H*} + p_y y^{H*} = (p_x + p_y)\exp\left[\frac{1}{2}(\bar{U} - \ln p_x - \ln p_y)\right]$

e) $\mathcal{L} = \ln x + \ln y + \lambda[W - p_x x - p_y y] \Rightarrow \quad x^* = W/2p_x$ and $y^* = W/2p_y. \Rightarrow V(P,W) = \ln W - 2\ln 2 - \ln p_x - \ln p_y$
$\mathcal{L} = p_x x + p_y y + \lambda[\bar{U} - \ln x - \ln y] \Rightarrow \quad x^{*H} = \exp[\frac{1}{2}(\bar{U} + \ln p_x - \ln p_y)]$ and $y^{*H} = \exp[\frac{1}{2}(\bar{U} + \ln p_y - \ln p_x)] \Rightarrow \quad E(P,\bar{U}) = p_x \exp[\frac{1}{2}(\bar{U} + \ln p_x - \ln p_y)] + p_y \exp[\frac{1}{2}(\bar{U} + \ln p_y - \ln p_x)].$
$V(P, E(P,\bar{U})) = \ln E(P,\bar{U}) - 2\ln 2 - \ln p_x - \ln p_y$
$= \ln\left(p_x \exp[\frac{1}{2}(\bar{U} + \ln p_x - \ln p_y)] + p_y \exp[\frac{1}{2}(\bar{U} + \ln p_y - \ln p_x)]\right) - 2\ln 2 - \ln p_x - \ln p_y.$
$= \bar{U}$

f) $x = x^H(P, V(P,w)) = exp[\frac{1}{2}(lnW - 2\ln 2 - \ln p_x - \ln p_y) - 2\ln 2 - \ln p_x - \ln p_y + \ln p_x - \ln p_y] = W/2p_x$
$x^H = E(P,\bar{U})/2p_x = \left[p_x \exp[\frac{1}{2}(\bar{U} + \ln p_x - \ln p_y)] + p_y \exp[\frac{1}{2}(\bar{U} + \ln p_y - \ln p_x)]\right]/2p_x = \exp[\frac{1}{2}(\bar{U} + \ln p_x - \ln p_y)]$

g) $x_M(P,W) = -\frac{\partial V(P,W)/\partial P_M}{\partial V(P,W)/\partial W} \Rightarrow \quad = -\frac{\partial\left[\ln\left(\frac{W^2}{4p_D p_M}\right)\right]/\partial P_M}{\partial\left[\ln\left(\frac{W^2}{4p_D p_M}\right)\right]/\partial W} \Rightarrow \quad = -\frac{\frac{4p_D p_M}{W^2}(-1)\frac{W^2}{4p_D p_M^2}}{\frac{4p_D p_M}{W^2}(2)\frac{W}{4p_D p_M}} \Rightarrow = \frac{W}{2p_M}$

h) $S\left(\bar{C}, w_G, w_B\right) = A\left(\frac{\bar{C}}{\alpha+\beta}\right)^{\alpha+\beta}\left(\frac{\alpha}{w_G}\right)^{\alpha}\left(\frac{\beta}{w_B}\right)^{\beta}$

$$z_i^* = -\left(\frac{\partial S(\bar{C},w_G,w_B)/\partial w_i}{\partial S(\bar{C},w_G,w_B)/\partial \bar{C}}\right)$$

$$z_G^* = -\left(\frac{\partial S(\bar{C},w_G,w_B)/\partial w_G}{\partial S(\bar{C},w_G,w_B)/\partial \bar{C}}\right) = -\left(\frac{A\left(\frac{\bar{C}}{\alpha+\beta}\right)^{\alpha+\beta}\left(\frac{\beta}{w_B}\right)^{\beta}\alpha^{\alpha}\left(-\alpha w_G^{-\alpha-1}\right)}{A(\alpha+\beta)\bar{C}^{\alpha+\beta-1}\left(\frac{1}{\alpha+\beta}\right)^{\alpha+\beta}\left(\frac{\alpha}{w_G}\right)^{\alpha}\left(\frac{\beta}{w_B}\right)^{\beta}}\right) = \left(\frac{\alpha}{\alpha+\beta}\right)\left(\frac{1}{w_G}\right)\bar{C}$$

$$z_B^* = -\left(\frac{\partial S(\bar{C},w_G,w_B)/\partial w_B}{\partial S(\bar{C},w_G,w_B)/\partial \bar{C}}\right) = -\left(\frac{A\left(\frac{\bar{C}}{\alpha+\beta}\right)^{\alpha+\beta}\left(\frac{\alpha}{w_G}\right)^{\alpha}\beta^{\beta}\left(-\beta w_B^{-\beta-1}\right)}{A(\alpha+\beta)\bar{C}^{\alpha+\beta-1}\left(\frac{1}{\alpha+\beta}\right)^{\alpha+\beta}\left(\frac{\alpha}{w_G}\right)^{\alpha}\left(\frac{\beta}{w_B}\right)^{\beta}}\right) = \left(\frac{\beta}{\alpha+\beta}\right)\left(\frac{1}{w_B}\right)\bar{C}$$

9.6.2 Practice Problems

1. Xiyangyang at Jia Chang's:

   (a) From the text in Section 9.1, suppose that Xiyangyang returned to Jia Chang's Homestyle Restaurant yesterday but only had \$18 with him. His utility is still $U = ln(D) + ln(M)$, the price of dumplings remains \$1, and the price of mooncakes remains \$2. Set up and solve his UMP.

   (b) Suppose that Xiyangyang left Jia Chang's yesterday less than satisfied because he did not get enough food. Xiyangyang feels he could have been 25% happier (i.e., recieved 25% more utility than he did yesterday.) Using your solutions from part (a), determine the level of utility that Xiyangyang wants, then solve his EMP to obtain that utility level.

   (c) Compare your answers for D and M from part a and part b. Are they the same? Why or why not?

2. **Expenditure minimization.** Reconsider problem 7 from Chapter 4. The next day, Tombo considers returning to the fair, but only if he can achieve even greater happiness. He decides that in order for it to be worth the long walk to the fair, his utility from consuming sushi and takoyako needs to reach at least 100. Let $\bar{U}$ represent this minimum utility level. What is the minimum amount of money he needs to bring with him to the fair to attain that goal? To set up this minimization problem, use a Lagrangian, where expenditures, $p_S S + p_T T$ are your objective function and the constraint is $\bar{U} = ST$ or $100 = ST$.

3. Through data analysis of consumer spending and satisfaction surveys at the Hong Lang supermarket chain, the average indirect utility function for goat meat, x, and sheep meat, y, is estimated to be

$$V(P,W) = \left(\frac{W}{p_x + p_y}\right)\left[\frac{p_x + p_y}{(p_x p_y)^{1/2}}\right].$$

   (a) Use Roy's identity to recover the Marshallian demand functions for goat and sheep meat.

(b) For the Hong Lang supermarket chain, the consultants have also found the average consumer expenditure function is

$$E(P, \bar{U}) \;=\; p_x \left(\frac{p_y}{p_x + p_y} \right)^2 \bar{U}^2 + p_y \left(\frac{p_x}{p_x + p_y} \right)^2 \bar{U}^2.$$

Use that equation to recover the average consumer's Hicksian demand functions.

4. Refer back to Section 9.4.2. Derive the Marshallian demand for mooncakes and the Hicksian demand for dumplings using Roy's identity and Shephard's lemma.

5. Returning to the production maximization problem in Section 9.5.1, from the two first-order conditions, work through the algebra to get the solutions for G^* and B^*:

$$\frac{\partial \mathcal{L}}{\partial G} \;=\; \alpha PAG^{\alpha-1}B^{\beta} - p_G = 0$$

$$\frac{\partial \mathcal{L}}{\partial B} \;=\; \beta PAG^{\alpha}B^{\beta-1} - p_B = 0.$$

6. Using the solutions for Paddi's PMP and CMP in Section 9.5.1, show that $S(w, C(w, \bar{Q})) = \bar{Q}$ and $C(w, S(w, \bar{C})) = \bar{C}$.

7. Using the solutions to the PMP and the CMP for Paddi's gingerbread problem, show that
i) $z(w, \bar{C}) = z^C(w, S(w, \bar{C}))$, and
j) $z^C(w, \bar{Q}) = z(w, C(w, \bar{Q}))$.

8. Complete consumer problem. Consider Weslie the Happy Goat who consumes two goods: grass, x, and mushrooms, y. Weslie has the following utility function $U = \frac{1}{3}\ln(x) + \frac{1}{4}\ln(y)$. His budget constraint is $W = p_x x + p_y y$.

   (a) Set up Weslie's UMP and solve for his Marshallian demand functions for grass and mushrooms.

   (b) Use your solutions to Weslie's UMP to derive his indirect utility function.

   (c) Use Roy's identity to recover the Marshallian demand functions from the indirect utility function you got in part b.

   (d) Set up Weslie's EMP and solve for his Hicksian (compensated) demand functions for grass and mushrooms.

   (e) Use your solutions to Weslie's EMP to derive his expenditure function.

   (f) Use Shephard's lemma to recover the Hicksian demand functions from the expenditure function you derived in part e.

   (g) Show that the following relationships hold with your solutions:

   i. $x^H(P, \bar{U}) = x(P, E(P, \bar{U}))$

   ii. $x(P, W) = x^H(P, V(P, W))$

   iii. $V(P, E(P, \bar{U})) = \bar{U}$

   iv. $E(P, V(P, W)) = W$

9. Complete producer problem.: Sparky's Pest Control factory makes hammers for killing cockroaches. Sparky buys steel, S, and hires labor, L, to forge cockroach–smashing hammers. The cost of a unit of steel is given by p_s while labor gets paid the wage rate w. Sparky's hammer production function is $\frac{1}{2}S + \ln(L)$.

   (a) Set up Sparky's PMP and solve for his input demand functions for steel and labor, assume he has a cost ceiling of $\bar{C}$.

   (b) Use your solutions to Sparky's CMP to derive his supply function.

   (c) Use Roy's identity to recover the input demand functions from the supply function you got in part b.

   (d) Set up Sparky's CMP and solve for his conditional input demand functions for steel and labor.

   (e) Use your solutions to Sparky's EMP to derive his cost function.

   (f) Use Shephard's lemma to recover the conditional factor demand functions from the cost function you derived in part e.

   (g) Show that the following relationships hold with your solutions:

      i. $z^C(w, \bar{Q}) = z(w, S(w, \bar{Q}))$

      ii. $z(w, \bar{C}) = z^C S(w, \bar{Q}))$

      iii. $S(w, C(w, \bar{Q})) = \bar{Q}$

      iv. $C(w, S(w, \bar{C})) = \bar{C}$

10. **Effort minimization.** (You might wish to review problem 4 from Chapter 4.) Paddie works for Professor Laincz producing math that Professor Laincz sells to students. Paddie does not like his job. Professor Laincz wants to maximize his profits according to the following profit function:

$$\Pi = AL^\alpha - wL,$$

where A and α are exogenous parameters and w is the wage rate paid to Paddie. Let $A = 26$, $\alpha = 1/2$, and the wage rate Professor Laincz pays Paddie is $w = \$1$ per hour.

   (a) Solve Professor Laincz's profit-maximization problem. How many hours a week should he make Paddie work?

   (b) Professor Laincz goes on vacation for a week, but he still expects Paddie to work to maximize his profits. Because of variable demand, sadly for Professor Laincz he cannot observe how hard Paddie works. He knows that if Paddie does his job he should at least make \$52 in revenue. Paddie secretly subcontracts with Jonie. He agrees to pay Jonie the same \$1 per hour that he earns, but he instructs Jonie to work until she has made \$52 in revenue. That is, Paddie minimizes wL subject to $AL^\alpha = 52$. How many hours does Jonie need to work? If Paddie reports that he worked that same as he did the previous week when Professor Laincz was here, how much does he make after he pays Jonie for the subcontracted work?

9.7 Math Appendix

Here, we list some key math definitions, theorems, and formulae in formal math for your reference.

Slutsky decomposition One very useful result from duality is the following:

$$\frac{\partial x_n^*(p, W)}{\partial p_m} = \frac{\partial x^H(p, V(p, W))}{\partial p_m} - \frac{\partial x_n^*(p, W)}{\partial W} x_m(p, W).$$

That result states that the marginal effect of a change in the price of some good, m, p_m, on the optimal demand for good n can be separated into a substitution effect and an income effect. The first term on the right-hand side of the equality is the **substitution effect**. It is the marginal change in the demand for good n from a marginal change in p_m *holding utility constant* which we get from the Hicksian (compensated) demand function. Note that this effect could be positive or negative depending on whether the goods are substitutes or complements. The second term is the **income effect** and shows how the optimal demand for good n changes with wealth times the level of the demand for good m.

Homogenous functions A function $f(x_1, x_2, ..., x_n)$ is said to be homogenous of degree h, where h is a whole number, if for any value of $\alpha > 0$ we have that:

$$f(\alpha x_1, \alpha x_2, ..., \alpha x_n) = \alpha^h f(x_1, x_2, ..., x_n).$$

For example, the production function $Y = K^{1/3}L^{2/3}$ is homogenous of degree 1. To see that, multiply K and L by α and evaluate

$$\begin{aligned}
&(\alpha K)^{1/3}(\alpha L)^{2/3} \\
=& (\alpha^{1/3} K^{1/3} \alpha^{2/3} L^{2/3} \\
=& \alpha^{1/3} \alpha^{2/3} K^{1/3} L^{2/3} \\
=& \alpha^1 K^{1/3} L^{2/3}.
\end{aligned}$$

Since we are left with α to the first power times the original function we can conclude that the function is homogenous of degree 1. Intuitively, that means if we double all the inputs (capital and labor), the output will double. More generally, if we change the inputs proportionately by a factor of α, the output will also change by a factor of α.

The functions we derived in the dual consumer problems are homogenous functions.

- The **Marshallian demand function** is homogenous of degree zero. That means if prices, P, and wealth, W, change by a factor of α, the optimal demand, $x(P, W)$, will not change.

- The **indirect utility function** is also homogenous of degree zero. If prices, P, and wealth, W, change by a factor of α, the utility, $V(P, W)$, the consumer derives will not change.

- The **Hicksian demand function** is also homogenous of degree zero. If prices, P, and the utility level, $\bar{U}$, change by a factor of α, the Hicksian demand functions, $x^{*H}(P, \bar{U})$, will not change.

- The **expenditure function** is homogenous of degree 1 in prices, P. If all the prices, P, change by a factor of α, then the amount required to achieve the desired utility level, $\bar{U}$, as expressed by the expenditure fuction, $E(P, \bar{U})$, will also increase by a factor of α.

10   Comparative Statics and the Envelope Theorem

If you recall, in Chapter 1 we discussed the distinction between exogenous and endogenous variables and have tried to be clear throughout our examples as to which is which. Hopefully, by now you are getting a feel for that. We now turn to the analysis of the relationship between the two.

Comparative statics is the term for analyzing how the **endogenous variables** change when the **exogenous variables** change. When we solve for an optimal or equilibrium value of an endogenous variable, the solution is determined by the exogenous variables and parameters. That is, the solution needs to be written as a function of the exogenous variables and

> **Definition**
>
> **Endogenous variable:** A variable that is determined within the model. Choice variables are endogenous variables, *but not all* endogenous variables are choice variables.
>
> **Exogenous variable:** A variable that is outside the agent's control and not determined within the model.

paramters *only*. We cannot have, for example, the quantity demanded be a function of the quantity supplied in a market; they are typically both endogenous variables.

Once we have our endogenous variable expressed as a function of exogenous variables, we can conduct comparative statics to see how important forces affect choices and behavior. We compare states of the world. For example, would consumers use the Shanghai Metro more or less if incomes rose? Under what conditions? One can think of it as a before–and–after comparison, or a comparison of two different universes (car traffic increases the length of time to commute to work by driving), or the comparison between what would or wouldn't happen if a policy was able to change some exogenous value (the government increases import tariffs on cars). These are often important questions, and we address the logic here and one particular result in comparative statics known as the envelope theorem.

We also examine *implicit* differentiation, which is how we can find derivatives when we cannot isolate the endogenous variable as a function of exogenous variables. Our first–order condition(s) *imply* that a function exists even if we can not write it explicitly. It turns out we can take the derivative and perform the comparative statics anyway.

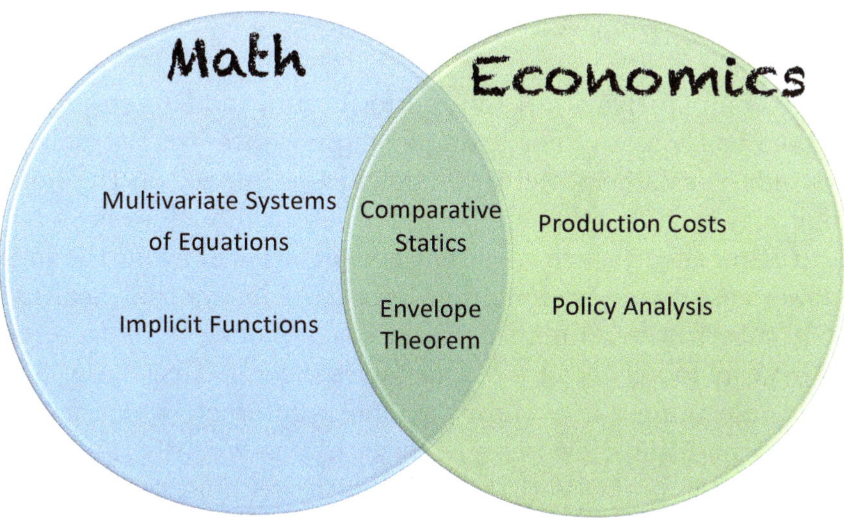

10.1 Comparative Statics

Comparative statics is the term for analyzing how the endogenous variables change when the exogenous variables change. An example might be ice cream sales. Ice cream sales at Mei's Multi–dimensional Ice Cream store often depend on the average daily temperature. The temperature is exogenous and is not affected or caused in anyway by her ice cream sales. However, her ice cream sales are positively related to the average daily temperature. If we let Q be quantity of ice cream and T be temperature, we should have a function like $Q = Q(T)$, which says that ice cream sales are a function of temperature. Furthermore, suppose that Mei has solved her profit–maximization problem to determine Q^*, where $Q^*(..., T, ...)$ is a function of several exogenous variables, such as the wages for paying workers, the cost of the ice cream she purchases wholesale, the average income in the neighborhood, and, of course, the daily outdoor temperature,

T. The comparative statics say that $Q_T^*(...., T, ...) = \frac{\partial Q^*}{\partial T} > 0$; optimal ice cream sales are strictly increasing with temperature. In other words, holding all else constant, a higher temperature means the optimal choice of ice cream sales increases. These kinds of qualitative statements are often the result of investigating the implications of a math model.

In many cases, the comparative statics we perform on a model's results are the key prediction that we take to the data. For example, suppose we ask the policy question, "If the sales tax on soda is raised in a city, what will the impact on the health among children living in that city be?" Further suppose that in our model we have two competing forces, the substitution effect which says that consumers will avoid the higher priced good and buy the lower cost goods, and another force, which says that healthy foods are often treated as poor substitutes, so higher priced sugary foods leads to a reduction in the consumption of healthy foods because of a negative income effect. A well–structured model can tell us multiple things: how those two forces interact, which force will dominate, and under what conditions. Moreover, as with any scientific approach, we can translate those results into hypotheses that can be measured and tested with data.

10.1.1 Example

Let's start with an example that only requires some algebra and no calculus. Consider a simple model of sales of Radon Ramen in Osaka. The demand and supply are given by the following linear model:

$$
\begin{aligned}
Q^D &= \alpha - \beta p + \gamma y \\
Q^S &= \omega + \phi p,
\end{aligned}
$$

where p is the price of Radon Ramen and y is the average income of the households in Osaka. Average income, y, is taken to be exogenous. The parameters α, β, γ, ω, and ϕ are all strictly positive. Together the two equations represent your basic supply and demand graph from principles of microeconomics. You can construct that graph by solving for price, p, in each equation and then graphing them.

However, as things stand we have three unknowns, Q^D, Q^S, and p, and only two equations. In equilibrium, supply equals demand; that is, the market clears. We then also have $Q^S = Q^D$, which provides us with a third equation. Assign Q^{**} as the equilibrium quantity, so $Q^{**} = Q^S = Q^D$.

Important distinction: The double asterisks on Q here represent *equilibrium*, where things are in balance and not the idea of *optimal*. Those two concepts are distinct. Sometimes the *equilibrium* is also the optimal quantity/choice, but they do not always overlap. In fact, many interesting policy questions arise when they do not coincide. For example, suppose the Shiva Chemical Corporation maximizes profits by producing 100,000 tons of what it calls red dust, and it can sell that product on the market for $P = \$100$ dollars per ton. That quantity and price are the *optimal* level of output and price from the firm's point of view. Those are also the *equilibrium* levels if supply equals demand. However, suppose that red dust has a negative externality and causes severe cases of arithmophobia that costs society an estimated \$25 per ton sold. From society's point of view, the negative mental health cost of the red dust is not being accounted for when Shiva corporation makes its decision. So, the *socially optimal* level (the level that takes into account everything: the firm's profits, consumer welfare, and the health costs) will be some lower level of quantity. Therefore, the government taxes red dust to change the *equilibrium* supply of red dust and move it to the socially *optimal* level. Thus, the equilibrium and socially optimal levels differ without the policy intervention. It is economists' and/or policymakers' job to investigate mechanisms and incentives to create that alignment and raise welfare.

Back to the Radon Ramen example. Solving with algebra we get the following equilibrium expressions:

$$p^{**} = \frac{\alpha - \omega + \gamma y}{\beta + \phi}$$

$$Q^{**} = \frac{\alpha\phi + \beta\omega + \gamma\phi y}{\beta + \phi}.$$

From those results, one can ask how a change in income, y, affects the equilibrium price and quantity. To get those results, take the partial derivatives. Why partial derivatives? Because we are asking the question how the equilibrium (or optimal) values change when **one** exogenous variable changes, holding all the other exogenous variables constant. We get:

$$\frac{\partial p^{**}}{\partial y} = \frac{\gamma}{\beta + \phi} > 0$$

$$\frac{\partial Q^{**}}{\partial y} = \frac{\gamma\phi}{\beta + \phi} > 0.$$

These are the comparative static results. Now that we have expressed the endogenous variables (Q, p) as functions of the exogenous variables, y, and the parameters (α, β, γ, ω, and ϕ), we ask how changes to the exogenous variables change the equilibrium. Since all the parameters are positive, both derivatives can be signed as positive. That means both p^{**} and Q^{**} are increasing functions of y.

Therefore, the comparative static results are that an increase in average income in Osaka, y, will lead to an increase in the quantity sold and price of Radon Ramen.

10.1.2 Another Example

Now let us look at an example that involves optimization. Mei lives all alone on the island of Hachijo. She picks passion fruit for consumption, C, and that requires work, L. She spends her leisure time sleeping and playing Taiko drums loudly, D, because, being the only one on the island, she can. Her utility function, time constraint, and passion fruit production function are given by

$$
\begin{aligned}
U(C, D) &= C^{1/2} + \beta D^{1/2} \\
D + L &= 1 \\
C &= \gamma L.
\end{aligned}
$$

The parameter γ measures how efficiently she is able to gather passion fruit. β is strictly positive and provides a relative weight on drum playing verus eating passion fruit. We set the time available in the day to 1 so that we can interpret the results as fractions or percentages of time. Her friend Pelops offers her a moped, which would enable her to gather passion fruit much faster. But she wonders if she would spend more time gathering passion fruit since she can get it more easily or would she spend less time gathering passion fruit because the moped makes her more efficient and allows her to increasing time drumming. That is, if γ rises, does Mei's optimal level of L^* increase or decrease? That is the comparative static answer we seek.

Setting up her problem by substituting the constraint and production function and deriving the FOC we get:

$$
U(L) = (\gamma L)^{1/2} + \beta(1 - L)^{1/2}
$$
$$
\frac{DU}{dL} = \frac{1}{2}\gamma^{1/2}L^{-1/2} - \frac{\beta}{2}(1 - L)^{-1/2} = 0.
$$

After some algebra we get

$$
L^* = \frac{\gamma}{\beta^2 + \gamma}.
$$

That is the *optimal* amount of time Mei spends gathering passion fruit. Therefore, $D^* = \frac{\beta^2}{\beta^2 + \gamma}$ and $C^* = \frac{\gamma^2}{\beta^2 + \gamma}$.

So, if she takes the moped and γ rises, what happens? Take the partial derivatives of the optimal results to find out:

$$
\begin{aligned}
\frac{\partial L^*}{\partial \gamma} &= \frac{\beta^2}{(\beta^2 + \gamma)^2} > 0 \\
\frac{\partial D^*}{\partial \gamma} &= \frac{-\beta^2}{(\beta^2 + \gamma)^2} < 0 \\
\frac{\partial C^*}{\partial \gamma} &= \frac{2\beta^2\gamma + \gamma^2}{(\beta^2 + \gamma)^2} > 0.
\end{aligned}
$$

Those comparative statics say the Mei will increase the amount of time she spends gathering passion fruit, L^*, and therefore decrease the time she spends playing drums, D^*, but she will now get a higher consumption level, C^*.

Let's change the scenario slightly. Suppose the king of Hachijo, who does not actually live there, imposes a tax on Mei's collection of passion fruit. She must give him τ percent of all the passion fruit she collects. Let's answer two questions here: First, how does the tax affect Mei's optimal choices.; Second, if the king wanted to maximize the tax revenue, what tax rate should he set?

Mei's production function effectively changes to $C = (1 - \tau)\gamma L$, where $1 - \tau$ shows how much of her efforts remain for her to consume, while $\tau\gamma L$ goes to the king as tax revenue. Now her problem becomes

$$
\begin{aligned}
U(L) &= ((1 - \tau)\gamma L)^{1/2} + \beta(1 - L)^{1/2} \\
\frac{DU}{dL} &= \frac{1}{2}((1 - \tau)\gamma)^{1/2}L^{-1/2} - \frac{\beta}{2}(1 - L)^{-1/2} = 0.
\end{aligned}
$$

After some algebra we get

$$
L^* = \frac{(1 - \tau)\gamma}{\beta^2 + (1 - \tau)\gamma}.
$$

From that solution, we find that $D^* = \frac{\beta^2}{\beta^2 + (1-\tau)\gamma}$ and $C^* = \frac{[(1-\tau)\gamma]^2}{\beta^2 + (1-\tau)\gamma}$.

Here, to make taking the derivatives a little bit easier, and less prone to a sign error, we will take the partial derivatives of the optimal choices with respect to the quantity $1 - \tau$ (instead of just τ), which represents the fraction of passion fruit that Mei *keeps*. Therefore, an increase in the tax rate, τ, means the fraction she gets to keep declines. The signs we get we taking the derivative using $(1 - \tau)$ will be the opposite of the result we are looking for with the original question:

$$
\begin{aligned}
\frac{\partial L^*}{\partial(1 - \tau)} &= \frac{\beta^2\gamma}{(\beta^2 + (1 - \tau)\gamma)^2} > 0 \\
\frac{\partial D^*}{\partial(1 - \tau)} &= \frac{-\beta^2\gamma}{(\beta^2 + (1 - \tau)\gamma)^2} < 0 \\
\frac{\partial C^*}{\partial(1 - \tau)} &= \frac{(2\beta^2 + (1 - \tau)\gamma) + (1 - \tau)\gamma^2}{(\beta^2 + (1 - \tau)\gamma)^2} > 0.
\end{aligned}
$$

Keeping in mind the quick discussion about $1 - \tau$, we conclude that an *increase* in the tax rate would decrease Mei's optimal time spent collecting passion fruit, L^*, and her overall consumption level of passion fruit, C^*, but increase the amount of time she spends playing drums, D^*.

We can also determine the impact of raising the tax rate on the tax revenue the king gets. He gets

$$
\text{Tax revenue} \quad = \quad \tau\gamma L^* = \tau\gamma\frac{(1 - \tau)\gamma}{\beta^2 + (1 - \tau)\gamma}.
$$

I have deliberately chosen not to combine the γ's in the numerator. I want to clearly distinguish between two effects of the tax rate. The first τ in the expression represents the tax rate on the revenue that Mei generates. The other τ's in the expression are part of L^*, so changes in the

two τ's there show how Mei's optimal choice of labor supply changes with the tax rate. We already figured out the answer to the latter above when we took $\partial L^*/\partial(1-\tau)$. However, here we are taking the derivative with respect to τ, and it will generate the opposite sign but otherwise remain the same (compare $\frac{d}{d(1-\tau)}(1-\tau) = 1$ and $\frac{d}{d\tau}(1-\tau) = -1$). In sum, there is the effect of changing the tax rate, *holding Mei's labor supply constant*, plus the effect of Mei changing her labor supply because of her reduced income, *holding the tax rate constant*.

That partial derivative is

$$\frac{\partial \text{Tax revenue}}{\partial \tau} = \gamma \frac{(1-\tau)\gamma}{\beta^2 + (1-\tau)\gamma} - \tau\gamma \frac{\beta^2 \gamma}{(\beta^2 + (1-\tau)\gamma)^2} \lesseqgtr 0,$$

where the two effects described are there separated by the minus sign. The first effect is the positive effect on the tax revenue the king collects. Holding Mei's labor supply constant, and therefore also holding her passion fruit production constant, an increase in the tax rate leads to higher revenues for the king. However, as the king raises the tax rate that discourages Mei from working and she reduces her labor supply and passion fruit production, as shown in the second term. The overall result is that the effect of an increase in the tax rate has an ambiguous effect on tax revenue. It depends on which of the two effects is stronger.

To give that some more intuition, suppose that $\beta = 1$ and $\gamma = 1$. Putting those values for the parameters into our result we have:

$$\frac{\partial \text{Tax revenue}}{\partial \tau} = \frac{(1-\tau)}{1 + (1-\tau)} - \tau \frac{1}{(1 + (1-\tau))^2} \lesseqgtr 0$$

$$\frac{\partial \text{Tax revenue}}{\partial \tau} = \frac{(1-\tau) + (1-\tau)^2 - \tau}{(1 + (1-\tau))^2} \lesseqgtr 0.$$

After combining the two fractions, the sign of the result will depend only on the numerator:

$$(1-\tau) + (1-\tau)^2 - \tau \lesseqgtr 0$$
$$\tau^2 - 4\tau + 2 \lesseqgtr 0.$$

Setting that equal to zero, the quadratic formula then yields:

$$\tau = 2 \pm \sqrt{2}$$
$$\tau = 0.59 \text{ or } \tau = 3.41.$$

Since the tax rate can only be between 0 and 1, the result is that for $\tau < 0.59$, the first effect dominates and tax revenues will rise with a higher tax rate, but for tax rates already at or above 0.59, tax revenues will decline with further tax rate increases because the second effect will dominate. Tax revenues are maximized at $\tau = 59\%$! Poor Mei.

10.1.3 Harder Example

Let's work through a somewhat more complex example to gain a deeper understanding. The Otaki factory has a monopoly on robots, but the government taxes the revenue generated at the rate of τ, where $0 < \tau < 1$. The government asks you, as their consultant, what would happen to robot output and tax revenue if they raised the tax rate. The Otaki factory faces the following maximization problem:

$$\max_{Q} \Pi(Q; \tau) = (1 - \tau)(A - BQ)Q - CQ^2.$$

A, B, and C are all strictly positive parameters and $P = A - BQ$ is the implied inverse demand curve.[14] Note as always that the firm chooses Q but not τ, so they treat the tax rate as exogenous. The *government* chooses τ but not Q. So, for the firm, the first-order condition is:

$$(1 - \tau)(A - 2BQ) - 2CQ = 0,$$

which yields the following profit-maximizing solution:

$$Q^* = \frac{A(1 - \tau) - 2C}{2B(1 - \tau)}.$$

To answer the first question, "How would robot output be affected by a change in τ?" we take the derivative of the solution with respect to τ to get

$$\frac{dQ^*}{d\tau} = \frac{-4BC}{(2B(1 - \tau))^2}.$$

From that, we can get our first result. Since B and C are strictly positive, $0 < \tau < 1$, and the denominator is squared, we can deduce that the numerator is unambiguously negative and the denominator is unambiguously positive. That makes the entire term negative. Therefore, an increase in the revenue tax rate, τ, will lead to a reduction in the quantity of robots.

What about tax revenue? Taking our optimal solution for Q^* and combining it with

---

[14]Recall that the inverse demand function writes price P on the left-hand side as a function of quantity demanded, Q^D. However, the causal relationship goes the other way: The demand function would write Q^D as a function of price, P.

the inverse demand function, we have that the total revenue for the firm is

$$
\begin{aligned}
\text{Revenue} \;=\; & P^*Q^* = (A - BQ^*)Q^* \\
=\; & \left(A - B\left(\frac{A(1-\tau) - 2C}{2B(1-\tau)} \right) \right) \left(\frac{A(1-\tau) - 2C}{2B(1-\tau)} \right) \\
=\; & A\left(\frac{A(1-\tau) - 2C}{2B(1-\tau)} \right) - B\left(\frac{A(1-\tau) - 2C}{2B(1-\tau)} \right)^2 .
\end{aligned}
$$

That is, total revenue and the fraction τ is tax revenue:

$$
\text{Tax revenue} \;=\; \tau\left[A\left(\frac{A(1-\tau) - 2C}{2B(1-\tau)} \right) - B\left(\frac{A(1-\tau) - 2C}{2B(1-\tau)} \right)^2 \right]
$$

$$
\begin{aligned}
\frac{d\text{Tax revenue}}{d\tau} \;=\; & \left[A\left(\frac{A(1-\tau) - 2C}{2B(1-\tau)} \right) - B\left(\frac{A(1-\tau) - 2C}{2B(1-\tau)} \right)^2 \right] \\
& - \tau\left[A\frac{dQ^*}{d\tau} - 2BQ^*\frac{dQ^*}{d\tau} \right] .
\end{aligned}
$$

To make interpretation easier, let's line that up with the tax revenue function and rewrite the line using Q^* and $\frac{dQ^*}{d\tau}$:

$$
\begin{aligned}
\text{Tax revenue} \;=\;& \tau P Q^* \\
\text{Tax revenue} \;=\;& \tau\left[A - BQ^*\right]Q^*
\end{aligned}
$$

$$
\frac{\partial(\text{Tax revenue})}{\partial\tau} \;=\; (A - BQ^*)Q^* - B\tau\frac{dQ^*}{d\tau}Q^* + \tau\left[A - BQ^*\right]\frac{\partial Q^*}{\partial\tau}
$$

From the line that says "Tax revenue" to the line that gives the marginal change in tax revenue with respect to a marginal change in the tax rate, τ, draw some lines to help you connect the terms. I deliberately left a space between those equations for you to do so. Draw an arrow from the tax rate, τ, in the first line to the $(A - BQ^*)Q^*$ in the line below. Next draw an arrow from $[A - BQ^*]$ down to $-B\tau\frac{\partial Q^*}{\partial\tau}Q^*$. Finally, draw an arrow from the final Q^* in the first line to $\tau[A - BQ^*]\frac{\partial Q^*}{\partial\tau}$. That should help you see where each part of the derivative comes from and understand how to interpret everything in economics terms.

The first term, $(A - BQ^*)Q^*$, is just the firm's optimal revenue level and comes from the first step in applying the product rule. Thus, the first term represents the marginal increase in tax revenue from holding the firm's revenue constant but increasing τ. The term is intuitively positive and would raise tax revenue.

The second term, $B\tau \frac{\partial Q^*}{\partial \tau} Q^*$, is positive overall because it has minus sign in front, multiplies B, τ, and Q^*, which are all positive, but also multiplies $\frac{\partial Q^*}{\partial \tau}$ which we established was negative. Thus, the minus sign and $\frac{\partial Q^*}{\partial \tau}$ result in a positive term, which increases tax revenue. The second term represents the change in *price*, holding everything else constant. The price rises with a fall in quantity, which leads to an increase in revenue. That might seem contradictory. The price is rising because there will be lower quantity, but the quantity we multiply by is Q^* and the price change comes through from $\frac{\partial Q^*}{\partial \tau}$.

Finally, the last term, $\tau\left[A - BQ^*\right] \frac{\partial Q^*}{\partial \tau}$, is the effect of the change in the optimal quantity, Q^*. Its the tax rate, τ, times the price, $[A - BQ^*]$, times the change in the optimal quantity that results from a change in the tax rate.

In these models, we are drawing out the effects of these exogenous variables on the endogenous variables when we allow our agents to make optimizing decisions. As you can imagine, these are critically important results. We can discuss all sorts of policy implications from conducting such analyses in these theoretical models and then take these results to real–world data to see if the predictions hold. Moreover, for well–established results, such analysis allows us to forecast and make predictions about the effects, and sometimes those effects are surprising or the result of unintended consequences.

10.2 Derivatives of Implicit Functions

In nearly all the derivatives we have taken so far, we have been using *explicit functions*. By "explicit" we mean that the y variable is isolated on the left-hand side entirely by itself, and the y also does not appear on the right-hand side. That is, y has been isolated and expressed in terms of x. For example,

$$y = 3x^2 + 4$$

is an explicit function. However,

$$y - 3x^2 - 4 = 0$$

is not explicit because y is not isolated. You could, of course, merely add $3x^2 + 4$ to both sides of the equation and make it explicit. In some cases that is not possible, such as in the following equation:

$$3y^2 - 4xy = 4 - x^2 y.$$

Such functions are called *implicit functions* because there still exists a relationship between x and y, even though we can't isolate the y variable. With an implicit function, we can still find $\frac{dy}{dx}$, or $\frac{\partial y}{\partial x}$ if it is a multivariate function, how the y variable changes with x (i.e., the slope).[15]

It turns out there is a pretty simple rule for finding the derivative in problems like the example called the *implicit function rule*. To apply the implicit function rule, there are two steps: First write the equation such that it is equal to zero. In the example, we would write it as

> **Definition**
>
> **Implicit and explicit functions.**: An explicit function is a function that can be written in the form $y = f(x)$, where y only appears on the left-hand side and there are no operations on y such as a square root, power, and so forth. An implicit function is a function, which is not expressed in the $y = f(x)$ form and often cannot be expressed as one.

$$3y^2 - 4xy + x^2 y - 4 = 0.$$

For notational convenience, we will call everything on the left-hand side $F(x, y)$. The first step is really finding $F(x, y) = 0$. The second step involves finding the derivatives of $F(x, y)$ with respect to both x and y. The derivative of y with respect to x is:

$$\frac{dy}{dx} = -\frac{F_x(x, y)}{F_y(x, y)}.$$

In words, the derivative of y with respect to x is the negative of the ratio of the partial derivative of $F(x, y)$ with respect to x to the partial derivative of $F(x, y)$ with respect to y.

Let's do a few examples. Suppose we have

$$y = 3x^2 + 4.$$

which is an explicit function, and from our most basic rules in Chapter 2 we know how

---

[15]This example is single–variable calculus, and we could have introduced this topic earlier. However, many of the interesting applications appear in a multivariate calculus context. Moreover, familiarity with the total derivative will be helpful in following the explanation.

to do this:

$$\frac{dy}{dx} = 6x.$$

Suppose we write it as an implicit function:

$$y - 3x^2 - 4 = 0.$$

If we apply the implicit function rule, we should get the same result:

$$\frac{dy}{dx} = -\frac{F_x(x, y)}{F_y(x, y)} = -\frac{-6x}{1} = 6x,$$

where the numerator, $-6x$, is the derivative of $y - 3x^2 - 4$ with respect to x, and the denominator, 1, is the derivative of $y - 3x^2 - 4$ with respect to y.

Now let's try this technique on the more complicated implicit function we had before:

$$3y^2 - 4xy = -x^2y + 4,$$

which, as a reminder, we must rewrite such that it is equal to zero:[16]

$$3y^2 - 4xy + x^2y - 4 = 0.$$

Applying the implicit function rule we have:

$$\frac{dy}{dx} = -\frac{F_x(x, y)}{F_y(x, y)} = -\frac{-4y + 2xy}{6y - 4x + x^2} = \frac{4y - 2xy}{6y - 4x + x^2}.$$

The interpretation here is exactly the same as before. The derivative shows us the marginal change in y that occurs because of a marginal change in x. In some ways this concept is even easier to intuitively understand with an implicit function. Again, reconsider the original equation when set equal to zero:

$$3y^2 - 4xy + x^2y - 4 = 0.$$

Suppose we increase x by a small amount. How much must y change in order to keep the equation true and equal to zero? By $\frac{dy}{dx}$.

You may have wondered, because the notation looks similar, if there is any relationship between this technique and the total derivatives we did in Chapter 6. The answer is yes. The implicit function theorem is a short–cut version of taking a total derivative on a single variable. However, it also allows us to compute that derivative when the function cannot be arranged into an explicit function, as we discussed at the start of this chapter. However, in essence, it is the same technique and the same application of taking derivatives.

---

[16]Whether we move everything to the left or right of the equals sign is irrelevant. We could also rearrange to get $-3y^2 + 4xy - x^2y + 4 = 0$. The implicit function rule will yield the same answer either way.

Let's use the total derivative approach on the example to illustrate. Starting from:

$$3y^2 - 4xy = 4 - x^2 y$$

we do not have to arrange it such that it equals zero. We could, but with total derivatives it does not matter. Taking the total derivative with respect to both x and y we get

$$6y\,dy - 4x\,dy - 4y\,dx = -2xy\,dx - x^2\,dy.$$

Remember, we are looking for the ratio dy/dx. We can get there by simply using algebra and solving. First, arrange the equation such that all the dy items are on the left-hand side and all the dx items are on the right-hand side.:

$$6y\,dy - 4x\,dy + x^2\,dy = -2xy\,dx + 4y\,dx.$$

Factor out the dx and dy:

$$\left(6y - 4x + x^2\right)dy = \left(-2xy + 4y\right)dx.$$

Now divide both sides by dx and by the factor multiplying dy, $(6y - 4x + x^2)$:

$$\frac{dy}{dx} = \frac{4y - 2xy}{6y - 4x + x^2}.$$

The answer is the same as before.

10.2.1 Examples With Models

Now, let us consider applying this technique to an economic model to show where and how it becomes useful. Suppose utility for Totoro and his budget constraint are given by

$$U(x_1, x_2) = \ln x_1 + \alpha \ln x_2$$
$$W = (1 + \tau)p_1 x_1 + p_2 x_2.$$

where x_1 is acorns and x_2 is umbrellas. $\alpha > 0$ is a preference parameter. In the forest where Totoro lives there is a sales tax on acorns of τ. Solving for x_2 and substituting the result into the objective function, Totoro's optimization problem is

$$Max_{x_1} U = \ln x_1 + \alpha \ln \left(\frac{W - (1 + \tau)p_1 x_1}{p_2}\right).$$

Taking the derivative with respect to the choice variable, x_1, and setting the result equal to zero to form the first-order condition, we have

$$\frac{dU}{dx_1} = \frac{1}{x_1} - \alpha\frac{(1+\tau)p_1}{W - (1+\tau)p_1x_1} = 0$$
$$\rightarrow \quad W - (1+\tau)p_1x_1 - \alpha(1+\tau)p_1x_1 = 0.$$

What if we want to know how x_1^* changes when W changes? The first-order condition is only satisfied by the optimal level of x_1, but the first-order condition is currently an implicit function. We can use the implicit function rule to get $\frac{dx_1^*}{dW}$. Of course, we could do what we did at the start of the Chapter and solve for x_1^* as an explicit function of W and the other exogenous variables and then take the derivative of that with respect to W. However, we want to illustrate the use of the implicit function theorem here. In the next example, we will not be able to solve for an explicit function, so we will have to use the implicit function to get the result.

Applying the implicit function theorem,

$$\frac{dx_1^*}{dW} = -\frac{1}{-(1+\tau)p_1 - \alpha(1+\tau)p_1}$$
$$= \frac{1}{(1+\alpha)(1+\tau)p_1} > 0,$$

which is positive since $\alpha > 0$, $\tau \geq 0$, and $p_1 > 0$. So, the demand for x_1 increases with wealth, making acorns a normal good for Totoro.

Let's use this technique on a general function. Suppose that the Nekobus company faces the following profit–maximization problem:

$$max_{K,L}\Pi = PF(K,L) - rK - wL - g[|K - \bar{K}|],$$

where P is price, $F(K,L)$ is the production function based on capital, K, and labor, L. Production has positive but diminishing returns to both capital and labor, which means that $F_K > 0$, $F_L > 0$, $F_{KK} < 0$, and $F_{LL} < 0$. That is, the production function is concave in both capital and labor. r is the rental rate of capital (the cost of using capital), and w is the wage rate for labor. The function $g[|K - \bar{K}|]$ represents an idea called *adjustment costs*. The idea is that the firm has $\bar{K}$ amount of capital. If it needs to make a change to that level of capital it can, but the adjustment comes at an additional cost (the cost of installing or removing machinery). The function has the following properties, $g(0) = 0$, $g' > 0$, and $g'' > 0$. That says the function is convex. Intuitively, when the choice of capital is different from existing capital $\bar{K}$, there are positive adjustment costs. As the difference between desired capital and installed capital grows, the adjustment costs grow at an increasing rate.

Taking the derivative with respect to K but holding L fixed, we get the following first-order condition:

$$\frac{\partial \Pi}{\partial K} = PF_K(K, L) - r - g'[|K - \bar{K}|] = 0$$

That equation implicitly defines a choice of K^*, but we cannot solve for it explicitly. However, we would like to know, for example, how a change in r affects the optimal choice of K^*, continuing to hold L fixed. To get that answer, we apply the implicit function theorem:

$$\frac{\partial K^*}{\partial r} = -\frac{-1}{PF_{KK}(K, L) - g''[|K - \bar{K}|]}$$
$$= \frac{1}{PF_{KK}(K, L) - g''[|K - \bar{K}|]}.$$

We can sign this expression since we know that $P > 0$, and from our assumptions we know that $F_{KK} < 0$ and $g'' > 0$. Therefore, the entire denominator is negative, and thus the whole expression is negative. That means that the optimal choice of capital is decreasing in the rental rate of capital.

In the example we just did, we explicitly held L constant. However, if K^* were to change and L is also an endogenous choice variable, then L^* would also change, which, in turn, would have feedback effects on K^*. This type of situation is one with a system of first-order conditions, and those conditions need to be accounted for simultaneously to derive the full effect of a change in one exogenous variable. We look at this type of problem next.

10.3 Multivariate Comparative Statics

Comparative statics, of course, extends to the multivariate case, but gets more complicated. In particular when using first-order conditions to derive comparative statics, one must be careful to account for the effects on *all* the optimality conditions when

an exogenous variable changes. If you have had some basic linear algebra, you may
have come across one of the key tools for comparative statics known as **Cramer's rule**.
Cramer's rule is explained in Chapter 12 and is a very powerful short–cut method for
handling large complex algebra problems with many variables. Here, we will do an example without any linear algebra. However, the reader is encourage to learn Cramer's
rule properly, as it is a commonly used technique in economic analysis.

Consider the basic profit–maximization problem for Yun's durian farm with two
inputs capital, K, and labor, L, but with a general production function, as follows:

$$\max_{K,L} \Pi \;=\; pf(K,L) - wL - rK$$

where

$$f_K > 0, f_L > 0$$
$$f_{KK} < 0, f_{LL} < 0.$$

The price of the durian he sells is p. The labor wage rate is w, and the rental rate of
capital is r. The production function is increasing in both K and L and strictly concave.
Taking the derivatives and forming the first-order conditions yields

$$\frac{\partial \Pi}{\partial K} \;=\; pf_K(K,L) - r = 0$$
$$\frac{\partial \Pi}{\partial L} \;=\; pf_L(K,L) - w = 0.$$

Suppose the question is, how would an increase in r affect the optimal choice of K?
That is, what is $\frac{\partial K^*}{\partial r}$? It might be tempting to use the implicit function theorem
on the first-order condition above that applied to choosing capital. One would get

$$\frac{\partial K^*}{\partial r} \;=\; \frac{1}{pf_{KK}(K,L)}.$$

That result is *wrong!* Why? Those two first-order conditions pin down the optimal
levels of the choice variables, K^* and L^*. When we write out the solutions properly, we
should have $K^* = K(p,w,r)$ and $L^* = L(p,w,r)$, where our solutions are functions of
all the exogenous variables. If an exogenous variable like r changes, it will affect both
K^* and L^*. So, a change in r not only affects K^* directly, but also indirectly through
its effect on L^*. That means we need to account for how *all* the first-order conditions
are affected. That is, K^* will change because r changes *and* will change because L^*
changes.

To account for how all the first–order conditions are affected when an exogenous
variable changes, we take the total derivatives of both with respect to the exogenous
variable of interest, r, in the example, and all the endogenous variables, K and L. We
would get

231

$$pf_{KK}(K,L)dK + pf_{KL}(K,L)dL - 1dr = 0$$
$$pf_{KL}(K,L)dK + pf_{LL}(K,L)dL = 0.$$

We can solve for $\frac{dK}{dr}$ by solving the one equation for dL, substituting into the other and rearranging. After a bit of algebra we have

$$\frac{dK}{dr} = \frac{f_{LL}(K,L)}{p(f_{KK}(K,L)f_{LL}(K,L) - f_{KL}(K,L)^2)}$$
$$\frac{dK}{dr} = \frac{f_{LL}}{p(f_{KK}f_{LL} - f_{KL}^2)},$$

where the second line drops the arguments (K,L) in the prodution function to make it easier to read. The first thing to note is the denominator contains $f_{KK}f_{LL} - f_{KL}^2$. Where have you seen that before? Try to answer before reading on.

That expression is one of the second-order conditions for an extreme point, as it is the saddle point condition. If we assume a maximum and an interior solution, that expression must be positive. p also appears in the denominator and must be positive, so the entire denominator is positive. The numerator is the second derivative of the production function with respect to L. We assumed that to be negative at the outset of the problem such that the production function is concave in L. Therefore, the entire expression for $\frac{dK}{dr} < 0$, which makes intuitive sense. An increase in the cost of capital will lead to a reduction in capital used.

Let's do the same for dL/dr. We now get

$$\frac{dL}{dr} = \frac{-f_{KL}}{p(f_{KK}f_{LL} - f_{KL}^2)}.$$

Notice the denominator is the same here, so we know the sign is positive for a profit-maximizing solution. The numerator, however, is the negative of the cross-derivative of the production function, which we did not make any assumptions about when we started. So, what is f_{KL}? Recall Young's theorem from Section 6.1.1. We arrive at the same result regardless of the order. So, if we had f_K and took the derivative with respect to L, we would interpret that as the marginal effect of L on the marginal productivity of K. Coming from the other direction, if we had f_L and took the derivative with respect to K, we would interpret that as the marginal effect of K on the marginal productivity of L. In economics terms, a positive value would represent *complements* in production; more of one input makes the other factor more productive. If the cross-derivative is negative then they are *substitutes* in production.

Returning to the result on $\frac{dL}{dr}$, the sign will be the opposite of the sign of f_{KL}. If $f_{KL} > 0$ and capital and labor are complements, then an increase in r will lead to a decrease in L. That makes intuitive sense. We know an increase in r leads to a reduction in K employed from our previous example. If the amount of K used falls, then the marginal productivity of labor will also fall, so Yun optimally chooses less labor. Alternatively, if $f_{KL} < 0$ and capital and labor are substitutes, then an increase in r will lead to a increase in L. Again, we know an increase in r leads to a reduction in K employed. If the amount of K used falls, that means the marginal productivity of labor will rise, so Yun optimally chooses more labor.

10.4 Envelope Theorem

One item we have been emphasizing is that when we are optimizing something in a model, the derivatives we take to form first-order conditions are always on the *choice* variables. We do that because the idea is the agents are doing the maximizing/minimizing by controlling these things. Agents cannot control exogenous variables and therefore cannot optimize with respect to them.

But what would it mean if we took the derivative of the objective function with respect to an exogenous variable?

10.4.1 Introductory Example

Suppose that the Otaki firm continues to produce robots (we'll forget about the taxes here). The Otaki firm faces following the short–run profit–maximization problem:

$$\max_{L} \Pi = P\alpha K^{1/3}L^{2/3} - wL.$$

Here, in the short run, physical capital, K, is fixed and cannot be changed. Thus, it is an exogenous variable in this problem. P is the price of robots and α is a productivity parameter for the Otaki firm. That leaves the choice variable as labor to employ, L. The cost of labor is given by the wage rate, w. We would take the derivative on the choice variable, L, to find the optimal choice, as follows:

$$\frac{d\Pi}{dL} = \frac{2}{3}P\alpha K^{1/3}L^{-1/3} - w = 0$$

$$L^* = \frac{8}{27}\left(\frac{P\alpha}{w}\right)^3 K.$$

and we solve for L^*, all in terms of exogenous variables and parameters as before.

But what if we took the derivative of the original maximization problem with respect to exogenous variables like K or w? For example,

$$\frac{\partial \Pi}{\partial K} = \frac{1}{3}P\alpha K^{-2/3}L^{2/3}$$

$$\frac{\partial \Pi}{\partial w} = -L.$$

What do those mean? Note that we do *not* set those derivatives equal to zero. Our firm owner is not optimizing by choosing these variables, so these cannot be used as first-order conditions.

The first expression might be read as the marginal change in profits with respect to a change in physical capital, and the second might be read as the marginal change in profits with respect to a change in the wage rate. Is that correct? No and sort of yes, but mostly no. You may be thinking, "*I find that answer vague and unconvincing,*" well, that is a vague, contradictory and confusing answer, so let's explore this more carefully.

The answer is no because if an exogenous variable changes the optimal value for the choice variable(s) will change, which will affect the profit level at the optimum. Notice that the two lines are partial derivatives in that they presume to hold everything else constant including labor, L. However, the firm would then change their optimal choice of labor as either K or w changed and you can see that directly in the solution for L^* and more formally by using our comparative statics:

$$\frac{\partial L^*}{\partial K} = \frac{8}{27}\left(\frac{P\alpha}{w}\right)^3, \qquad \frac{\partial L^*}{\partial w} = -\frac{8}{9}\left(\frac{(P\alpha)^3}{w^4}\right)K.$$

Again, the derivatives on the exogenous variables are not accurately interpreted as we wrote before because they affect the endogenous choice variables, which, in turn, affect the profits, but the naïve derivatives hold L constant.

You may now be thinking, why did you say "yes, but mostly no?" Seems like it should be just "no," right? Well, let's take those exogenous variable derivatives again but actually account for the optimal level of L by substituting in our solution for L^*. We get

$$\Pi^* = P\alpha K^{1/3}(L^*)^{2/3} - wL^*.$$

We formed a *value function*. We encountered this type of expression before when we looked at the indirect utility function in Chapter 9 when we discussed duality. The objective function is expressed entirely in terms of exogenous variables and parameters because we replaced the endogenous variables with their optimal solutions. Now when we take that derivative, we account for the fact that the optimal values of the choice variable will change when an exogenous variable changes. Let's do that:

$$\frac{\partial \Pi^*}{\partial K} = \frac{1}{3}P\alpha K^{-2/3}(L^*)^{2/3} + \frac{2}{3}P\alpha K^{1/3}(L^*)^{-1/3}\frac{\partial L^*}{\partial K} - w\frac{\partial L^*}{\partial K}.$$

Notice the application of chain rule in the second term. Now $\frac{\partial L^*}{\partial K}$ is exactly what we derived, but we shall leave it in this form momentarily, and rearranging you can see why:

$$\frac{\partial \Pi^*}{\partial K} = \frac{1}{3}P\alpha K^{-2/3}(L^*)^{2/3} + \left[\frac{2}{3}P\alpha K^{1/3}(L^*)^{-1/3} - w\right]\frac{\partial L^*}{\partial K}.$$

Examine the term inside the braces [ ]. It is the first-order condition from our original optimization problem! At the optimal level of labor, L^*, the first-order condition is zero, of course. The above expression reduces to:

$$\frac{\partial \Pi^*}{\partial K} = \frac{1}{3}P\alpha K^{-2/3}(L^*)^{2/3},$$

which is *almost* exactly the same as we got above when we naively took the derivative of Π with respect to the exogenous variable K. Almost.

Why almost? Study the expressions and find what is different before moving on. The difference is that in the naïve derivative we have the endogenous variable L, and in the expression we have the optimtized value of labor L^*. This example is the main result of the envelope theorem. The derivative with respect to an exogenous variable of the objective function gives the correct value *at the optimum*. That is, in this case, $\frac{d\Pi}{dK} = \frac{d\Pi^*}{dK}$ at the optimally chosen level of L^*, but not elsewhere.

You might be thinking to yourself, "um...so what?" Well, this result will make your modeling and math life a lot easier. You can identify the impact of an exogenous variable on the objective function by taking the derivative directly and noting that the result is correct at the optimized values of the endogenous variables. That means you do not have to go through a large chunk of calculus and algebra work to solve a problem and then find this effect. You can take the derivative directly, *so long as* you are careful with how exactly you interpret that result by remembering it only applies at the optimal choices of the endogenous variables.

10.4.2 General Case

Consider the following optimization problem:

$$y = f(x_1, x_2; \alpha),$$

where x_1 and x_2 are choice variables and α is exogenous. You can think of that function as containing the entire problem written in a general form, and it may or may not have constraints. This problem generalizes what we had before, though we could make it even more general and allow there to be N choice variables and M exogenous variables. However, the results will not change, and we'll just carry around lots of extra notation.

The necessary first-order conditions are

$$\frac{\partial y}{\partial x_1} = f_{x_1} = 0$$

$$\frac{\partial y}{\partial x_2} = f_{x_2} = 0.$$

Those two conditions define the optimal levels of x_1^* and x_2^*, even if we cannot solve for them explicitly. We write these solutions as $x_1^*(\alpha)$ and $x_2^*(\alpha)$. Those solutions show that the optimized levels of the endogenous choice variables are functions of the exogenous variable(s), α in this case. Placing these optimized values back into the objective function, we form the value function (an indirect objective function), which we will call $\omega(\alpha)$

$$\omega(\alpha) = f(x_1^*(\alpha), x_2^*(\alpha); \alpha).$$

How does α affect y? Or, equivalently, how does y vary with α?

$$\frac{\partial \omega(\alpha)}{\partial \alpha} = f_{x_1}(x_1^*(\alpha), x_2^*(\alpha); \alpha)\left(\frac{\partial x_1^*(\alpha)}{\partial \alpha}\right) + f_{x_2}(x_1^*(\alpha), x_2^*(\alpha); \alpha)\left(\frac{\partial x_2^*(\alpha)}{\partial \alpha}\right) + f_\alpha(x_1^*(\alpha), x_2^*(\alpha); \alpha)$$

But again, $f_{x_1}(x_1^*(\alpha), x_2^*(\alpha); \alpha)$ and $f_{x_2}(x_1^*(\alpha), x_2^*(\alpha); \alpha)$ are the first-order conditions, which must be equal to zero at the optimum. The entire expression reduces to

$$\frac{\partial \omega(\alpha)}{\partial \alpha} = f_\alpha(x_1^*(\alpha), x_2^*(\alpha); \alpha).$$

That is the result one would get if you took the derivative of the original function, $y = f(x_1, x_2; \alpha)$, with respect to α, except that the answer is at the optimum. Again, taking the derivative of the objective function with respect to an exogenous variable only yields the correct answer at the optimum. Otherwise, the result ignores the fact that as the exogenous variable changes, the endogenous choice variables adjust. To see this more clearly and visually we have one more example.

10.4.3   One More Example

Let's go back to Otaki's robots but consider the long-run problem where both capital and labor can be adjusted:

$$\max_{K,L} \Pi = P\alpha K^{1/3}L^{2/3} - rK - wL,$$

where we have added the rental rate of capital, r, times the amount of capital to account for the capital costs. Suppose we want to know how r affects profits, and we take the derivative of Π with respect to r. We get

$$\frac{\partial \Pi}{\partial r} = -K,$$

which makes sense at some level. If r increases, the marginal cost will depend directly on how many units of capital the firm has. But, at the same time, a change in r should induce a change in how much capital the firms uses.

Looking at the optimal solution, observe that, without even taking the first-order conditions, we can see that optimal solutions will have the form $L^*(r, w, p)$ and $K^*(r, w, p)$. The optimal solutions for the production inputs will be functions of the three exogenous variables, r, w, and p, so we can write the solutions as $L^*(r, w, p)$ and $K^*(r, w, p)$.

Forming the indirect profit function by substituting in the optimal solutions, we have

$$\Pi^* = P\alpha K^*(r, w, p)^{1/3}L^*(r, w, p)^{2/3} - rK^*(r, w, p) - wL^*(r, w, p).$$

Now, taking the derivative of Π^* with respect to r, we get

$$\frac{d\Pi^*}{dr} = P\alpha\frac{1}{3}K^*(r, w, p)^{-2/3}L^*(r, w, p)^{2/3}\frac{\partial K^*}{\partial r} + P\alpha\frac{2}{3}K^*(r, w, p)^{2/3}L^*(r, w, p)^{-1/3}\frac{\partial L^*}{\partial r}$$
$$- K^*(r, w, p) - r\frac{\partial K^*}{\partial r} - L^*(r, w, p) - w\frac{\partial L^*}{\partial r}.$$

Notice that after we rearrange we get back the original first-order conditions on K^* and L^*, as follows:

$$\frac{d\Pi^*}{dr} = \left[P\alpha\frac{1}{3}K^*(r, w, p)^{-2/3}L^*(r, w, p)^{2/3} - r\right]\frac{\partial K^*}{\partial r}$$
$$+ \left[P\alpha\frac{2}{3}K^*(r, w, p)^{2/3}L^*(r, w, p)^{-1/3} - w\right]\frac{\partial L^*}{\partial r}$$
$$- K^*(r, w, p).$$

The expressions in brackets are the first-order conditions. At the optimum they are equal to zero, so the entire expression reduces to

$$\frac{d\Pi^*}{dr} = -K^*(r, w, p).$$

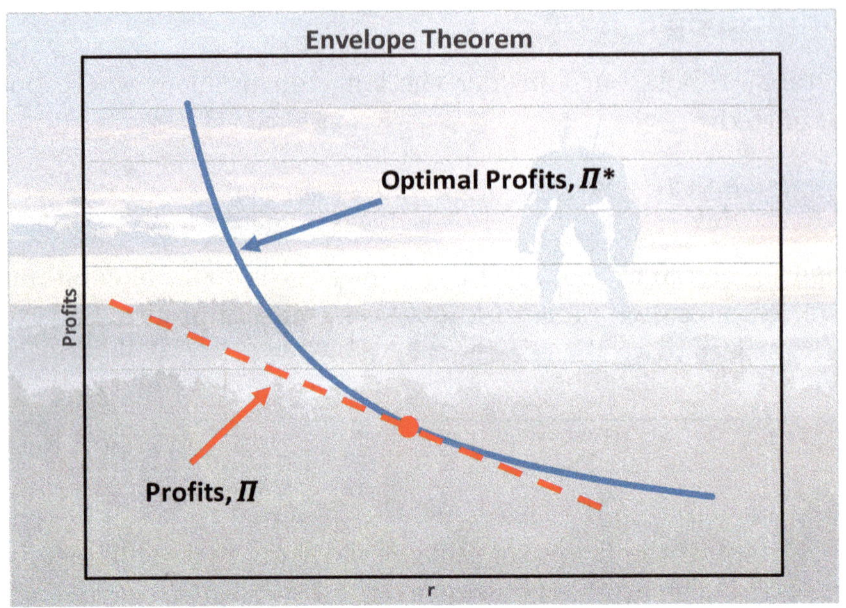

Figure 10.1: Profits and optimal profits as function of r.

Hence, at the optimum we have the same result as the naïve derivative. To illustrate, Figure 10.1 shows profits, Π, as a function of r, and optimal profits, Π^*, also as a function of r. The profit function itself, where the inputs are unchanging, is just a linear function, with a negative slope of $-K$. However, the optimal profits adjust both K^* and L^* as r changes, creating a convex function. The two have a tangency, and therefore the same slope, at the optimal solution to the profit–maximization problem.

The end result here is that taking the derivative of the original problem with respect to an exogenous variable does give us the correct result, but only at the optimized value. Once you move away from that level of r, the slopes are no longer equal.

10.5   Exercises

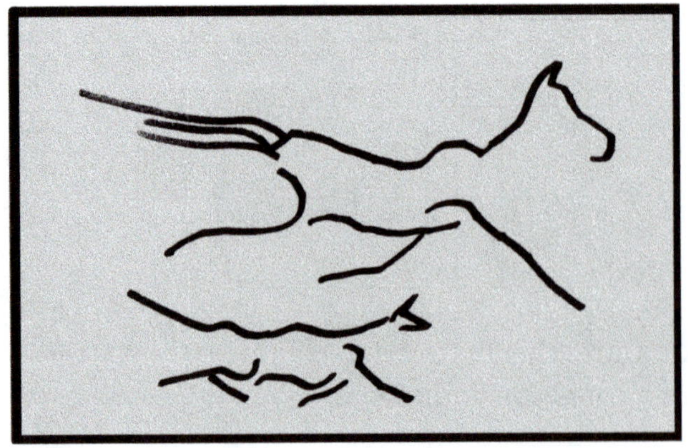

10.5.1 Quick Check Answers

a) The equlibrium price and quantity are $p^{**} = \frac{\alpha - \omega + \gamma y + \eta r}{\beta + \phi}$ and $Q^{**} = \frac{\beta \omega + \alpha \phi + \gamma \phi y - \eta \beta r}{\beta + \phi}$.
Thus, $\frac{\partial p^{**}}{\partial r} = \frac{\eta}{\beta + \gamma} > 0$ and $\frac{\partial Q^{**}}{\partial r} = \frac{-\eta \beta}{\beta + \gamma} < 0$.

b) $\frac{\partial}{\partial \beta} \tau \gamma \frac{(1-\tau)\gamma}{\beta^2 + (1-\tau)\gamma} = -\tau \gamma \frac{2(1-\tau)\gamma}{[\beta^2 + (1-\tau)\gamma]^2} < 0$. If Mei's preference for playing drums increases, she reduces her passion fruit picking, which reduces the tax revenue collected by the king.

c) $\frac{\partial}{\partial A} \frac{A(1-\tau) - 2C}{2B(1-\tau)} = \frac{1}{2B} > 0$ and $\frac{\partial}{\partial B} \frac{A(1-\tau) - 2C}{2B(1-\tau)} = \frac{-2(1-\tau)[A(1-\tau) - 2C]}{[2B(1-\tau)]^2} < 0$. An increase in the intercept of the inverse demand function increases the optimal quantity, but an decrease in the slope reduces the optimal quantity.

d) $\frac{dy}{dx} = 18x^2$

e) $\frac{dy}{dx} = \frac{y^2 - 2x}{1 - 2xy}$

f) $\frac{\partial x}{\partial(1+\tau)} = -\frac{x_1}{1+\tau}$ and $\frac{\partial x_1}{\partial p} = -\frac{x_1}{1+\tau} = -\frac{x_1}{p_1}$

g) The result is $\frac{dL}{dw} = \frac{f_{KK}}{p(f_{KK}f_{LL} - f_{KL}^2)} < 0$. We know its negative because the denominator is the second-order condition which must be positive for a maximum, and the numerator is $f_{KK} < 0$, which was assumed negative to give the production function diminishing returns to capital.

10.5.2 Practice Problems

1. Find dy/dx from the following using the implicit function rule:

   (a)
   $$y + 4x - 5 = 0$$

   (b)
   $$y + 8xy + 5 = 0$$

   (c)
   $$y^2 + 2xy^{1/2} + 5x - 99 = 0$$

   (d)
   $$2x^3 - 4x^2y + 6xy^2 = -8$$

   (e)
   $$4y^3 + 2xy^4 + 5x^2y + \ln x = 0$$

   (f) Find dK/dL from the following using the implicit function rule:
   $$\alpha K^{\alpha-1}L^\beta - r - \theta(K - \Gamma)^{\theta-1} = 0$$

   (g) Find dc_1/dy_2 from the following using the implicit function rule:
   $$\frac{1}{c_1} - \frac{\beta(1+r)}{y_2 + (1+r)(a + y_1 - c_1)} = 0$$

2. Consider the following general functions for the demand and supply of Mandaburg-

ers:

$$Q^D = \alpha + \beta P$$
$$Q^S = \gamma + \delta P$$

(a) Consider the four parameters, α, β, γ, and δ. Given what you know of supply and demand from principles of microeconomics, what restrictions can you assume for these parameters, and why? Make those assumptions for the rest of the problem.

(b) Solve for the equilibrium values of P^* and Q^* such that the market clears.

(c) Use partial derivatives to find the effects of changes in each of the parameters (α, β, γ, and δ) on the equilibrium values of P^* and Q^*, so eight derivatives in all.

3. Return to the tax example from Section 10.1.3. Note that the demand curve is our usual simple linear inverse demand function, $P = A - BQ$. A is the y-intercept and B is the slope.

(a) Find the effect of an increase in A on both the optimal quantity produced and tax revenue.

(b) Find the effect of an increase in B on both the optimal quantity produced and tax revenue.

(c) Bonus: Find the tax rate, τ^*, that maximizes government revenue. Determine how A and B affect τ^*.

4. For the Nekobus problem in Section 10.2.1, find $\partial K^*/\partial \bar{K}$ and $\partial K^*/\partial P$.

5. Return to the example from Section 10.3. Derive the following comparative static results from the model there for Yun's farm: $\frac{dL}{dw}$, $\frac{dK}{dw}$, $\frac{dK}{dp}$, and $\frac{dL}{dp}$.

6. Consider the following problem. Yun has a plot of land on which he grows coconuts and durian. Let C be the quantity of coconuts grown and D be the amount of durian he grows. Harvesting coconuts is costless as the ripe coconuts fall to the ground when ready. However, he must hire labor, L, at the wage rate of w to obtain the durian. If the total amount of land is N, then the production of coconuts is $C = C(cN)$, which is strictly concave, so $C_N > 0$ $C_{NN} < 0$. The choice variable c is the fraction of land devoted to coconuts and must lie between zero and one, $0 \leq c \leq 1$. The production function for durian is multivariate and given by $D = D((1-c)N, L)$ and is again concave in both land and labor, so $D_N > 0$, $D_{NN} < 0$, $D_L > 0$, and $D_{LL} < 0$. Let the prices of coconuts and durian be P_C and P_D, respectively. Yun's profit maximization problem is then:

$$\max_{c,L} \Pi = P_C C(cN) + P_D D((1-c)N, L) - wL.$$

(a) Use the approach of Section 10.3 to first get the two first-order conditions.

(b) Take the total derivatives of both first-order conditions with respect to the two endogenous choice variables, c and L, and the exogenous variables, P_C, P_D, and w.

(c) Set $dP_C = 0$ and $dP_D = 0$, so there are no changes in the prices of coconuts and durian. Then find the effect of a wage rate increase on the amount of labor hired, dL/dw.

(d) Set $dP_D = 0$ and $dw = 0$, so there are no changes in the price of durian or the wage rate. Then find the effect of a coconut price increase on the fraction of land devoted to coconuts, dc/dP_C.

(e) Set $dP_C = 0$ and $dw = 0$, so there are no changes in the price of coconuts or the wage rate. Then find the effect of a durian price increase on the fraction of land devoted to coconuts, dc/dP_D.

11   Integration

Integration is the opposite of taking derivatives. In fact, some textbooks refer to integration as anti-derivatives. So, yes, we're going to go in reverse. The good news is that having practiced going forward and taken lots of derivatives, we can invert the rules. The bad news is that reversing the process is not as easy as it sounds.

When do we use integration in economics? Mathematically one of the most useful things an integral does is find the area under a curve. For example, you probably have computed consumer welfare or producer surplus or a dead–weight loss. In most, if not all, those cases, you were finding the areas of familiar and regular geometric shapes –triangles, rectangles, squares, trapezoids– where you could use familiar area formulas from geometry. But what would you do if the shapes were actually *curved*? Integration will do that for you. If we need to find the area between the price and the demand curve in a supply–and–demand analysis, where the demand curve is not straight but strictly convex, we can use integration to perform the task. Another area where integrals appear is evaluating objective functions or constraints over time. Graphically, we often have time on the x-axis when we are thinking about how some economic variables (e.g. , GDP, CPI, unemployment) change over time. The area under the curve may have an interpretation and meaning we care about. Thus, integration of that nature is common in macroeconomics, where making comparisons of the economy over time becomes important for understanding and modeling. Furthermore, if you have taken a statistics course or an econometrics course, you may have encountered integrals as a way of computing a cumulative distribution function (CDF) from a probability distribution function (PDF).

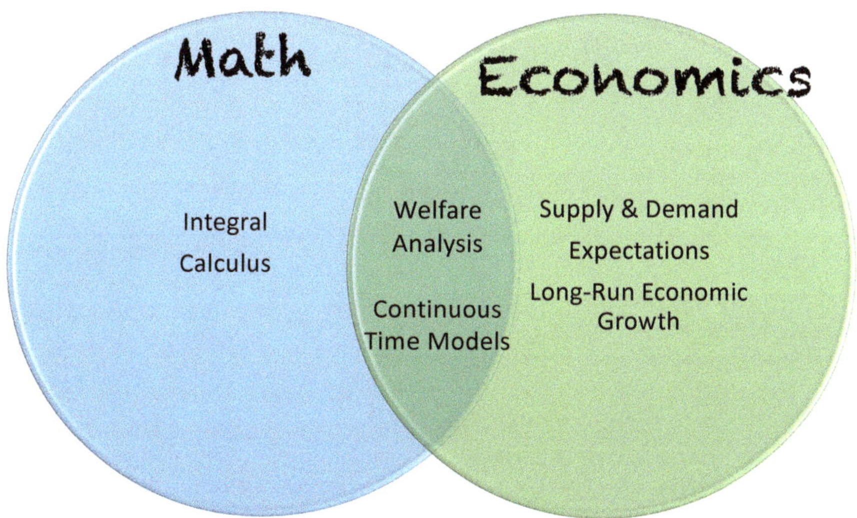

11.1 Indefinite Integrals

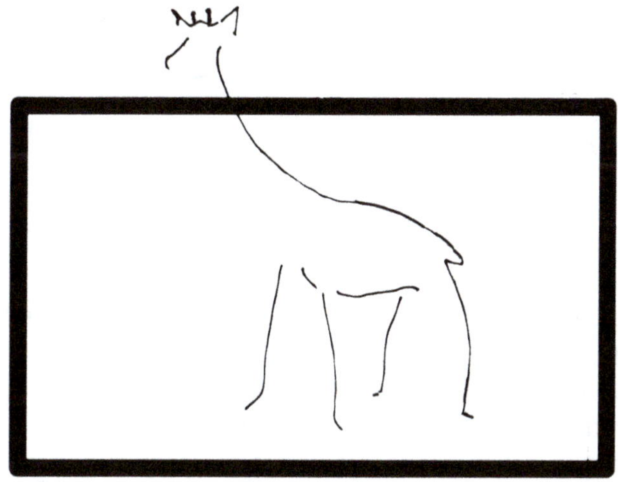

Indefinite??!! What does that mean? Well, among a few things, it means that in a few pages we will discuss *definite* integrals. What is the difference between *indefinite* and *definite* integrals? For the moment, we provide a brief explanation, but all will be more clear when we get to definite integrals. In short, if we are looking at the area under the curve generated by some function, for a definite integral we are looking at the area between some point a and some point b on the x-axis. That is, the integral has a defined lower limit a and a defined upper limit b. In contrast, for an indefinite integral, we do not have a lower or upper limit; we merely find the expression for the area under the entire curve.

Let's move to a simple example to illustrate. Consider the following function that describes the growth rate of some country; let's call it SSSP. In the economy of SSSP, the growth rate of the population is given by

$$\frac{dH(t)}{dt} = t^{-1/2}.$$

Figure 11.1 graphs the function.

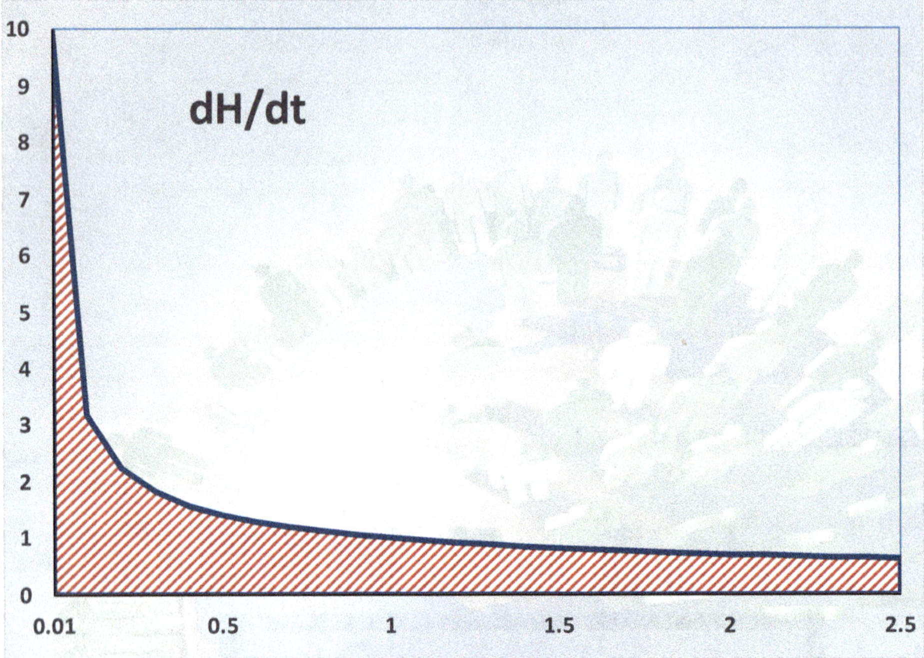

Figure 11.1: Integration: Area under the curve.

H is the population and t is time. We want to know the *level* of the population at some point in time; perhaps, for example, we want to forecast the future population in 10 years. To figure that out, we need to reverse the **power** derivative rule found in Chapter 2. For your convenience we reproduce the derivative rules summary table.

Table 11.1: Derivative Rules

Rule	Function	Derivative
Constant	$y = c$	$\frac{dy}{dx} = 0$
Linear	$y = bx$	$\frac{dy}{dx} = b$
Power	$y = x^n$	$\frac{dy}{dx} = nx^{n-1}$
Sum–difference	$y = f(x) + g(x)$	$\frac{dy}{dx} = f'(x) + g'(x)$
Product	$y = f(x)g(x)$	$\frac{dy}{dx} = f'(x)g(x) + f(x)g'(x)$
Quotient	$y = \frac{f(x)}{g(x)}$	$\frac{dy}{dx} = \frac{g(x)f'(x) - f(x)g'(x)}{[g(x)]^2}$
Chain	$y = f(g(x))$	$\frac{dy}{dx} = f'(g(x))g'(x)$

For $\frac{dH}{dt} = t^{-1/2}$, applying the reverse of the power rule *and* the reverse of the constant rule, we would have

$$H(t) \;=\; 2t^{1/2} + C.$$

Before addressing the constant rule and why there is a C in the result, make sure you can see how we get $2t^{1/2}$. Take the derivative of $H(t)$ and verify that you get the original dH/dt. Also observe that we write $H(t)$ to indicate that the population level, H, is a function of time, t.

Where does the C come from? Suppose that $H(t) = 2t^{1/2} + 5$. Take the derivative with respect to t, and you will get dH/dt. Now suppose $H(t) = 2t^{1/2} + 7,777,777.77$. Take the derivative again, and you get the same result. Or suppose $H(t) = 2t^{1/2} - 1,966$. Taking the derivative again yields the same answer. All those results follow from the **constant** rule for derivatives, which says the derivative of any constant number is zero. As a reminder from Chapter 2, that should make sense. If the derivative is how a function changes, then a constant number represents no change, so the derivative is zero.

The take–away is that when we integrate and reverse the derivative process, we need to account for the fact that there may be a constant term. That constant term could be *any* number, and we do not have any information (so far) about what it is. Therefore, traditionally that constant is labeled with the capital letter C to represent an *arbitrary constant*. It is called the *constant of integration*. C could be any number, anything at all. If we were to ignore the constant's existence, it would be as if we just assumed that $C = 0$.

How does that affect our results? Is it important? Consider the following diagram of our resulting population equation, $H(t)$, where we let C take on three possible values: 0, 2, and 5. Imagine the y-axis and C are in units of billions. The exact value of C is going to make an enormous difference in the population of our fictional economy, SSSP.

Can we ever get the exact value of C? Yes, we can, but we need an additional piece of information. Notice in the diagram how the differing values of C shift the entire curve in a parallel manner. The curves never intersect. Each value of C puts the population function on a different path, and those paths never cross. Thus all we need to know is one point where the true curve passes through. If we know just a single point, we can figure out which curve we have and what the exact value of C is.

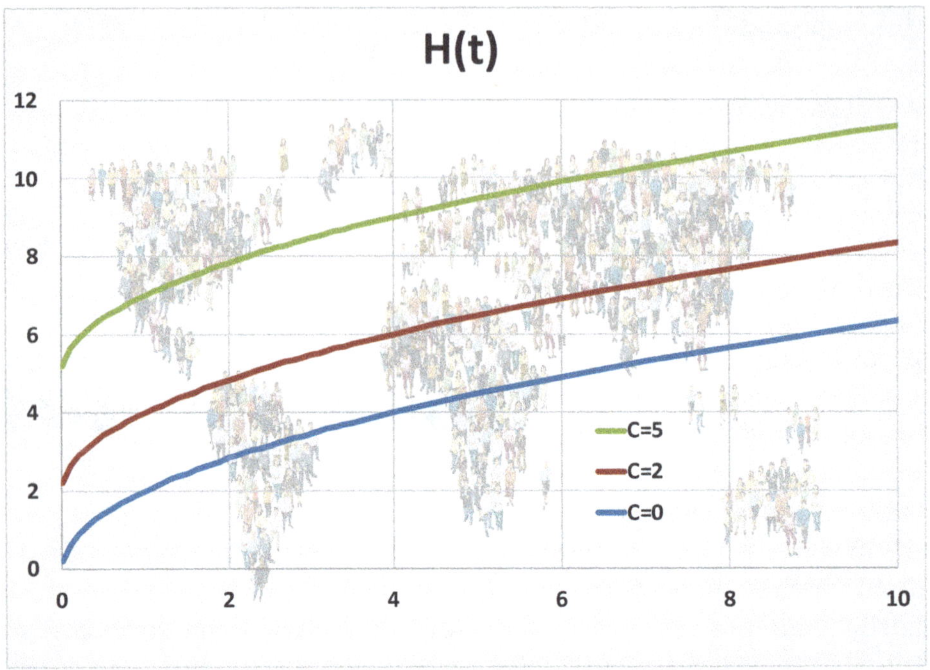

Figure 11.2: Integration: Area under the curve.

Suppose in our SSSP example that we know the starting value of H at time $t = 0$ is 2. In effect, we know the starting population. To find C we substitute $t = 0$ and $H(t) = H(0) = 2$ into our integration result as follows:

$$
\begin{aligned}
H(t) &= 2t^{1/2} + C \\
H(0) &= 2(0)^{1/2} + C \\
2 &= 2(0)^{1/2} + C \\
2 &= 2(0) + C \\
2 &= 0 + C \\
C &= 2.
\end{aligned}
$$

Therefore, the function is

$$
H(t) = 2t^{1/2} + 2.
$$

In this example, we knew the starting point. In other situations we may know the end point and use that to pin down the arbitrary constant. But any point will do. It does not have to be at either end. If we do not have an endpoint or some other point, the resulting function from the integration must include the arbitrary constant.

Notation

The standard notation used for an indefinite integral is as follows:

$$
\int f(x)dx = F(x) + C.
$$

Let's pull that apart and identify the components to better understand what that means.

$$\underbrace{\int}_{\text{Integral sign}} \underbrace{f(x)}_{\text{Integrand}} \underbrace{dx}_{\text{Differential}} = \underbrace{F(x)}_{\text{Integrated function}} + \underbrace{C}_{\text{Arbitrary constant}}$$

$\int$ is an operator indicating we are integrating what follows. $f(x)$ is the function that we are integrating and it is referred to as the "integrand." dx we have seen many times before in the context of derivatives. It is the "differential of x," meaning a marginal change in x, and written here it indicates which variable we are integrating. That becomes extra important in multivariate cases where we could integrate over more than one variable (like reversing the derivative rules for multivariate derivatives). $F(x)$ is the resulting function of x after carrying out the integration plus C, the constant of integration we discussed. Furthermore, note how the original function is indicated with a lower case f while the integrated function uses an upper case F. That is standard practice such that one can keep track of the relationship among the functions.

To illustrate the point about dx being important in the multivariate case, consider the following problem:

$$\int f(x,y)dx = \int \left(2xy + x^2 - y^2\right) dx.$$

In what we have, the operation to be performed is integrating with respect to x. The process for doing so is akin to taking partial derivatives, as we learned in Chapter 6. When we integrate with respect to x we treat the other variables, y in this case, as constants. For this example, we would get

$$\int f(x,y)dx = \int \left(2xy + x^2 - y^2\right) dx$$
$$= x^2 y + \frac{1}{3}x^3 - xy^2 + C.$$

What if we wanted to integrate over both x and y? First, the notation would look like this:

$$\int_y \int_x f(x,y)dxdy = \int_y \int_x \left(2xy + x^2 - y^2\right) dxdy$$

and then we would go one variable at a time, reversing the process of taking the cross-derivative. For our purposes, we will not worry about multivariate integration but move on to the rules of integration and definite integrals. We will define the integral more precisely after we have covered the rules of integration. Overall, if one is comfortable with integration in the single–variable case, the multivariate case is not significantly different. But it is important, as the previous example tries to convey, to recognize when the problem is multivariate and what the notation signifies.

11.2 Rules of Integration

11.2.1 Basic Rules

Here, we quickly cover the basic rules, which largely amount to reversing the rules for derivatives while making sure to keep track of the constant of integration. Following Table 11.2, we briefly discuss each.

Table 11.2: Integration Rules

Rule	function	Integral
Power	$\int x^n dx$	$= \frac{x^{n+1}}{n} + C$
Sum	$\int [f(x) + g(x)]\, dx$	$= \int f(x)dx + \int g(x)dx$
Multiple	$\int k f(x)dx$	$= k \int f(x)dx$
Exponential	$\int e^x dx$	$= e^x + C$
Exponential function	$\int f'(x)e^{f(x)}dx$	$= e^{f(x)} + C$
Natural log	$\int \frac{1}{x}dx$	$= \ln|x| + C$
Natural log function	$\int \frac{f'(x)}{f(x)}dx$	$= \ln(f(x)) + C,\ f(x) > 0$
Exponent	$\int a^x dx$	$= \frac{a^x}{\ln a} + C,\ a > 0$

Power rule. The power rule, which we used in the previous examples, is the most straightforward, as it reverses the most familiar derivative rule. Note the addition of the constant of integration, C, here and throughout.

Sum rule. The sum rule is quite helpful. The sum rule allows us to take more

> **Quick Check**
>
> Use the power rule to integrate the following functions. Don't forget the constant of integration!
> a) $5x^4$
> b) αx^β

complicated integrals and simplify them in many cases. For example, suppose we have

$$\int f(x) \;=\; \int (5x^4 + 4x^3 - 2x^2 + 100x - 42)dx.$$

The sum rule allows us to separate each of the terms out and perform integrations on each part. In effect, the sum rule allows us to rewrite the problem as

$$\int f(x) \;=\; \int 5x^4 dx + \int 4x^3 dx - \int 2x^2 dx + \int 100x dx - \int 42 dx.$$

Each piece merely requires a straightforward application of the power rule, so we would have

$$\int f(x) \;=\; x^5 + C_1 + x^4 + C_2 - \frac{1}{3}x^3 + C_3 + 50x^2 + C_4 - 42x + C_5.$$

Notice that although each integration effectively generates a constant of integration, since they are all *arbitrary* we can combine them all into one constant without changing the result at all. No matter what each C_i is, it will get added to all the others to form one constant. Thus, we would just write:

$$\int f(x) \;=\; x^5 + x^4 - \frac{1}{3}x^3 + 50x^2 - 42x + C.$$

Multiple rule. The multiple rule is also helpful in simplifying integrals. What it effectively says is we can factor out any constant from the integration. That can be useful in a couple of ways. First, it can just reduce the size of the numbers we are dealing with so that the integration is a bit easier. Here is a quick example to illustrate

$$\int f(x) \;=\; \int 198x^2 dx$$
$$=\; 66 \int 3x^2 dx$$
$$=\; 66x^3 + C.$$

By factoring out a multiple, the part of the integrand remaining is a straightforward application of the power rule.

Second, the multiple rule can often be used to manipulate an integral and put it into a form such that we can apply other rules. to complete the integration. We show some examples of that when we discuss the natural log and exponential functions. Here is another quick example where the multiple rule is quite helpful in simplifying a problem that initially seems difficult:

$$\int f(x) \;=\; \int 3\left(\frac{5ln(y)}{11 + y}\right)^{\alpha} x^2 dx.$$

Notice that everything multiplying x^2 is either a constant term, a variable that we are not integrating, y, or a parameter, α. All those items can be treated as constants, so we can rewrite our problem using the multiple rule as

$$\int f(x) = 3\left(\frac{5ln(y)}{11+y}\right)^{\alpha} \int x^2 dx$$

$$= 3\left(\frac{5ln(y)}{11+y}\right)^{\alpha} \frac{1}{3} x^3$$

$$= \left(\frac{5ln(y)}{11+y}\right)^{\alpha} x^3.$$

Exponential and exponential function rules. The exponential rule may be the easiest because it is the counterpart to the rule for taking the derivative of e^x. Since the slope of e^x is e^x, the integral of e^x is also e^x but with the added constant of integration, C.

However, the exponential function rule is a bit trickier. In order to carry out this operation either we must recognize that the function multiplying $e^{f(x)}$ is the derivative of the exponent *or* we algebraically manipulate the expression to get it into that form so that we can carry out the integration. Here is an example:

$$\int f(x) = \int 18xe^{x^2} dx$$

$$= 9 \int 2xe^{x^2} dx$$

$$= 9e^{x^2} + C.$$

By using the multiple rule, we can factor out the 9, leaving $2x$ multiplying e^{x^2}. We do that because $2x$ is the derivative of x^2. Thus, the second line has the integral in the form of $\int f'(x)e^{f(x)}dx$, so we can use the exponential rule listed. To check your work, take the derivative of your result. You should get back to the original function as follows:

$$\frac{df(x)}{dx} = \frac{d}{dx}(9e^{x^2} + C)$$

$$= 9(2x)e^{x^2}$$

$$= 18xe^{x^2}.$$

Natural log and natural log function rules. The natural log rule, as shown in the table, has a unique feature in that we have to use the absolute value sign because the natural log is not defined for zero or negative numbers. The integration only makes mathematical sense if the ln function operates on positive numbers. The same applies to the natural log function rule, where we write it such that $f(x) > 0$. Moreover, similar to the exponential function just discussed, in order to use the rule we need to recognize

that the fraction has the derivative in the numerator and the original function in the
denominator. Alternatively, we would have to manipulate the integrand to get it in that
form. Here is an example:

$$\int f(x)dx \;=\; \int \left(\frac{2}{x}\right)\left(\frac{12x^2+1}{8x^2+2}\right)dx$$
$$=\; \int \left(\frac{24x^2+2}{8x^3+2x}\right)dx.$$

In the first line, we cannot apply the natural log function rule, and it may not be
immediately obvious, but after multiplying the fractions, the numerator is indeed the
derivative of the denominator, $\frac{df(x)}{dx}(8x^3+2x)=24x^2+2$. We can apply the rule to get

$$\int f(x)dx \;=\; \int \left(\frac{24x^2+2}{8x^3+2x}\right)dx$$
$$=\; \ln(8x^3+2x)+C.$$

As always, we can double–check our work by taking the derivative of the result to verify
we did it right:

$$\frac{df(x)}{dx} \;=\; \frac{d}{dx}\ln(8x^3+2x)+C$$
$$=\; \frac{1}{8x^3+2x}(24x^2+2)$$
$$=\; \frac{24x+2}{8x^3+2x} = \left(\frac{2}{x}\,\frac{12x^2+1}{8x^2+2}\right).$$

Exponent rule. That brings us to the last of the rules listed, the exponent rule. To
be honest, in most intermediate economics applications and models, the need for this
rule is exteremely rare. However, we add it here for completeness and to make sure you
are alert to the difference between this type of function and what you may have become
used to at this point. If you are given the following problem:

$$\frac{df(x)}{dx} \;=\; \frac{d}{dx}5^x$$

The answer to that problem is

$$\frac{df(x)}{dx} \;=\; ln(5)5^x.$$

It is *not*

$$\frac{df(x)}{dx} = x5^{x-1}.$$

If you recall the rule for taking the derivative of the exponential function, $\frac{d}{dx}e^x = e^x$, that is a special case of the exponent rule. If we apply the exponent rule we would have $\frac{d}{dx}e^x = ln(e)e^x$, but $ln(e) = 1$, leaving us with our familar and simple result: $\frac{d}{dx}e^x = e^x$.

If we have to integrate with the variable of interest in the numerator but the base is something *other* than e, we need to account for this natural log term:

$$\int a^x dx \;\; = \;\; \frac{a^x}{\ln a} + C, a > 0.$$

11.3 Integration Tricks

At this point, you may be looking at integration and thinking, "This is getting messy." In general, integration is harder than taking derivatives. It often involves manipulating the problem such that we can apply a rule we know. What we present now are two common tricks for solving problems that as initially written cannot be integrated. The idea is to manipulate them into forms such that we can complete the integration.

It should also be noted that we cannot always perform an integration and get a *closed form solution*. What does that mean? Technically, it means you cannot "solve" it in a finite number of standard math steps. Practically, it means you can't use pencil and paper to figure out the solution. Often we say, we can't solve it analytically. For example, the following function and its graph should be familiar to anyone who has taken basic statistics or introductory econometrics:

$$f(x) \;=\; \frac{1}{2\pi\sigma^2}e^{-(x-\mu)^2/2\sigma^2}.$$

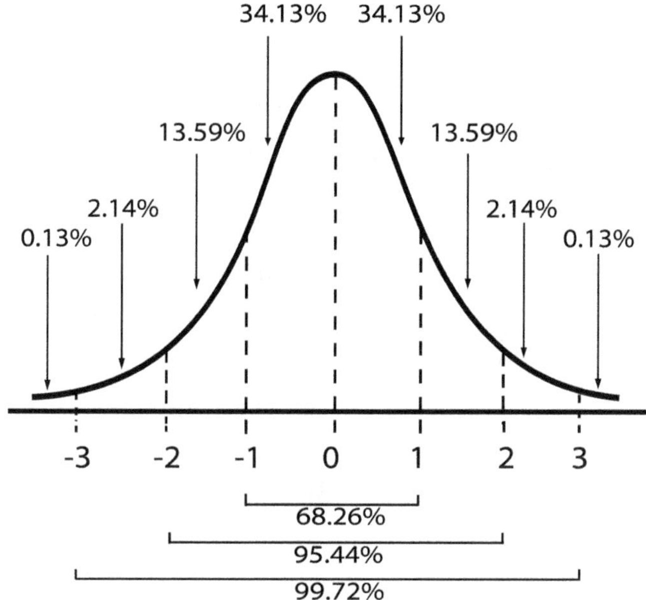

Figure 11.3: Integration: Normal distribution.

It is the probability distribution function (PDF) for the normal distribution. In that function, π is the π you associate with calculating the area or diameter of a circle, (3.14159....), μ is the average of the distribution, and σ^2 is the variance of the distribution.[17] Without going into a long tangential discourse on statistics (which would amount to more chapters), it tells you the probability of x equaling some particular value; call it x^*. It is often useful to find the cumulative distribution function (CDF), which is the integral of the pdf. If the value at any point in the pdf is the probability, then by finding the area under the curve up to x^* we can find the probability of x being less than or equal to x^*. That is

$$\int^{x^*} f(x)dx \;=\; \int^{x^*} \frac{1}{2\pi\sigma^2}e^{-(x-\mu)^2/2\sigma^2}dx.$$

However, you cannot perform that integration with pencil (or pen) and paper.[18] It needs

---

[17]If you have never taken statistics and are not sure what a distribution is, think of it as the results of selecting a bunch of random numbers. The distribution is the pattern those numbers form. Sometimes each number has an equal chance to be selected as any other, and that would be a uniform distribution. Lottery drawings are uniform distributions, as every combination has an equal probability of occuring. In a normal distribution, as shown in the figure, numbers near the average, are more likely to be drawn and the further away from the average, the less likely a number will be selected.

[18]Yours truly and my college study buddy once spent an entire weekend trying to do that exact

to be solved in another manner. Thus, if you have taken statistics you almost always find a table with a large amount of numbers for the normal distribution and possibly more tables for other distributions. These are the results of performing the integration numerically, where we compute actual values separately at different parts of the function (distribution). Numerical integration is well beyond the scope of this textbook.

That preface/warning aside, there are two commonly used "tricks" in our tool bag to manipulate an integration problem from something we cannot do directly into a different integration problem that we can do analytically. These techniques are *integration by substitution* and *integration by parts*, and we turn to them now.

11.3.1 Integration by Substitution, aka Change of Variable

Integration by substitution is also sometimes called change of variable. Both are descriptive of how the technique works. When faced with a complex integration to perform that our rules do not cover, we can sometimes find a substitution that allows us to complete the integration. That sounds great, but let's start with a relatively easy example to illustrate. Suppose we have the following problem:

$$\int f(x)dx \;=\; \int 2x(x^2+1)dx.$$

As written, we cannot solve this problem directly, but we could multiply the $2x$ through and get to something we could integrate, as follows:

$$\int f(x)dx \;=\; \int (2x^3 + 2x)dx.$$

This expression can be handled with our power and sum rules We get

$$\int f(x)dx \;=\; \frac{1}{2}x^4 + x^2 + C.$$

But, let's suppose that we couldn't do that initial multiplication. There is another way. Suppose we do the following:

$$\text{Let } u = x^2 + 1.$$

Then, if we take derivatives on either side we get

$$du = 2xdx.$$

Looking back at our original problem, we now have a way to replace everything in terms of x with an expression in terms of the variable u. That includes the dx term. Making

---

integration analytically when it came up on an assignment. We totally wasted the weekend, and our professor found this experience highly entertaining and used our efforts as a teaching moment for the entire class :(.

those replacements, we do the following:

$$\int f(x)dx \;=\; \int 2x(x^2+1)dx$$

$$=\; \int \overbrace{(x^2+1)}^{u}\overbrace{2xdx}^{du}$$

$$=\; \int udu,$$

where in the second line, the integrand is manipulated slightly such that exactly where and how the substitution takes place is easier to see. Now we have a very simple integration to perform, and we get

$$\int udu \;=\; \frac{1}{2}u^2+C.$$

As a final step, replace u with x^2+1 to get

$$=\; \frac{1}{2}(x^2+1)^2+C$$

$$=\; \frac{1}{2}x^4+2x^2+\frac{1}{2}+C$$

$$=\; \frac{1}{2}x^4+2x^2+C,$$

where in the last line we just merged the $+\frac{1}{2}$ with the arbitrary constant.

The substitution is meant to take something complicated or just not integratable based on our rules and convert it to something we can do. In the example, we could complete the integration either way. Now let's look at an example in which we need to use a substitution to get the answer. Consider

$$\int f(x)dx \;=\; \frac{4x}{x^2+1}dx.$$

We do not have an integration rule to handle that directly, but we can make a substitution to complete the integration. Let $u = x^2+1$, then $du = 2xdx$:

$$\int f(x)dx \;=\; \int \frac{4x}{x^2+1}dx$$

$$=\; \int \frac{2du}{u}$$

$$=\; 2\int \frac{1}{u}du$$

$$=\; 2\ln|u|+C$$

$$=\; 2\ln|x^2+1|+C$$

$$=\; 2\ln(x^2+1)+C.$$

Once we get the expression to $\int \frac{1}{u}du$, we can apply the natural log rule. Notice that we can drop the absolute value sign here because $(x^2 + 1)$ can never be zero or negative. Let's verify that we did it correctly by taking the derivative of our result:

$$
\begin{aligned}
\frac{d}{dx}F(x) &= \frac{d}{dx}2\ln(x^2 + 1) + C \\
&= 2\frac{1}{x^2 + 1}2x \\
&= \frac{4x}{x^2 + 1}.
\end{aligned}
$$

We will see some more examples of integration by substitution when we look at definite integrals in the next subsection.

11.3.2 Integration by Parts

The second technique, integration by parts, is a little trickier. This technique works when we can break the integrand into two separate pieces. Through some clever rearrangement, we can go from an integral that cannot be solved analytically to one that can. To see how this works, please follow the next few steps of logic carefully. It may seem random and disconnected at first, but hopefully it will become clear rather quickly why we do these steps.

Suppose we wish to carry out an integration over a function of x. To make the notation (hopefully) easier to follow, let $w = f(x)$ and $z = g(x)$. Consider for a moment the product wz, and suppose we took a total derivative here:

$$
d(wz) = wdz + zdw,
$$

where that second line is just an application of the *product rule* of derivatives. Now, one algebra step gets us to

$$
wdz = d(wz) - zdw.
$$

Suppose we want to integrate that. Then, we have

$$
\begin{aligned}
\int wdz &= \int d(wz) - \int zdw \\
\int wdz &= wz - \int zdw
\end{aligned}
$$

$$
\underbrace{\int wdz}_{\text{Original problem / Can't integrate}} = \underbrace{wz - \int zdw}_{\text{New problem / Can integrate}} .
$$

Moving from one line to the next in the two equations, all we did was reverse the derivative $d(wz)$ with an integral. That gets us back to wz.

What was the point of that? The point is that suppose we can separate our integration problem into the two parts on the left-hand side, w and dz, but we can't integrate that expression. That manipulation tells us that instead of integrating $\int wdz$, we would get the same answer if we evaluate $wz - \int zdw$. We effectively exchange trying to carrying out the integration $\int wdz$ for the integration $\int zdw$.

Let's go through an example:

$$\int x(x+1)^{1/2}dx$$

Step 1: *Let one portion of the integrand be equal to z and the other portion be equal to* dw:

$$\text{Let } z = x \text{ and } dw = (x+1)^{1/2}dx.$$

Sometimes it requires a bit of trial and error to figure out the correct assignments of z and dw. After a bit of practice, the patterns that work do become clearer and easier to identify.

Step 2: *From z find dz and from dw find w.*
Now we have to take one derivative and perform one integration. Note that when choosing dw in step 1, we should be sure to assign dw to something we can integrate:

$$
\begin{aligned}
z &= x \\
dz &= 1dx \\
&\text{and} \\
dw &= (x+1)^{1/2}dx \\
\int dw &= \int (x+1)^{1/2}dx \\
w &= \frac{2}{3}(x+1)^{3/2},
\end{aligned}
$$

where to get that last line we used integration by substitution. Specifically, let $u = x+1$; then $du = dx$ and the integration can be written as $\int u^{1/2}du$, which is straightforward to integrate.

Step 3: *Rewrite the problem using $\int wdz = wz - \int zdw$:*

$$
\begin{aligned}
wdz &= \int x(x+1)^{1/2}dx \\
wdz &= wz - \int zdw
\end{aligned}
$$

$$\overbrace{\int x(x+1)^{1/2}dx}^{wdz} = \overbrace{\frac{2}{3}(x+1)^{3/2}x}^{wz} - \overbrace{\int \frac{2}{3}(x+1)^{3/2}dx}^{zdw}.$$

We traded doing the original integral $\int x(x+1)^{1/2}dx$, which we could not do directly with our existing rules, for the this integral $\int \frac{2}{3}(x+1)^{3/2}dx$, which we *can* do with our rules.

Step 4: *Evaluate the new integral.*
Applying integration by substitution again, let $u = x+1$, and $du = dx$, so we have

$$\int x(x+1)^{1/2} = \frac{2}{3}(x+1)^{3/2}x - \int \frac{2}{3}(u)^{3/2}du$$

$$= \frac{2}{3}(x+1)^{3/2}x - \frac{4}{15}(u)^{5/2}du$$

$$= \frac{2}{3}(x+1)^{3/2}x - \frac{4}{15}(x+1)^{5/2} + C.$$

One note about the constant of integration, before we move on to another example. You may have noticed that when initially applying integration by parts in step 2 to go from dw to w we integrated but did not add C. However, at the very end in step 4, we did add C. Why? For the original problem there will be a constant of integration, and, as always, it is arbitrary. It could be any number. We just need to make sure it is included at the end. If we included it when going from dw to w that would be fine, but we would be carrying that term around, making our notation more complicated and gaining nothing.

Here's one more example of integration by parts on a problem that seems easy at first:

$$\int \ln x\, dx.$$

Wait. Didn't we have a rule for that? No, no we didn't, though you may find it listed in some calculus textbooks as a rule in an expanded list. It turns out the answer is not as straightforward as you might think it would be.

In order to solve this problem, we use integration by parts. Wait!! We need two parts. I only see one $\ln x$. What gives? *The second part is dx.* So,...

Step 1:

$$w = \ln x, \text{ and } dz = dx.$$

Step 2:

$$dw = \frac{1}{x}dx, \text{ and } z = x.$$

Step 3:

$$\int w\,dz = wz - \int z\,dw$$

$$\int \ln x\,dx = (\ln x)(x) - \int (x)\frac{1}{x}\,dx$$

$$\int \ln x\,dx = x\ln x - \int 1\,dx.$$

Step 4:

$$\int \ln x\,dx = x\ln x - x + C$$

$$\int \ln x\,dx = x(\ln x - 1) + C.$$

That is the answer. It seems more complex than it should be, but to fully convince yourself, take the derivative and see that you get back to $\ln x$.

11.4 Definite Integrals

11.4.1 Definition of Definite Integral

A definite integral has a defined starting point and ending point along the dimension of the variable we are integrating over. See Figure 11.4. Imagine we have some function for the curve. We learned how to find the integral for the entire function or the area under the whole curve. With a definite integral, we are interested in the area under a specific part of the curve, that part between the starting and ending points and below the function and above the x-axis. We will call these points, a and b, the *lower limit* and the *upper limit*, respectively. So, in Figure 11.4 we would use a definite integral to find the area under the curve *between a and b*.

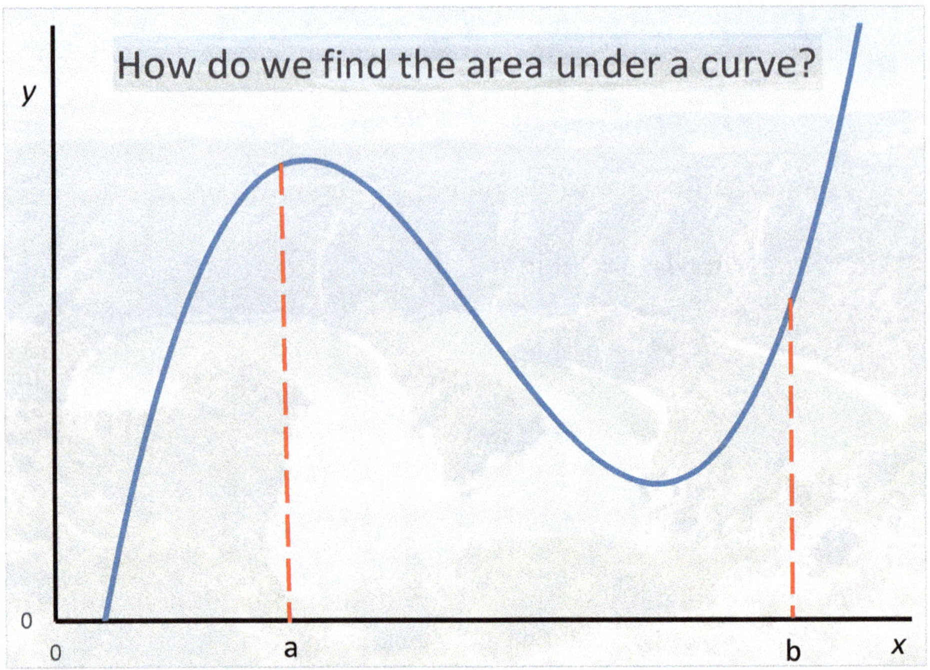

Figure 11.4: Integration: Area under the curve.

How do we do that? Effectively, we find the area under the entire curve up to point b, the upper limit, then subtract the area under the curve up to point a, the lower limit. That leaves the area between points a and b.

First, we have some additions to our standard notation to introduce:

$$\overbrace{\int_a^b}^{\text{Integral from } a \text{ to } b} f(x)dx = \overbrace{F(b)}^{\text{Area up to } b} - \overbrace{F(a)}^{\text{Area up to } a}$$

$$= F(x)\big|_a^b.$$

When we have a definite integral, the limits of integration are written above and below the integration sign itself, $\int_a^b$, with the lower limit at the bottom and the upper limit at the top. On the right-hand side, recall that the capital F indicates the integrated function. When we write $F(b)$, we mean, after finding $F(x)$ through integration, let $x = b$ and plug that value into $F(x)$. We say "evaluate $F(x)$ at b." The right-hand side is the difference between the integrated function evaluated at point b and the integrated function evaluated at point a. The last line, $F(x)\big|_a^b$, is an alternative and more compact notation for the same thing. It reads the function $F(x)$ evaluated from a to b.

Further, note that there is no constant of integration. Where did it go? It canceled out with itself. Let's go step-by-step with an easy example so you can see how that happens. Consider the following:

$$\int_1^2 2x\,dx \;=\;$$

$$
\begin{aligned}
&= \left[x^2 + C\right]\big|_1^2 \\
&= \left[2^2 + C\right] - \left[1^{1/2} + C\right] \\
&= (4 + C) - (1 + C) \\
&= 3.
\end{aligned}
$$

The C's cancel. That will always occur in a definite integral. As a result, when evaluating a definite integral we never bother to write out the constants of integration since they will disappear. Effectively, since the constant is the same number, it will add the same amount to the area up to both points a and b so that area contributed by C cancels out when we look at the difference in areas.

Before going any further, let's think a bit about how and why integration is the area under the curve. This sidestep will help us understand the definite integral.

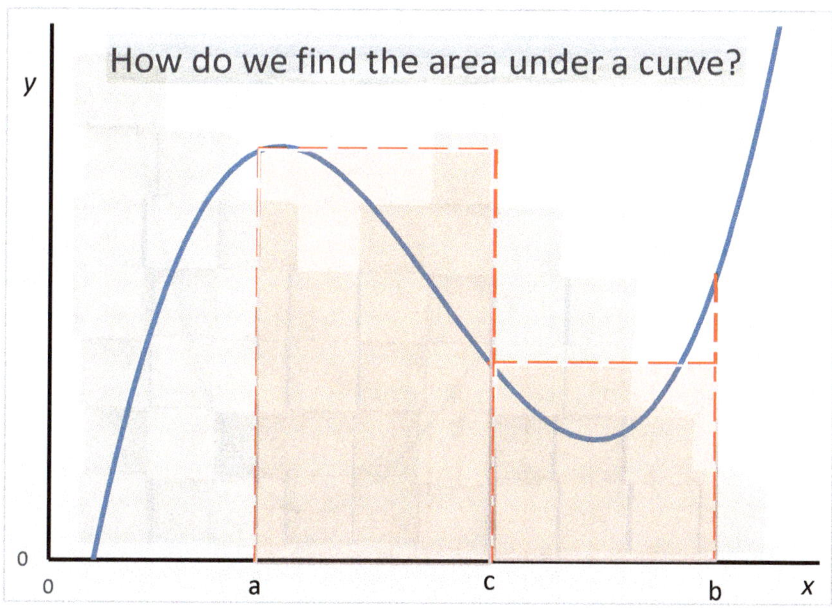

Figure 11.5: Integration: First approximation.

If you recall our discussion of the derivative back in Chapter 2, in order to find the tangent line or the slope of a curve, we used an approximation. We drew a line between two points on the curve and used that as an approximation of the slope at the initial point. The distance between the two points we labeled h. However, visual inspection made it obvious the line was not completely accurate. There was an error. We drew another line using a second point on the curve that was closer to the first, effectively using a smaller h. We still had an error, so we kept moving the point closer until *in the limit* there was no distance between the points, $\lim h \to 0$, just dx, an infinitessimally small change along the x-axis, or a marginal change in x.

To find the area under the curve, integral calculus effectively uses the same trick, but instead of creating a line and looking at the slope, we will create a series of rectangles and compute the combined area. See Figure 11.5.

We have a function with some curves, $y = f(x)$, and we are interested in finding the area under the curve between points a and b. Suppose we divide the distance between a and b in half and use that midpoint to create two rectangles. Call that midpoint, point c. We can use our elementary geometry now to get the areas of the two rectangles and add them together. That would give us an approximation of the area under the curve as follows:

$$Area \;=\; f(a)(c-a) + f(c)(b-c),$$

where $f(a)$ and $f(c)$ represent the heights of the rectangles and $c - a$ and $b - c$ are the widths of each rectangle. Adding those areas together gives us the total area of the approximation. Visual inspection of the figure makes it obvious the approximation is crude. There are large chunks of the rectangles outside the actual area we want to measure.

What can we do? Use more and smaller rectangles. See Figure 11.6.

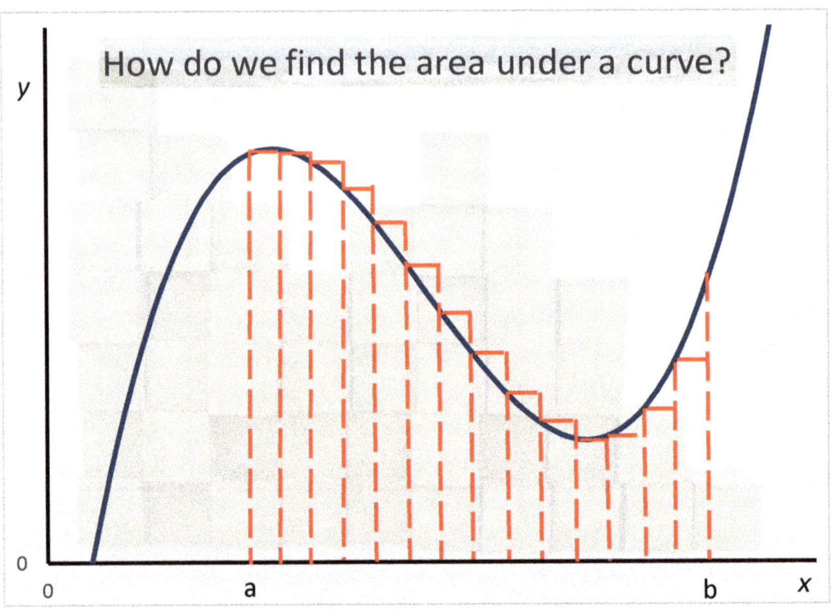

Figure 11.6: Integration: Second approximation.

Now we have a larger series of rectangles. Suppose we call the number of rectangles n. The base length of each is a difference in the value along the x-axis, so let's call that Δx to represent the difference in x. We will make our rectangles of equal width so that Δx is the same for each rectangle. We can write the total area covered by the rectangles as

$$Area \;=\; \sum_{i=1}^{n} f(x_i)\Delta x,$$

where i is our counter variable that goes from 1 to n, the total number of rectangles. Thus, $f(x_i)$ is the height of each rectangle. The base for each rectangle is Δx representing the change in x value. This new approximation is better than the first one. It's less crude and more closely follows the actual curve. But it's still just an approximation, and there are obvious errors in the approximation from looking at the figure.

To do even better, let's further shrink the base of the rectangles. The rectangles are getting thinner and thinner and thinner, until we reach the limit, $\lim \Delta x \to 0$. Wait. That's familar; it is the marginal change in x, $dx = \lim \Delta x \to 0$. We write

$$\int_a^b f(x)dx \;=\; \lim_{n\to\infty} \sum_{i=1}^{n} f(x_i)\Delta x,$$

where instead of using $\lim \Delta x \to 0$, on the right-hand side, we use $\lim_{n\to\infty}$. That represents creating up to an infinite number of rectangles. In order to do that, the width (the base) of each rectangle would have to reach the limit of a marginal change in x. The equation is the effective definition of the definite integral.

Integration is really just an infinite sum of rectangles. In fact, the symbol for integration $\int$ is the cursive version of $\sum$ for summation! What are you doing when integrating? *Adding* all the space under the curve to get the area. Fortunately, though, the rules of integration are fantastic time savers such that we do not actually have to carry out an infinite sum. However, the idea of summing up the area is a useful way to think about integrals and will make understanding the properties of definite integrals, discussed in the next subsection, a bit more intuitive.

Let's do a quick example before exploring those useful properties of definite integrals:

$$
\begin{aligned}
\int_1^4 &= 3x^{1/2}dx \\
&= \left(\frac{2}{3}\right) 3x^{1/2+1}\Big|_1^4 \\
&= 2x^{3/2}\Big|_1^4 \\
&= 2(4)^{3/2} - 2(1)^{3/2} \\
&= 2(8) - 2(1) \\
&= 16 - 2 \\
&= 14.
\end{aligned}
$$

This definite integral is fairly straightforward. The reader is encouraged to graph the orginal function $f(x) = 3x^{1/2}$, make rectangles as before between $a = 1$ and $b = 4$, and estimate the area. How close do you get?

11.4.2 Properties of Definite Integrals

Property I:

$$\int_a^b f(x)dx \; = \; -\int_b^a f(x)dx$$

If we reverse the limits of integration and go from b to a instead of from a to b, we get the negative of the original value. Why? Think about the area interpretation and look at Figure 11.7. When we integrate from a to b and calculate $F(b) - F(a)$, we get the area under the curve up to point b and then subtract off the area up to point a. If we reverse the limits we are really just reversing that subtraction operation. We take the area up to point a and then subtract the area up to point b, $F(a) - F(b)$.

In Figure 11.7 the definite integral is the difference between the areas shown. The area up to point a is the cross-hatched region. The area up to point b is the cross-hatched region and the region with the diagonal lines. To get the area between a and b, we need to subtract the cross–hatched area. If we change the limits, we just takie the difference between the area up to a and the area up to b.

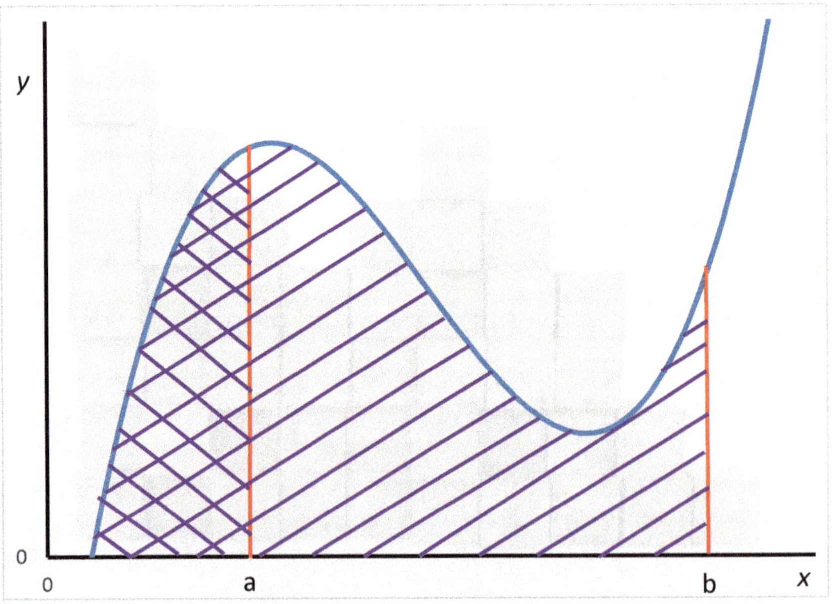

Figure 11.7: Property I of definite integrals.

Property II:

$$\int_a^a f(x)dx \;\;=\;\; 0$$

Property II is quite intuitive. Basically, it just says the area up to some point a minus the area up to *the exact same point* is zero.

Property III:

$$\int_a^d f(x)dx \;\;=\;\; \int_a^b f(x)dx + \int_b^c f(x)dx + \int_c^d f(x)dx$$
$$\text{for} \qquad a < b < c < d$$

Figure 11.8 below illustrates property III, which says we can take an integration from one point, a, to another, d, and break it up into a series of smaller areas.

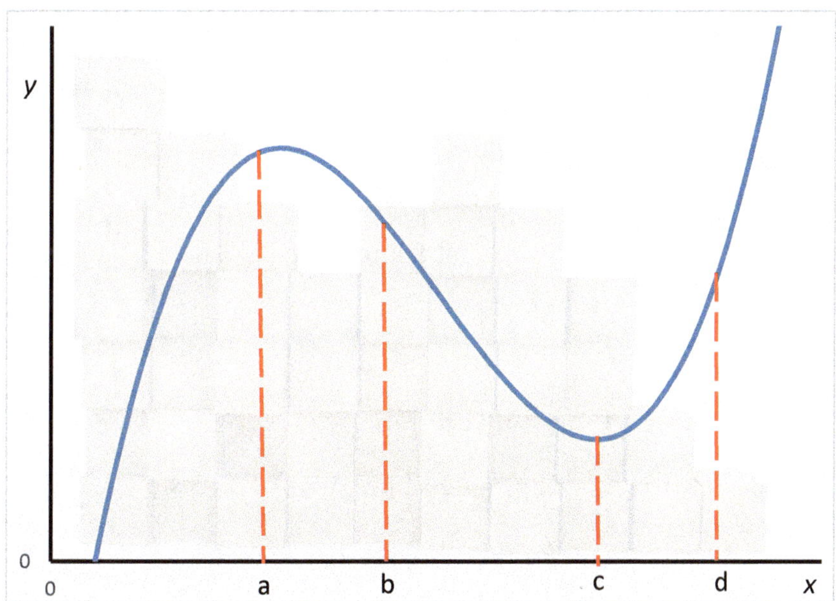

Figure 11.8: Property III of definite integrals.

There are cases where that might be easier to do because the function changes. For example, suppose that Hayata's company produces a quantity of Q beta capsules by employing labor, L. Hayata's production function is given by

$$Q(L) = \begin{cases} L^{1/2} & L \leq 9 \\ 3 + 0.1(L-3)^{1/2} & L > 9. \end{cases}$$

If you needed to find the integral for L between 0 and 10, you might find it challenging to write because of the piecewise nature of the production function. Property III says we can do separate integrals and then add the results:

$$\int_0^{10} Q dL \;=\; \int_0^9 L^{1/2} dL + \int_9^{10} 3 + 0.1(L-3)^{1/2} dL.$$

Warning: As a final note, be careful when using integration by substitution (change of variable) and definite integrals. When using integration by substitution and converting x to u make sure you change the limits of integration or convert back into x before evaluating. An example should be sufficient to make this point clear:

$$\int_1^2 f(x) \;=\; \int_1^2 \frac{5x}{x^2+1} dx$$
$$\text{Let}: \qquad u = x^2 + 1, \qquad du = 2x dx.$$

Since $(5/2)du = 5x dx$, that would allow us to replace the integrand with $(5/2u)du$. But now our variable is u and not x, so we can do the following:

$$= \frac{5}{2} \int \frac{1}{u} du$$
$$= \frac{5}{2} \ln|u|.$$

But in order to evaluate it from 1 to 2, we must remember to convert back into x terms:

$$= \frac{5}{2} \int \frac{1}{u} du$$
$$= \frac{5}{2} \ln(x^2+1)\big|_1^2$$
$$= \frac{5}{2} \ln(5) - \frac{5}{2} \ln(2)$$
$$= \frac{5}{2} \ln(5) - \frac{5}{2} \ln(2)$$
$$= \frac{5}{2} \ln(5/2).$$

Alternatively, we could have converted the upper and lower limits of integration from x to u terms, such that since we let $u = x^2 + 1$, then the lower limit is $(1)^2 + 1 = 2$ and the upper limit is $(2)^2 + 1 = 5$. Then when we integrate,

$$
\begin{aligned}
&= \frac{5}{2} \int_2^5 \frac{1}{u} du \\
&= \frac{5}{2} \ln(u)|_2^5 \\
&= \frac{5}{2} \ln(5) - \frac{5}{2} \ln(2) \\
&= \frac{5}{2} \ln(5/2).
\end{aligned}
$$

The same result as before.

A common mistake students make is they use the original limits in the result with the changed variable. They might do the following:

$$
\begin{aligned}
&= \frac{5}{2} \ln|u||_1^2 \\
&= \frac{5}{2} \ln(2) - \frac{5}{2} \ln(1) \\
&= \frac{5}{2} \ln(2) - 0 \\
&= \frac{5}{2} \ln(2),
\end{aligned}
$$

which is wrong! Do not do that!

11.5   Application to Consumer Welfare

In many principles of economics classes, you learn that consumer surplus can be measured by the area below the demand curve and above the price. That is, consumer surplus is the difference in dollars that a consumer was *willing to pay* versus what they *actually paid*. In exercises to illustrate that concept, the demand curve is nearly always linear.

That is true, because it means finding the area in those graphs requires only a straight-forward application of the formulas for the areas of regular geometric shapes like squares, rectangles, and triangles. Suppose the demand curve is not a straight line. That means we need to find the area under a curve. With the integration techniques we just covered, you can now derive these results.

Let us return to the example from Chapter 9 involving Xiyangyang and his demand for dumplings and mooncakes. Xiyangyang's utility is given by $U = ln(D) + ln(M)$. W is wealth, and the prices are denoted p_D and p_M. From solving Xiyangyang's utility maximization problem, the Marshallian demand functions we got were

Consumer surplus: The gain to consumers as measured by the difference between their willingness to pay (i.e., the maximum they would pay for a good or service and the price that they actually pay on the market).

$$D^*(P, W) \;=\; \frac{W}{2p_D}, \qquad M^*(P, W) = \frac{W}{2p_M}.$$

The expenditure minimization problem yielded the following Hicksian demand functions:

$$D^{*H} \;=\; \left(\frac{p_M}{p_D} e^{\bar{U}} \right)^{1/2}, \qquad M^{*H} = \left(\frac{p_D}{p_M} e^{\bar{U}} \right)^{1/2}.$$

Suppose that the price of dumplings and mooncakes is one, $p_D = p_M = 1$, and the profit–maximizing owner of the dumplings and mooncakes shop, Hong Tai Lang, one day decides to double the price of dumplings to $p_D = 2$. What will happen to Xiyangyang's welfare?

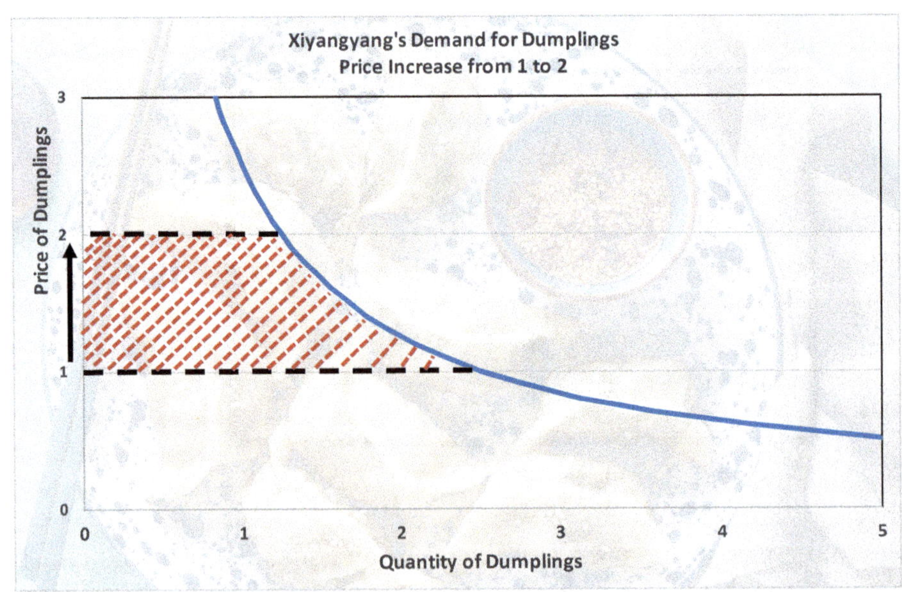

Figure 11.9: Xiyangyang's demand for dumplings.

As the figure shows, Xiyangyang loses that loosely drawn area with dashed lines from his consumer surplus. How much is that? It is the area under the demand curve between the price of 1 and the price of 2. Did you notice that the variable we are integrating is on the y-axis and not the x-axis, as we usually draw an integration problem? Recall that our Marshallian demand function, $D^*(P, W) = \frac{W}{2p_D}$, gives us quantity as a function of price, but we traditionally represent that with prices on the y-axis and quantities on the x-axis. You can always redraw the diagram reversing the axes and integrate the prices on the x-axis and the answer you get will, of course, be the same.

Moreover, this integral is a definite integral, where we are going from a price of 1 to a price of 2. The area we are finding is the negative change in consumer surplus for Xiyangyang. Setting that up we have:

$$\Delta CS = \int_1^2 \frac{W}{2p_D} dp_D.$$

Carrying out the integration in steps, we get the following result:

$$\begin{aligned}
\Delta CS &= \frac{W}{2} \int_1^2 \frac{1}{p_D} dp_D \\
&= \frac{W}{2} \ln(p_D)|_1^2 \\
&= \frac{W}{2} (\ln(2) - \ln(1)) \\
&= \frac{W}{2} \ln(2).
\end{aligned}$$

So, Xiyangyang's welfare declines by that amount. Note that the area lost under the Marshallian demand curve yields the consumer surplus lost due to both the *substitution effect*, switching to more of the lower cost good, and the *income effect*, demand falling as Xiyangyang is made effectively poorer with the price increase. To fix ideas, let $W = 12$. Then, initially Xiyangyang's demand for dumplings and mooncakes would be $D^* = 6$ and $M^* = 6$ before Hong Tai Lang's price increase and $D^* = 3$ and $M^* = 6$ after the price increase: Demand for dumplings falls through both the income and substitution effects. However, the substitution and income effects for mooncakes exactly cancel each other out (this result is a property of the Cobb-Douglas utility function).

Let's look at the Hicksian demand functions for the same price increase. The Hicksian demand expresses the demand as a function of prices and utility. So, it will tell us the substitution effect of the price increase and allow the income level to adjust to maintain the required utility level. The area under the Hicksian demand curve is given by

$$
\begin{aligned}
D^{*H} &= \int_1^2 \left(\frac{p_M}{p_D}e^{\bar{U}}\right)^{1/2} dp_D \\
&= \left(p_M e^{\bar{U}}\right)^{1/2} \int_1^2 \left(\frac{1}{p_D}\right)^{1/2} dp_D \\
&= \left(p_M e^{\bar{U}}\right)^{1/2} 2p_D^{1/2}\big|_1^2 \\
&= \left(p_M e^{\bar{U}}\right)^{1/2} 2(2^{1/2} - 1^{1/2}) \\
&= \left(p_M e^{\bar{U}}\right)^{1/2} 2(0.414) \\
&= 4.972.
\end{aligned}
$$

That is how much our consumer, Xiyangyang, would need to be compensated with in order to achieve the same utility level following the price increase.

To see that result more clearly, compare the Hicksian demands under the old and new price. Using the utility derived at the optimum under the UMP, $\bar{U} = 2ln(6) = 3.584$. When $p_D = p_M = 1$, we have $D^{*H} = 6$ and $M^{*H} = 6$. If p_D increases to 2, then $D^{*H} = 4.243$ and $M^{*H} = 8.485$. In order to achieve those levels of consumption, and maintain the same utility level, the consumer would need a wealth level of $W = 16.972$, which is the same as the orginal 12 plus the compensated amount, 4.972, that we derived earlier!

11.6   Exercises

11.6.1   Quick Check Answers

a) $x^5 + C$

b) $\frac{\alpha}{\beta+1}x^{\beta+1} + C$

c) $\int \frac{\alpha}{\beta+1}x^{\beta+1}dx \Rightarrow \frac{\alpha}{\beta+1}\int x^{\beta+1}dx \Rightarrow \frac{\alpha}{\beta+1}\frac{1}{\beta+2}x^{\beta+2} + C \Rightarrow \frac{\alpha}{(\beta+1)(\beta+2)}x^{\beta+2} + C$

d) $\int \left(3x^7 + 5x^{1/2} - \gamma x^{-2}\right) dx \Rightarrow \int 3x^7 dx + \int 5x^{1/2}dx - \int \gamma x^{-2}dx \Rightarrow \frac{3}{8}x^8 + \frac{10}{3}x^{3/2} + \gamma x^{-1} +$

C

e) $\beta \ln x + C$

f) $e^{x^2} + C$

g) $\int \frac{3x^2+6}{x^3+6x}dx \Rightarrow$ Let $u = x^3 + 6x \Rightarrow du = (3x^2 + 6)dx \Rightarrow \int \frac{1}{u}du = ln|u| + C = ln|x^3 + 6x| + C$

h) $\int xe^x dx \Rightarrow$ Let $w = x$ and $dz = e^x dx \Rightarrow dw = dx$ and $z = e^x \Rightarrow \int xe^x dx = xe^x - \int e^x dx = xe^x - e^x + C = e^x(x-1) + C$

i) $\int_1^4 x^2 dx \Rightarrow \frac{1}{3}x^3|_1^4 = \frac{64}{3} - \frac{1}{3} = 21$

j) $\int_0^{16} x^{1/2} - x^{1/4}dx \Rightarrow \frac{2}{3}x^{3/2} - \frac{4}{5}x^{5/4}|_0^{16} = \frac{128}{3} - \frac{128}{5} = \frac{256}{15}$

11.6.2   Practice Problems

1. Evaluate the following indefinite integrals:

   (a)
   $$\int \frac{dx}{xlnx}$$

   (b)
   $$\int (x^2 + x + \sqrt{x})dx$$

   (c)
   $$\int \frac{dx}{2x - 4}$$

   (d)
   $$\int \frac{ax^2}{\sqrt{x^3 + 8}}dx$$

   (e)
   $$\int \frac{ln(2x + a)}{2x + a}dx \qquad for \qquad a > 0$$

2. Use integration by substitution to solve the following problems:

   (a) $\int \frac{2x}{x^2+1}dx$

   (b) $\int \frac{5x}{x^2+1}dx$

   (c) $\int (x^2 + 2x + 4)^{1/2}(x + 1)dx$

3. Use integration by parts to solve the following problems:

   (a) $\int x^2 e^{2x}dx$

   (b) $\int xlnxdx$

   (c) $\int (x^2 + 5)^{-1}x^3 dx$

4. Evaluate the following definite integrals:

(a)

$$\int_0^2 x^3 dx$$

(b)

$$\int_0^1 2e^x dx$$

(c)

$$\int_0^3 \frac{dx}{1+x}$$

(d)

$$\int_1^e \frac{dx}{x}$$

(e)

$$\int_{-5}^5 (x^3 + 3x) dx$$

(f)

$$\int_1^z \frac{dx}{2x - 1}$$

5. Referring to the example in Section 11.5, suppose Tai Hui Long drops the price of mooncakes from $p_M = 1$ to $p_M = 0.5$. Assume Xiyangyang has $W = 12$ and the price of dumplings is $p_D = 1$; find the consumer surplus as measured under the Marshallian demand curve from the price change. Find the change under the Hicksian demand curve.

6. Again referring to the example in Section 11.5, suppose that Xiyangyang's wealth is $W = 12$ and the intial prices are $p_D = 2$ and $p_M = 1$. Calculate and compare the welfare impact under two scenarios: (a) Hui Tai Long raises the price of dumplings to $p_D = 3$, or (b) Hui Tai Long raises the price of mooncakes to 2. Which would Xiyangyang prefer?

11.7   Math Appendix

Here we list some key math definitions, theorems, and formulae in formal math for your reference.

Liebniz rule: In some applications, we need to take the derivatives of expressions that are in integral form. The rule for performing these integrations is known as Liebniz rule:

$$\frac{\partial F(z)}{\partial z} = \frac{\partial}{\partial z} \int_{a(z)}^{b(z)} f(z;t)dt$$

$$= \int_{a(z)}^{b(z)} f_z(z;t)dt + f(z;b(z)) - f(z;a(z)).$$

The intuition for the result here is perhaps easier to grasp if you refer to Figure 42 illustrating the definite integral. There are three parts in that expression that correspond to the variable z's appearance in three different places in the original expression. Start with the z inside the integrand. The result is straightforward in that one should take the derivative of the integrand with respect to z and then integrate. Visually, in Figure 11.4 imagine z as marginally raising or lowering the entire curve, which would change the size of the area under the curve.

The second and third components appear when z enters as a variable in either or both the lower and upper limits. In the case of the upper limit of integration, $b(z)$, if z increases the function $b(z)$, it moves the upper limit to the right. That increases the area under the curve. In contrast if $a(z)$ is increasing in z, that moves the lower limit to the right, effectively decreasing the area under the curve, so it enters negatively here.

12   Linear Algebra and Cramer's Rule

In this final chapter, we will go through a brief introduction to linear algebra. In addition, some intermediate–level economics textbooks use another method for finding comparative statics solutions (as we did in Chapter 10) when there are two (or more) equations. More advanced textbooks that incorporate linear algebra almost invariably use the method we will introduce here. The method is called Cramer's rule, but in order to understand how to apply Cramer's rule it is necessary to cover some basic linear algebra (the terms "linear algebra" and "matrix algebra" are interchangeable). If you have never taken linear algebra before; do not fear, it is very easy largely because it is just algebra written in a different form. The hardest part about linear algebra is learning the terminology because there is *a lot* of it.

In what follows, the first few parts cover the key linear algebra basics often used at the intermediate economics level. They are not hard at all. The final section goes over Cramer's rule. It is useful for finding derivatives when there are multiple interdependent equations involved. That situation often arises when you have an agent making many choices. Each choice generates a separate first-order condition in a multivariate maximization (or minimization) problem. Solving that system of equations can be really tedious if, for example, there were 5 or 10 or 50 different choices made. But Cramer's rule helps us reduce the problem to something more manageable by allowing us to isolate and solve for a single variable at time.

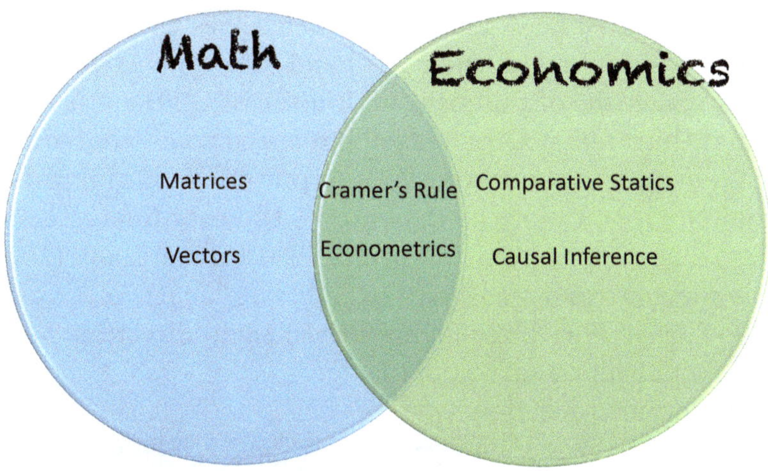

12.1　Matrix Basics

12.1.1　Matrix Representation

Matrices do many things but are frequently used to represent systems of equations. For example, consider the following two simple algebraic equations:

$$2x + 3y = 15$$
$$x - 2y = 4.$$

There is another way to represent that system of equations:

$$\begin{bmatrix} 2 & 3 \\ 1 & -2 \end{bmatrix} \begin{bmatrix} x \\ y \end{bmatrix} = \begin{bmatrix} 15 \\ 4 \end{bmatrix}$$

The first set of numbers inside the braces [ ] are the coefficients on the x's and y's arranged in a particular order. The x and the y themselves are in the second set of braces. We have one equals sign, and on the right-hand side we have the constant terms in another set of braces. Each set of numbers with a set of brackets is referred to as a **matrix**. The second and third matrices can also be referred to as **vectors** because they only have one column. Vectors often have the interpretation of giving both quantity and direction. For example, if you plot the **d** vector at the coordinate (15,4) on the x-y plane, the direction is a line drawn from the origin $(0,0$ to the point $(15,4)$, with the idea that something is moving from $(0,0)$ to $(15,4)$. The length of that vector is the quantity (or speed). A shorter line, but moving in the same direction, would be going "slower." A longer line would represent "going faster."

To further reduce notation, each matrix can be assigned a letter to represent them. In the example, we assign the following:

$$\mathbf{A} = \begin{bmatrix} 2 & 3 \\ 1 & -2 \end{bmatrix}, \mathbf{x} = \begin{bmatrix} x \\ y \end{bmatrix}, \text{ and } \mathbf{d} = \begin{bmatrix} 15 \\ 4 \end{bmatrix}.$$

Note that the **bold font** is frequently used to distinguish a matrix or a vector from a single variable (single number). Single variables are referred to as scalars. Each element within a matrix or vector is a scalar. Also, capital letters are used to represent a matrix with more than one row and more than one column, whereas bold lower–case letters typically represent vectors. We can then write the whole system more compactly as

$$\mathbf{Ax} = \mathbf{d}.$$

This equation reads as the **A** matrix times the **x** vector equals the **d** vector. (We will get to matrix multiplication shortly.) For the simple two–equation, two–unknown system we started with, the change in notation to $\mathbf{Ax} = \mathbf{d}$ makes little difference in terms of the effort required to write or type it out. However, imagine an algebraic system with 100 equations and 100 unknowns. Writing everything out would be tedious, to say the least. Using the matrices we could represent the system simply as $\mathbf{Ax} = \mathbf{d}$. In econometrics or data science, where each observation can be represented as a separate equation, it is not uncommon to have 10,000 or even 10,000,000 observations. For this reason, econometrics is almost always presented in matrix notation.

Suppose you have data on salaries of graduates from your university 10 years after they finished college. You have data on their salaries, majors, sex, race, age, addresses, whether they obtained their graduate degree, and what kind of graduate degree. You are asked to figure out the value of getting a master's degree from that data, how much the salary increases with a master's degree holding all the other factors constant. Further, suppose your data set covers 10,000 alumni. Technically, each observation (each person's set of data) is a separate equation, so that would mean writing out 10,000 equations without matrix notation. Tedious, indeed.

12.1.2 Dimension

The **dimension** of a matrix is the number of rows and columns. The previous **A** matrix has a dimension of 2 by 2. The **x** and **d** matrices are 2x1, where the number of rows is always written first and the number of columns is written second. The following matrices' dimensions are 4x2, 2x3, and 4x3, respectively:

$$\begin{bmatrix} 2 & 3 \\ 1 & -2 \\ -2 & 3 \\ 1 & 0 \end{bmatrix} \qquad \begin{bmatrix} 2 & 3 & -5 \\ 1 & -2 & 8 \end{bmatrix} \qquad \begin{bmatrix} 2 & 3 & -5 \\ 1 & -2 & 8 \\ -2 & 3 & 2 \\ 1 & 0 & 10 \end{bmatrix}.$$

When a matrix has an equal number of rows and columns, it is called a **square matrix**. The **A** matrix in the Section 12.1.1 is a square 2x2 matrix.

12.1.3 Transpose

If we take a matrix and flip its rows and columns, that is the **transpose** of the original matrix. For example, consider the following three matrices:

$$\mathbf{A} = \begin{bmatrix} 2 & 3 \\ 1 & -2 \\ -2 & 3 \\ 1 & 0 \end{bmatrix} \qquad \mathbf{B} = \begin{bmatrix} 2 & 3 & -5 \\ 1 & -2 & 8 \\ -2 & 3 & 2 \end{bmatrix} \qquad \mathbf{C} = \begin{bmatrix} 1 & 2 & 3 \\ 0 & 1 & 2 \\ 0 & 0 & 1 \end{bmatrix}.$$

The transpose of **A** notationally can be written as $\mathbf{A}^T$ or $\mathbf{A}'$. They mean exactly the same thing. The transposes for **A**, **B**, and **C** would be, respectively

$$\mathbf{A}^T = \begin{bmatrix} 2 & 1 & -2 & 1 \\ 3 & -2 & 3 & 0 \end{bmatrix} \qquad \mathbf{B}^T = \begin{bmatrix} 2 & 1 & -2 \\ 3 & -2 & 3 \\ -5 & 8 & 2 \end{bmatrix} \qquad \mathbf{C}^T = \begin{bmatrix} 1 & 0 & 0 \\ 2 & 1 & 0 \\ 3 & 2 & 1 \end{bmatrix}.$$

Observe how the dimensions flip in the matrix **A** example. **A** is 4x2, but $\mathbf{A}^T$ is 2x4. Matrices **B** and **C** are square matrices. Notice how these matrices "flip" along the diagonal from top left to the bottom right. That diagonal $(2, -2, 2)$ in **B** is referred to as the **principal diagonal**. In the **C** matrix, the principal diagonal is $(1, 1, 1)$.

12.2 Math Operations With Matrices

12.2.1 Addition and Subtraction

The dimensions matter for performing matrix operations such as addition, subtraction, and multiplication. For addition and subtraction, the matrices must have the exact same dimensions. Adding matrices is really straightforward. Simply add the individual elements that are in the same row and column to form a new matrix with the same dimensions. For example, suppose we have two matrices $\mathbf{A}$ and $\mathbf{B}$:

$$\mathbf{A} = \begin{bmatrix} 2 & 3 \\ 1 & 7 \\ -2 & 3 \end{bmatrix} \qquad \mathbf{B} = \begin{bmatrix} 5 & 8 \\ 3 & -2 \\ 10 & 6 \end{bmatrix}.$$

Adding the two matrices together we get

$$\begin{bmatrix} 2 & 3 \\ 1 & 7 \\ -2 & 3 \end{bmatrix} + \begin{bmatrix} 5 & 8 \\ 3 & -2 \\ 10 & 6 \end{bmatrix} = \begin{bmatrix} 7 & 11 \\ 4 & 5 \\ 8 & 9 \end{bmatrix}.$$

Subtraction is essentially the same. For $\mathbf{B}$-$\mathbf{A}$ we have

$$\begin{bmatrix} 5 & 8 \\ 3 & -2 \\ 10 & 6 \end{bmatrix} - \begin{bmatrix} 2 & 3 \\ 1 & 7 \\ -2 & 3 \end{bmatrix} = \begin{bmatrix} 3 & 5 \\ 2 & -9 \\ 12 & 3 \end{bmatrix}.$$

12.2.2 Multiplication

There are two main types of multiplication operations involving matrices. In the first type, **scalar multiplication**, we simply multiply one number (a scalar) times a matrix. In the second type, **matrix multiplication**, we multiply two full matrices times each other.

Scalar Multiplication

In scalar multiplication, we multiply a single number times an entire matrix. For example, suppose we multiply 3 times the **A** matrix from before as follows:

$$3 \begin{bmatrix} 2 & 3 \\ 1 & 7 \\ -2 & 3 \end{bmatrix} = ?$$

The process is as straightforward as one could ask. Merely multiply each element by the scalar, 3, to get

$$3 \begin{bmatrix} 2 & 3 \\ 1 & 7 \\ -2 & 3 \end{bmatrix} = \begin{bmatrix} 6 & 9 \\ 3 & 21 \\ -6 & 9 \end{bmatrix}.$$

Matrix Multiplication

While addition, subtraction, and scalar multiplication are nicely straightforward, matrix multiplication is not so obvious. To multiply two matrices, we multiply the elements in the *rows* of the first matrix by the elements in the *columns* of the second matrix and add these products together. This process is probably best illustrated through an example. Consider the following **A** and **B** matrices

$$\mathbf{A} = \begin{bmatrix} 2 & 3 \\ 1 & 7 \\ -2 & 3 \end{bmatrix} \qquad \mathbf{B} = \begin{bmatrix} 5 & 3 & 10 \\ 8 & -2 & 6 \end{bmatrix},$$

and suppose we want to multiply **A** by **B**; we will call the resulting matrix **C**:

$$\overbrace{A}^{3x2} \ \overbrace{B}^{2x3} = \overbrace{C}^{3x3}.$$

Note the dimensions of each matrix. **A** is 3x2, and **B** is 2x3. The first important observation is that the number of columns in **A** is the same as the number of rows in **B**. Since we are multiplying a 3x**2** times a **2**x3 matrix, the two 2's are referred to as the **inner** dimensions. The fact that matrix **A** has 2 columns, and is the same as the number of rows in matrix **B**, is necessary for us to be able to multiply the matrices. When two matrices can be multiplied they are said to be **conformable**. The **outer** dimensions are the 3 rows from the **A** matrix and the 3 columns from the **B** matrix. The outer dimensions determine the dimensions of the matrix that results from the multiplication. The answer, matrix **C**, will be a matrix of dimension 3x3 because the outer dimensions of **A** and **B** are 3 and 3, respectively.

Let's carry out the multiplication step by step, starting from

$$\begin{bmatrix} 2 & 3 \\ 1 & 7 \\ -2 & 3 \end{bmatrix} \begin{bmatrix} 5 & 3 & 10 \\ 8 & -2 & 6 \end{bmatrix} = \begin{bmatrix} c_{11} & c_{12} & c_{13} \\ c_{21} & c_{22} & c_{23} \\ c_{31} & c_{32} & c_{33} \end{bmatrix}.$$

> **Definition**
>
> **Conformable**: Matrices are said to be conformable for matrix multiplication when the number of columns of the first matrix is equal to the number of rows of the second matrix.

Note that in the **C** matrix I have represented each individual element as c_{ij} with two subscript numbers, i and j. The first number, i, refers to the row, and the second number, j, refers to the column. For example, c_{23} means the element in the **C** matrix in the second row and third column.

The element in the first row and first column of $\mathbf{C}$, c_{11}, is found by multiplying the first row of $\mathbf{A}$ by the first column of $\mathbf{B}$. How do you do that? Multiply the first element of the first row in the $\mathbf{A}$ matrix by the first element of the first column in the $\mathbf{B}$ matrix *and then add* the second element of the first row in the $\mathbf{A}$ matrix times the second element of the first column in the $\mathbf{B}$ matrix. Here it is written out using the numbers:

$$c_{11} = 2 \times 5 + 3 \times 8 = 10 + 24 = 34.$$

To get the next element of $\mathbf{C}$, c_{12}, we carry out the same type of operation. c_{12} comes from multiplying the first row of $\mathbf{A}$ by the second column of $\mathbf{B}$ as follows:

$$c_{12} = 2 \times 3 + 3 \times (-2) = 6 + (-6) = 0.$$

Continuing in that manner,

$$c_{13} = 2 \times 10 + 3 \times 6 = 20 + 18 = 38.$$

At this point, we have the first row of the $\mathbf{C}$ matrix,

$$\begin{bmatrix} 2 & 3 \\ 1 & 7 \\ -2 & 3 \end{bmatrix} \begin{bmatrix} 5 & 3 & 10 \\ 8 & -2 & 6 \end{bmatrix} = \begin{bmatrix} 34 & 0 & 38 \\ c_{21} & c_{22} & c_{23} \\ c_{31} & c_{32} & c_{33} \end{bmatrix}.$$

Moving on the to the second row of the $\mathbf{C}$ matrix, we use the second row of the first matrix, $\mathbf{A}$, and move through the three columns of the second matrix, $\mathbf{B}$:

$$
\begin{aligned}
c_{21} &= 1 \times 5 + 7 \times 8 = 5 + 56 = 61 \\
c_{22} &= 1 \times 3 + 7 \times (-2) = 3 + (-14) = -11 \\
c_{23} &= 1 \times 10 + 7 \times 6 = 10 + 42 = 52.
\end{aligned}
$$

The third row would be

$$
\begin{aligned}
c_{31} &= (-2) \times 5 + 3 \times 8 = -10 + 24 = 14 \\
c_{32} &= (-2) \times 3 + 3 \times (-2) = -6 + (-6) = -12 \\
c_{33} &= (-2) \times 10 + 3 \times 6 = -20 + 18 = -2,
\end{aligned}
$$

and the final result is

$$\begin{bmatrix} 2 & 3 \\ 1 & 7 \\ -2 & 3 \end{bmatrix} \begin{bmatrix} 5 & 3 & 10 \\ 8 & -2 & 6 \end{bmatrix} = \begin{bmatrix} 34 & 0 & 38 \\ 61 & -11 & 52 \\ 14 & -12 & -2 \end{bmatrix}.$$

In this example, where we carried out $\mathbf{A} \times \mathbf{B}$, we could also have carried out $\mathbf{B} \times \mathbf{A}$. In that case we would have

$$\begin{bmatrix} 5 & 3 & 10 \\ 8 & -2 & 6 \end{bmatrix} \begin{bmatrix} 2 & 3 \\ 1 & 7 \\ -2 & 3 \end{bmatrix} = \begin{bmatrix} -7 & 66 \\ 2 & 28 \end{bmatrix}.$$

Note that the answer is completely different from $\mathbf{A} \times \mathbf{B}$. Unlike multiplication with scalars, matrices do not exhibit the *commutative* property of multiplication. That is, with regular numbers $5 \times 7 = 7 \times 5$, the order does not matter. But with matrices it does.

The following shows the steps taken in the previous example in a more general way to help you see how it is done. Pay careful attention to the subscripts on each element:

$$\begin{bmatrix} a_{11} & a_{12} \\ a_{21} & a_{22} \\ a_{31} & a_{32} \end{bmatrix} \begin{bmatrix} b_{11} & b_{12} & b_{13} \\ b_{21} & b_{22} & b_{23} \end{bmatrix} = \begin{bmatrix} c_{11} & c_{12} & c_{13} \\ c_{21} & c_{22} & c_{23} \\ c_{31} & c_{32} & c_{33} \end{bmatrix}$$

$$\begin{bmatrix} a_{11}b_{11} + a_{12}b_{21} & a_{11}b_{12} + a_{12}b_{22} & a_{11}b_{13} + a_{12}b_{23} \\ a_{21}b_{11} + a_{22}b_{21} & a_{21}b_{12} + a_{22}b_{22} & a_{21}b_{13} + a_{22}b_{23} \\ a_{31}b_{11} + a_{32}b_{21} & a_{31}b_{12} + a_{32}b_{22} & a_{31}b_{13} + a_{32}b_{23} \end{bmatrix} = \begin{bmatrix} c_{11} & c_{12} & c_{13} \\ c_{21} & c_{22} & c_{23} \\ c_{31} & c_{32} & c_{33} \end{bmatrix}.$$

Let's do one more example. Reconsider the first system of equations we introduced in Section 12.1.1:

$$\begin{matrix} A & x & = & d \\ \begin{bmatrix} 2 & 3 \\ 1 & -2 \end{bmatrix} & \begin{bmatrix} x \\ y \end{bmatrix} & = & \begin{bmatrix} 15 \\ 4 \end{bmatrix} \end{matrix}.$$

Here, we multiply the $\mathbf{A}$ matrix times the $\mathbf{x}$ vector, and we get back the original two equation system as follows:

$$\begin{bmatrix} 2x + 3y \\ 1x - 2y \end{bmatrix} = \begin{bmatrix} 15 \\ 4 \end{bmatrix}.$$

12.2.3   Squaring a Matrix

Suppose we want to square the $\mathbf{A}$ matrix from above (i.e., multiply it by itself). Observe that we can't do the following:

$$\begin{bmatrix} 2 & 3 \\ 1 & 7 \\ -2 & 3 \end{bmatrix} \begin{bmatrix} 2 & 3 \\ 1 & 7 \\ -2 & 3 \end{bmatrix} = ?$$

> **Quick Check**
>
> Using the two matrices,
> $$A = \begin{bmatrix} 2 & 3 \\ 1 & -2 \end{bmatrix},\ B = \begin{bmatrix} 4 & 1 \\ 1 & 0 \end{bmatrix},$$
> c) Add the matrices.
> d) Multiply the matrices.

Where is the problem? The matrices are not *conformable* for this operation. The first matrix has dimensions 3x2, and the second matrix is also 3x2. The inner dimensions do not match, $2 \neq 3$.

When we square a scalar, we multiply that number by itself (e.g., 5 squared is $5 \times 5 = 25$). When we square a matrix, however, we first need to take the *transpose* of the matrix and then multiply the transpose by the original matrix. Taking the transpose first, we would have

$$\begin{bmatrix} 2 & 1 & -2 \\ 3 & 7 & 3 \end{bmatrix} \begin{bmatrix} 2 & 3 \\ 1 & 7 \\ -2 & 3 \end{bmatrix} = \begin{bmatrix} 9 & 7 \\ 7 & 67 \end{bmatrix}.$$

Notice that the result is a *square* 2x2 matrix. Squaring a matrix of any dimensions will result in a *square* matrix. We can square a vector, too, with the same result. For example, suppose we have the following *row* vector (we call it a row vector, as it has one row and multiple columns):

$$v_1 \;=\; [1 \;\; 2 \;\; 3],$$

and we want to find v_1^2. Taking the transpose and then multiplying, we get

$$v_1' v_1 \;=\; \begin{bmatrix} 1 \\ 2 \\ 3 \end{bmatrix} [1 \;\; 2 \;\; 3] = \begin{bmatrix} 1 & 2 & 3 \\ 2 & 4 & 6 \\ 3 & 6 & 9 \end{bmatrix}.$$

This operation results in a 3x3 square matrix. If we started with the *column* vector or reversed the order of the multiplication, we would have

$$v_1 v_1' \;=\; [1 \;\; 2 \;\; 3] \begin{bmatrix} 1 \\ 2 \\ 3 \end{bmatrix} = [14] = 14,$$

which is a 1x1 matrix or just a scalar.

We have now covered adding matrices, subtracting matrices, and multiplying matrices. What about dividing by matrices? To do that we need to learn how to invert a matrix, which we do in the next section.

12.3 Inverse and Determinant

12.3.1 Inverse

This section takes you through how to find the inverse of a 2x2 matrix. The **inverse** of any number is its reciprocal. For example, the inverse of 5 is $\frac{1}{5}$, and the inverse of $\frac{3}{8}$ is $\frac{8}{3}$. The definition of the inverse is that, when multiplied by the original number, the result is 1. In effect, finding an inverse is a way of dividing. Suppose you want to divide 26 by 5. One way of doing that operation is to multiply 26 by the inverse of 5, $\frac{1}{5}$, and carry out $26 \times \frac{1}{5}$.

But what is the equivalent of 1 in matrix terms? That is a special matrix called the **identity matrix**. For any square matrix of dimension n, the identity matrix is also a square matrix, where all the diagonal elements are 1's and all the other elements are zeros.

For example, the following are identity matrices of dimensions 2 and 3, respectively:

$$I_2 = \begin{bmatrix} 1 & 0 \\ 0 & 1 \end{bmatrix}, \qquad I_3 = \begin{bmatrix} 1 & 0 & 0 \\ 0 & 1 & 0 \\ 0 & 0 & 1 \end{bmatrix}.$$

Test the identity matrices by multiplying them by any of the matrices we saw before Remember that the dimensions must match such that you multiply 2x2 matrices by the 2x2 identity matrix and 3x3 matrices by the 3x3 identity matrix.

Thus, when we want the inverse of a matrix, we are really looking for a matrix that, when we multiply by the original, gives us an identity matrix like those shown. In what follows, we are only concerned with the inverses of 2x2 square matrices. Moreover, we just give you the formula for finding the inverse rather than going through the full derivation, which you can find in any linear algebra text. The reason is that the derivation and the procedure for finding the inverses of larger dimension matrices, while not difficult mathwise, do get long and tedious. More to the issue at hand, we will not be dealing with higher dimension matrices where such procedures would be necessary. However, if you are curious, the appendix to the chapter presents how to find the determinant and inverse of larger matrices.

Suppose we have a 2x2 matrix $\mathbf{A}$, as follows:

$$A = \begin{bmatrix} a & b \\ c & d \end{bmatrix}$$

, where I am using the notation of a, b, c, and d to represent the four scalar elements. The reason for doing so is simply that it is easier to remember the inverse formula.

The notation of the inverse of $\mathbf{A}$ is $\mathbf{A}^{-1}$,1 which should make sense since it is the same as $5^{-1} = \frac{1}{5}$, for example. The inverse of $\mathbf{A}$ is given by

$$A^{-1} = \frac{1}{ad - bc} \begin{bmatrix} d & -b \\ -c & a \end{bmatrix}.$$

To be sure that really is the inverse, we will carry out the matrix multiplication and verify that it results in a 2x2 identity matrix:

$$A^{-1}A = I_2$$

$$\underbrace{\frac{1}{ad - bc} \begin{bmatrix} d & -b \\ -c & a \end{bmatrix}}_{A^{-1}} \underbrace{\begin{bmatrix} a & b \\ c & d \end{bmatrix}}_{A} = \underbrace{\begin{bmatrix} 1 & 0 \\ 0 & 1 \end{bmatrix}}_{I}.$$

The fraction in front, $1/(ad - bc)$, is a scalar. Carrying out the matrix multiplication, we have

$$\frac{1}{ad-bc}\begin{bmatrix} ad-bc & db-bd \\ -ca+ac & -cb+ad \end{bmatrix} = \begin{bmatrix} 1 & 0 \\ 0 & 1 \end{bmatrix}$$

$$\begin{bmatrix} \frac{ad-bc}{ad-bc} & 0 \\ 0 & \frac{-cb+ad}{ad-bc} \end{bmatrix} = \begin{bmatrix} 1 & 0 \\ 0 & 1 \end{bmatrix}$$

$$\begin{bmatrix} 1 & 0 \\ 0 & 1 \end{bmatrix} = \begin{bmatrix} 1 & 0 \\ 0 & 1 \end{bmatrix}.$$

Here's a numerical example. Given matrix $\mathbf{A}$,

$$\mathbf{A} = \begin{bmatrix} 2 & 3 \\ 1 & -2 \end{bmatrix},$$

the inverse is

$$\mathbf{A}^{-1} = \frac{1}{-4-3}\begin{bmatrix} -2 & -3 \\ -1 & 2 \end{bmatrix} = \begin{bmatrix} 2/7 & 3/7 \\ 1/7 & -2/7 \end{bmatrix}.$$

Use matrix multiplication, $A^{-1}A = I$, to verify that it is indeed the inverse.

In addition, note that there are cases when the inverse of a matrix *does not exist*. That will happen whenever $ad - bc = 0$, leaving the fraction undefined. Why does this happen? To explain, here's a simple example:

$$\mathbf{A} = \begin{bmatrix} 5 & 2 \\ 5 & 2 \end{bmatrix}.$$

Notice that the numbers across the rows are the same. Thus, when finding $ad - bc$, we get $10 - 10 = 0$. Since the matrix represents the coefficients in a two–equation algebra system, we could write it as

$$5x + 2y = d_1$$
$$5x + 2y = d_2.$$

There are two possibilities here. Suppose d_1 and d_2 are exactly the same. For simplicity, suppose they are both 1. Then we have two equations that are exactly the same. Try to solve those algebraically for x and y. You can't, because there are an infinite number of x's and y's that solve the system. Alternatively, suppose $d_1 \neq d_2$, such as $d_1 = 1$ and $d_2 = 2$. You still can't solve the system, because there does not exist an x and y that will make both equations true simultaneously.

The problem is pretty straightforward. Both equations describe the exact same line. They are said to be **linearly dependent**. Note that the elements in the matrix do not have to be exactly the same for there to be linear dependence. For example, suppose the matrix is

$$\mathbf{A} = \begin{bmatrix} 5 & 2 \\ 10 & 4 \end{bmatrix}.$$

The inverse does not exist since $5*4 - 10*2 = 0$. The lines described by both equations lie on top of each other in the case where $d_1 = d_2$ or never intersect because they are parallel lines when $d_1 \neq d_2$. For there to be an inverse and solution to the two–equation system, the lines need to be
textitlinearly independent. That is, the lines must not lie on top of each other, but intersect at a single point. When the lines *do* intersect at a point, the system of equations has a unique solution for x and y, $ad - bc \neq 0$, and the inverse exists.

12.3.2   Determinant

To find the inverse, we had the term $ad - bc$ in the denominator of the inverse formula. That term for our matrices has a special name, the **determinant**, and it exhibits a large number of properties. The basic idea of the determinant is that it shows the area or volume defined by the linear equations in the matrix. The formula for the determinant of a 2x2 matrix is $ad - bc$.[19] The notation for the determinant is essentially the same as the notation for absolute value. To write the determinant of matrix $\mathbf{A}$ we would write

$$|\mathbf{A}| = \begin{vmatrix} a_{11} & a_{12} \\ a_{21} & a_{22} \end{vmatrix} = a_{11}a_{22} - a_{21}a_{12},$$

where I have used a_{ij}, indicating the row and column instead of a, b, c, and d.

To get a better feel for what that means, consider the following example with graphs:

$$\mathbf{A} = \begin{bmatrix} 4 & -1 \\ 2 & 1 \end{bmatrix}.$$

If we graph that matrix considering each row as a vector (i.e., a ray from the origin to the point indicated), we would have the following graph.

---

[19]The determinant formula for matrices larger than 2x2 is similar but more complicated. We will not go into that here, but you can find it in the appendix to the chapter.

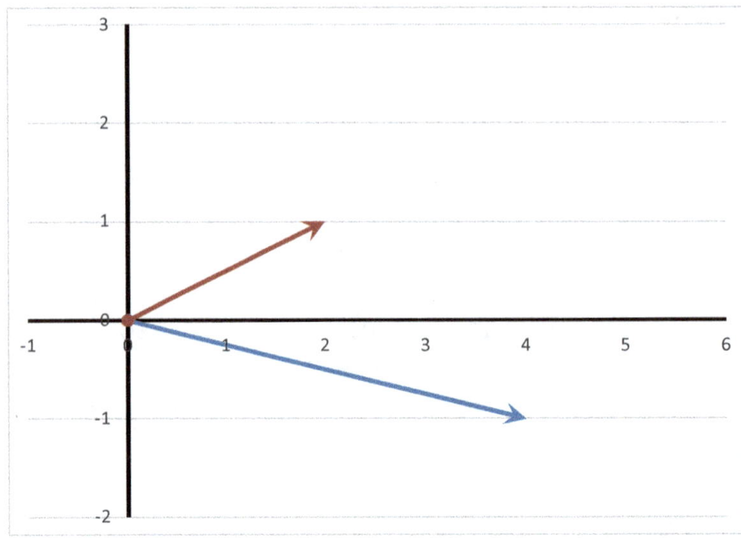

Figure 12.2: Two vectors of matrix **A**.

Now suppose we used these lines to form a parallelogram by taking each ray and starting it from the endpoint of the other ray:

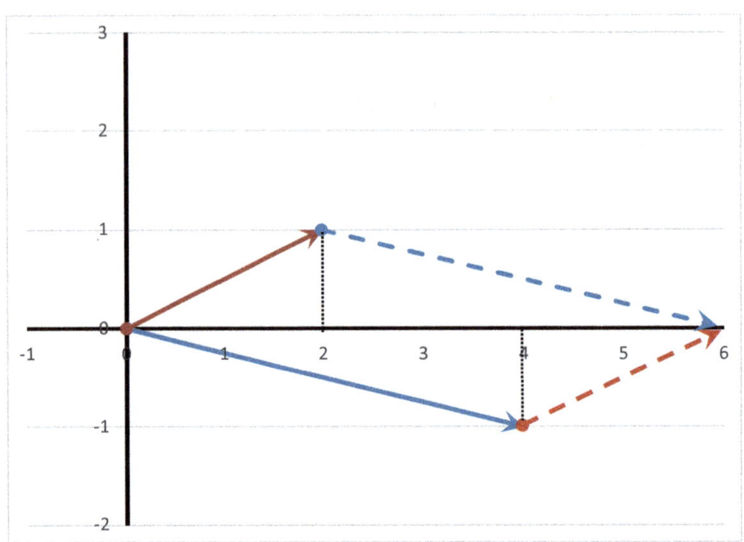

Figure 12.3: Parallelogram formed by matrix **A**.

What is the area of that parallelogram? We could use the basic area of a triangle formula to figure it out. The triangle determined by each ray relative to the origin is pretty simple; we just take 1/2 times the base and the height. Thus, for the upper triangle and lower triangle we would have

$$Area_1 = 1/2 \times (2) \times (1) = 1/2$$
$$Area_2 = 1/2 \times (4) \times (1) = 2.$$

Adding those together gives us 5/2. For the entire parallelogram, each triangle appears twice, so if we double the area of the two triangles, we will have the area of the parallelogram: $2\times(5/2)=5$. That is the value of the determinant when we apply the formula $|A| = ad - bc$:

$$|\mathbf{A}| = \begin{vmatrix} 4 & -1 \\ 2 & 1 \end{vmatrix} = 4 \times 1 - 2 \times (-1) = 5.$$

Consider the following example from earlier when the inverse does not exist:

$$\begin{bmatrix} 5 & 2 \\ 10 & 4 \end{bmatrix} \begin{bmatrix} x \\ y \end{bmatrix} = \begin{bmatrix} 1 \\ 1 \end{bmatrix}.$$

If you graph the two lines described by the equations, the lines lie right on top of each other. Thus, the value of the determinant is zero because there is no area covered by the two lines. Alternatively, suppose the **d** vector changes such that we have

Quick Check

Find the determinants of the following:

g) $A = \begin{bmatrix} 2 & 3 \\ 1 & -2 \end{bmatrix}$ h) B$= \begin{bmatrix} -3 & 3 \\ 1 & -4 \end{bmatrix}$

$$\begin{bmatrix} 5 & 2 \\ 10 & 4 \end{bmatrix} \begin{bmatrix} x \\ y \end{bmatrix} = \begin{bmatrix} 1 \\ 2 \end{bmatrix}.$$

Now the lines do not lie on top of each other, but the lines are parallel. If you try to form a parallelogram, it will fail since you can never close the shape.

12.4 Cramer's Rule

12.4.1 Cramer's Rule Basics

Cramer's rule provides a quick methodology for solving a system of equations for a single element. In what follows in this section, you could probably solve the equations faster using your basic algebra rather than Cramer's rule. However, when we have two equations derived from an economic model, Cramer's rule becomes more convenient when compared to the method we used in Chapter 10.

First, consider the following basic representation of a system of equations as matrices introduced at the beginning of this chapter on linear algebra:

$$\mathbf{Ax} = \mathbf{d}.$$

If we are interested in solving for the objects in the $\mathbf{x}$ vector, we can actually get the solution in one quick step: Multiply both sides by $\mathbf{A}^{-1}$, the inverse of $\mathbf{A}$. What we get is

$$\mathbf{A}^{-1}\mathbf{Ax} = \mathbf{A}^{-1}\mathbf{d}.$$

Look at the left-hand side. We have $\mathbf{A}^{-1}\mathbf{A}$, which is equal to the identity matrix.[20] What we get is

$$\mathbf{Ix} = \mathbf{A}^{-1}\mathbf{d}.$$

Since the identity matrix times any other matrix is just like multiplying by 1 we have

$$\mathbf{x} = \mathbf{A}^{-1}\mathbf{d},$$

which solves for $\mathbf{x}$. Great, that was easy. To find the answers for the $\mathbf{x}$ vector, we just need to multipy the inverse of $\mathbf{A}$ times the $\mathbf{d}$ vector. The following is an example:

$$\begin{bmatrix} 2 & 3 \\ 1 & -2 \end{bmatrix} \begin{bmatrix} x_1 \\ x_2 \end{bmatrix} = \begin{bmatrix} 15 \\ 4 \end{bmatrix}.$$

To find x_1 and x_2, we first find the inverse of the $\mathbf{A}$ matrix and then mutliply it by the $\mathbf{d}$ vector. The inverse of the $\mathbf{A}$ matrix, which we found before, is

$$\mathbf{A}^{-1} = \begin{bmatrix} 2/7 & 3/7 \\ 1/7 & -2/7 \end{bmatrix}.$$

Multiplying by the $\mathbf{d}$ vector, we have

$$\mathbf{x} = \begin{bmatrix} x_1 \\ x_2 \end{bmatrix} = \begin{bmatrix} 2/7 & 3/7 \\ 1/7 & -2/7 \end{bmatrix} \begin{bmatrix} 15 \\ 4 \end{bmatrix} = \begin{bmatrix} 30/7 + 12/7 \\ 15/7 - 8/7 \end{bmatrix} = \begin{bmatrix} 6 \\ 1 \end{bmatrix}.$$

Thus, in our system of equations, $x_1 = 6$ and $x_2 = 1$.

Cramer's rule cuts out one step in order to solve for either the x_1 or the x_2. In the general two–equation system, written out with more detail here,

$$\begin{bmatrix} a_{11} & a_{12} \\ a_{21} & a_{22} \end{bmatrix} \begin{bmatrix} x_1 \\ x_2 \end{bmatrix} = \begin{bmatrix} d_1 \\ d_2 \end{bmatrix},$$

---

[20]Keep in mind that the order of multiplication matters with matrices, so the $\mathbf{A}^{-1}$ must be on the left-side for both sides of the equals signs. That is, you cannot write: $\mathbf{A}^{-1}\mathbf{Ax} = \mathbf{dA}^{-1}$.

it turns out that we can find x_1 and x_2 by taking a couple, simple determinants instead of going through the full process of finding the inverse and then multiplying by the **d** vector. The formulas for x_1 and x_2 are

$$x_1 = \frac{\begin{vmatrix} d_1 & a_{12} \\ d_2 & a_{22} \end{vmatrix}}{\begin{vmatrix} a_{11} & a_{12} \\ a_{21} & a_{22} \end{vmatrix}} \text{ and } x_2 = \frac{\begin{vmatrix} a_{11} & d_1 \\ a_{21} & d_2 \end{vmatrix}}{\begin{vmatrix} a_{11} & a_{12} \\ a_{21} & a_{22} \end{vmatrix}}.$$

The denominator in each is the determinant of the **A** matrix. The numerator, however, is the determinant that comes from replacing a column of the **A** matrix with the **d** vector. For x_1, the *first* element in the **x** vector, we replace the *first* column of the **A** matrix with the **d** vector. For x_2, the *second* element in the **x** vector, we replace the *second* column of the **A** matrix with the **d** vector.

Going back to the same example, for x_1 we would have

$$x_1 = \frac{\begin{vmatrix} 15 & 3 \\ 4 & -2 \end{vmatrix}}{\begin{vmatrix} 2 & 3 \\ 1 & -2 \end{vmatrix}} = \frac{15 \times (-2) - 4 \times 3}{2 \times (-2) - 3 \times (1)} = \frac{-42}{-7} = 6.$$

Notice the first column of the determinant in the numerator; it is the **d** vector. For x_2 we would have

$$x_2 = \frac{\begin{vmatrix} 2 & 15 \\ 1 & 4 \end{vmatrix}}{\begin{vmatrix} 2 & 3 \\ 1 & -2 \end{vmatrix}} = \frac{2 \times 4 - 1 \times 15}{2 \times (-2) - 3 \times (1)} = \frac{-7}{-7} = 1.$$

Here is a second example, where we will go from the two-equation system through to the solution. Suppose we have

$$5x_1 + 4x_2 = 35$$
$$6x_1 - 2x_2 = 8.$$

Representing that in matrix form we have

$$Ax = d$$
$$\begin{bmatrix} 5 & 4 \\ 6 & -2 \end{bmatrix} \begin{bmatrix} x_1 \\ x_2 \end{bmatrix} = \begin{bmatrix} 35 \\ 8 \end{bmatrix}.$$

To find x_1, apply Cramer's rule

$$x_1 = \frac{\begin{vmatrix} d_1 & a_{12} \\ d_2 & a_{22} \end{vmatrix}}{\begin{vmatrix} a_{11} & a_{12} \\ a_{21} & a_{22} \end{vmatrix}} = \frac{\begin{vmatrix} 35 & 4 \\ 8 & -2 \end{vmatrix}}{\begin{vmatrix} 5 & 4 \\ 6 & -2 \end{vmatrix}} = \frac{-102}{-34} = 3.$$

For x_2, apply Cramer's rule:

$$x_2 = \frac{\begin{vmatrix} a_{11} & d_1 \\ a_{21} & d_2 \end{vmatrix}}{\begin{vmatrix} a_{11} & a_{12} \\ a_{21} & a_{22} \end{vmatrix}} = \frac{\begin{vmatrix} 5 & 35 \\ 6 & 8 \end{vmatrix}}{\begin{vmatrix} 5 & 4 \\ 6 & -2 \end{vmatrix}} = \frac{-170}{-34} = 5.$$

All that seems like extra tedious work to solve a simple, linear two–equation system, and in these example cases it is. However, when we apply it to our models, it makes life easier, as we will see. It is especially useful if we have a system of four or more equations but are interested

in the solution to just one particular x, because instead of having to solve for everything, we have a quick method for finding the answer to the variable of interest.

12.4.2 Cramer's Rule Applied to Derivatives

If you've made it this far, wow. Now we put it all together; in this section we will use derivatives to find a maximum (minimum), total derivatives, matrices, and Cramer's rule. Let the fun begin.

Consider the Ultra firm's profit maximization problem in producing beta capsules. Ultra chooses how much physical capital, K, to rent and how much labor, L, to hire. K and L are the endogenous choice variables. Ultra's production function is $F(K, L)$, where $F_K > 0$, $F_{KK} < 0$, $F_L > 0$, and $F_{LL} < 0$. All that means is that output is concave in both capital and labor (i.e., there are diminishing marginal returns to both capital and labor). We will also assume that $F_{KL} > 0$, the cross–derivative, is positive, which means that the marginal product of capital increases with more labor (and that the marginal product of labor increases with more capital). Ultra sells the capsules at the price P, rents the capital at the rate r, and pays labor the wage rate w. The firm takes P, w, and r as given, so they are the exogenous variables.

Writing out the profit maximization problem we have

$$max_{K,L}\Pi = PF(K, L) - rK - wL.$$

To find the choices of capital and labor that maximize profits, we need to take the first-order condition for both K and L, as follows:

$$\frac{\partial \Pi}{\partial K} = PF_K - r = 0$$

$$\frac{\partial \Pi}{\partial L} = PF_L - w = 0.$$

In words, the first one reads, the price times the marginal product of capital is the marginal benefit, which must equal the marginal cost of capital, r. The second reads, the price times the marginal product of labor is the marginal benefit, which must equal the marginal cost of labor, w.

Now suppose we want to know how an increase in wages, w, will affect Ultra's choice of K. That is, we want to find $\frac{dK}{dw}$. To find the answer we need to account for *both* first-order conditions. Obviously, if wages increase we would expect the firm to use less labor, as the second of the two first-order conditions suggests. However, if the firm uses less labor, that affects the marginal product of capital, which changes the value of F_K in the first first-order condition.

To find that derivative from a system of equations, we will go through the following steps:

1. Take the total derivatives of each first-order condition.

2. Arrange the equations such that the derivatives of the endogenous choice variables, dK and dL, are on the left-hand side and the derivatives of the exogenous variables (e.g., dw, dP, and dr) are on the right-hand side.

3. Put the resulting system of equations into matrix form.

4. Apply Cramer's rule to find the derivatives of interest.

Step 1: Total derivatives

Starting with the first condition,

$$\frac{d\Pi}{dK} = PF_K - r = 0,$$

take the derivative with respect to K, L, and w:

$$PF_{KK}dK + PF_{KL}dL = 0.$$

Notice, there is no w in the equation, so dw does not appear. The second condition is

$$\frac{d\Pi}{dL} = PF_L - w = 0,$$

and the total derivative is

$$PF_{LK}dK + PF_{LL}dL - dw = 0.$$

Step 2: Arrange the equations

Her,e we keep the choice variables (dL, dK) on the left-hand side and move the exogenous variables (dw) to the right-hand side. *Watch your negative signs when you do so!* In this case, it is pretty straightforward:

$$
\begin{aligned}
PF_{KK}dK + PF_{KL}dL &= 0 \\
PF_{LK}dK + PF_{LL}dL &= dw.
\end{aligned}
\tag{2}
$$

Step 3: Matrix Form

Here, we treat the dK and dL (i.e., the endogenous variables) like the x_1 and x_2 in the matrices of the previous subsection, so they go in the $\mathbf{x}$ vector. Everything that multiplies them are the coefficients that go into the $\mathbf{A}$ matrix. What are on the right-hand sides of the equations are the elements of the $\mathbf{d}$ vector. Thus, we get

$$\begin{bmatrix} PF_{KK} & PF_{KL} \\ PF_{LK} & PF_{LL} \end{bmatrix} \begin{bmatrix} dK \\ dL \end{bmatrix} = \begin{bmatrix} 0 \\ 1 \end{bmatrix}.dw$$

Notice that I factored out the dw on the right-hand side. To see why, for just a moment let's write the system using $Ax = d$ notation such that we have

$$\mathbf{Ax} = \mathbf{d}dw,$$

where the dw still multiplies the $\mathbf{d}$ vector. From Cramer's rule we know that the first element in the $\mathbf{x}$ vector (which is dK in this case) would be

$$dK = \frac{\begin{vmatrix} d_1 & a_{12} \\ d_2 & a_{22} \end{vmatrix}}{\begin{vmatrix} a_{11} & a_{12} \\ a_{21} & a_{22} \end{vmatrix}} dw.$$

Recall from Chapter 6 we treat dw as a scalar. Then we can divide both sides by dw to get

$$\frac{dK}{dw} = \frac{\begin{vmatrix} d_1 & a_{12} \\ d_2 & a_{22} \end{vmatrix}}{\begin{vmatrix} a_{11} & a_{12} \\ a_{21} & a_{22} \end{vmatrix}},$$

and that, dK/dw, is the relationship we are looking for! How capital changes with the wage rate.

Step 4: Apply Cramer's rule

Now we just follow the formula for Cramer's rule, putting in the values for the $\mathbf{A}$ matrix and the $\mathbf{d}$ vector:

$$\frac{dK}{dw} = \frac{\begin{vmatrix} 0 & PF_{KL} \\ 1 & PF_{LL} \end{vmatrix}}{\begin{vmatrix} PF_{KK} & PF_{KL} \\ PF_{LK} & PF_{LL} \end{vmatrix}}$$

$$\frac{dK}{dw} = \frac{-PF_{KL}}{P^2 F_{KK} F_{LL} - (PF_{LK})^2}.$$

What does that mean in economics terms? We need to figure out the sign of the fraction to determine whether capital increases or decreases with the wage rate. How do we do that? The denominator here is part of the second–order conditions for a two–variable system and is always positive whenever we are looking for a maximum or minimum. Refer back to Chapter 5 on second–order conditions and the saddle point condition. So, the denominator is positive.

Therefore, all we need is the sign of the numerator to determine the sign of $\frac{dK}{dw}$. From our assumptions at the start, we had that $F_{KL} > 0$. This assumption combined with the minus sign in front, means that the entire derivative is negative. In words, an increase in the wage rate will lead to a fall in the amount of capital that Ultra will rent. Notice that, if we had assumed the opposite, such that $F_{KL} < 0$, then the marginal product of capital falls with more labor, and we would have the opposite result (i.e., an increase in wages would increase capital chosen by firms).

12.4.3 Specific Example

Here, we will redo the problem in the previous section, but instead of using $F(K, L)$ as the production function, we will use a more specific example, where $F(K, L) = K^{1/3}L^{2/3}$. Furthermore, instead of looking for how capital responds to wages dK/dw, we will find how labor demand responds to the real interest rate, dL/dr. Now the profit–maximization problem is

$$max_{K,L}\Pi = PK^{1/3}L^{2/3} - rK - wL.$$

To find the choices of capital and labor that maximize profits, we need to take the first-order condition for both K and L, as follows:

$$\frac{d\Pi}{dK} = \frac{1}{3}PK^{-2/3}L^{2/3} - r = 0$$

$$\frac{d\Pi}{dL} = \frac{2}{3}PK^{1/3}L^{-1/3} - w = 0. \tag{3}$$

Step 1: Total derivatives

Taking total derivatives with respect to the choice variables, L and K, and the exogenous variable, r, we have

$$\frac{-2}{9}PK^{-5/3}L^{2/3}dK + \frac{2}{9}PK^{-2/3}L^{-1/3}dL - dr = 0$$

$$\frac{2}{9}PK^{-2/3}L^{-1/3}dK + \frac{-2}{9}PK^{1/3}L^{-4/3}dL = 0. \tag{4}$$

Step 2: Arrange the equations

Rearranging,

$$\frac{-2}{9}PK^{-5/3}L^{2/3}dK + \frac{2}{9}PK^{-2/3}L^{-1/3}dL = dr$$

$$\frac{2}{9}PK^{-2/3}L^{-1/3}dK + \frac{-2}{9}PK^{1/3}L^{-4/3}dL = 0. \tag{5}$$

Step 3: Matrix form

Arranging into matrix form,

$$\begin{bmatrix} \frac{-2}{9}PK^{-5/3}L^{2/3} & \frac{2}{9}PK^{-2/3}L^{-1/3} \\ \frac{2}{9}PK^{-2/3}L^{-1/3} & \frac{-2}{9}PK^{1/3}L^{-4/3} \end{bmatrix} \begin{bmatrix} dK \\ dL \end{bmatrix} = \begin{bmatrix} 1 \\ 0 \end{bmatrix} dr.$$

Step 4: Apply Cramer's rule

Note that since dL is the second element in the $\mathbf{x}$ vector, we replace the second column of the $\mathbf{A}$ matrix to find dL/dr:

$$\frac{dL}{dr} = \frac{\begin{vmatrix} \frac{-2}{9}PK^{-5/3}L^{2/3} & 1 \\ \frac{2}{9}PK^{-2/3}L^{-1/3} & 0 \end{vmatrix}}{\begin{vmatrix} \frac{-2}{9}PK^{-5/3}L^{2/3} & \frac{2}{9}PK^{-2/3}L^{-1/3} \\ \frac{2}{9}PK^{-2/3}L^{-1/3} & \frac{-2}{9}PK^{1/3}L^{-4/3} \end{vmatrix}}$$

$$\frac{dL}{dr} = \frac{\frac{-2}{9}PK^{-2/3}L^{-1/3}}{\frac{4}{81}P^2K^{-4/3}L^{-2/3} - \left(\frac{-2}{9}PK^{-2/3}L^{-1/3}\right)^2}.$$

The denominator must be positive for profit maximization. The numerator is negative because all the variables are positive, but there is a negative sign out front. That means increases in the real interest rate will cause firms to demand less labor (i.e., $dL/dr < 0$).

Quick Check

i) From the example in Section 12.5.3, find dK/dr.

12.4.4   Another Example

As a final example, we go through a model from macroeconomics. In this model, we have a consumer, Hayata, who works for the Science Patrol (SP). Hayata has the following utility function, where he chooses consumption, c, and leisure, l, subject to a budget constraint, as follows:

$$\max_{c,l} U(c, l)$$

$$\text{s.t.}$$

$$c = w(H - l) + \pi - T.$$

The Science Patrol pays Hayata the real wage of w. H is Hayata's total time endowment, and l is leisure time such that $H - l$ is time spent working. π is profits from the firms where Hayata holds stock, and T is the lump-sum level of taxes the government imposes on consumers, including Hayata. We make the following assumptions about the utility function: $U_c > 0$, means consumption has a positive marginal utility; $U_{cc} < 0$, but with diminishing returns; U_l, leisure time, also has positive marginal utility; $U_{ll} < 0$, but again with diminishing marginal utility; and $U_{cl} > 0$, which can be interpreted in two equivalent ways: (1) The marginal utility of consumption is increasing with leisure time; or (2) The marginal utility of leisure is increasing with consumption.

Setting up this problem using a Lagrangian (you can also do this problem by substituting the constraint directly into the utility function),

$$\mathcal{L} = U(c, l) + \lambda \left[w(H - l) + \pi - T - c \right].$$

The first-order conditions (FOCs) come from taking the derivative of the Langrangian with respect to the choice variables for consumers and with respect to the Lagrangian multiplier, λ, which as before just reproduces the original budget constraint. Consumers choose consumption, c, and leisure time, l. We have

$$\frac{\partial \mathcal{L}}{\partial c} = U_c(c, l) - \lambda = 0$$
$$\frac{\partial \mathcal{L}}{\partial l} = U_l(c, l) - \lambda w = 0$$
$$\frac{\partial \mathcal{L}}{\partial \lambda} = w(H - l) + \pi - T - c = 0.$$

The notation $U_c(c, l)$ indicates the partial derivative of the utility function, U, with respect to c, consumption. In what follows, after writing $U_c(c, l)$ out in full for now, but we will shorten that to just U_c and drop the argument notation, (c, l), to keep our equations from getting too cluttered and hard to read. However, even though we don't write them out, it should be understood that the arguments are still there.

We have three equations and three unknowns. However, it is very easy to reduce it to two equations and two unknowns. We can eliminate λ by dividing the first FOC by the second FOC as follows:

$$U_c(c, l) - \lambda = 0$$
$$U_l(c, l) - \lambda w = 0.$$

These equations can be rewritten as

$$U_c(c, l) = \lambda$$
$$U_l(c, l) = \lambda w,$$

and we divide one by the other:

$$\frac{U_c(c, l)}{U_l(c, l)} = \frac{1}{w},$$

or after inverting:

$$\frac{U_l(c, l)}{U_c(c, l)} = w,$$

which reads, the marginal rate of substitution (MRS, see Chapter 6) between leisure and consumption equals the real wage rate when utility is maximized. Our two–equation system is therefore the budget constraint:

$$w(h - l) + \pi - T - c = 0,$$

and the combined first-order conditions from above, rewritten slightly here as

$$U_l\left(c,l\right) - wU_c\left(c,l\right) = 0.$$

Step 1: Total derivatives

We have our two equations, and now we totally differentiate with respect to the choice variables $(c,\, l)$ and the exogenous variables (w, π, T). Note that you can select only one exogenous variable and follow the Cramer's rule procedure as described previously. To save extra work, we take the derivatives on *all* the exogenous variables; and we will see how to handle this process in what follows. Totally differentiating we get

$$(h-l)\,dw - w\,dl + d\pi - dT - dc \;=\; 0$$
$$U_{lc}\left(c,l\right)dc + U_{ll}\left(c,l\right)dl - wU_{cc}\left(c,l\right)dc - wU_{cl}\left(c,l\right)dl - U_c\left(c,l\right)dw \;=\; 0.$$

Step 2: Arrange the equations

Rearranging such that we have the derivatives of the endogenous variables on the left-hand side and the derivatives of the exogenous variables on the right–hand side, we have

$$-dc - w\,dl \;=\; -(h-l)\,dw - d\pi + dT$$
$$U_{lc}\left(c,l\right)dc - wU_{cc}\left(c,l\right)dc + U_{ll}\left(c,l\right)dl - wU_{cl}\left(c,l\right)dl \;=\; U_c\left(c,l\right)dw.$$

Factoring out dc and dl in the second of the two equations and dropping the notation (c,l) to make it easier to read,

$$-dc - w\,dl \;=\; -(h-l)\,dw - d\pi + dT$$
$$[U_{lc} - wU_{cc}]\,dc + [U_{ll} - wU_{cl}]\,dl \;=\; U_c\,dw.$$

Step 3: Matrix form

Now we can put this system into matrix form:

$$\begin{bmatrix} -1 & -w \\ U_{lc} - wU_{cc} & U_{ll} - wU_{cl} \end{bmatrix} \begin{bmatrix} dc \\ dl \end{bmatrix} = \begin{bmatrix} -(h-l)\,dw - d\pi + dT \\ U_c\,dw \end{bmatrix}.$$

Step 4: Apply Cramer's rule

We are ready to apply Cramer's rule to find the effects of the exogenous variables on the endogenous variables. Here, we will look at the effect of changing the tax level, T, on the optimal choices of consumption, c, and leisure, l. The exercises at the end of the chapter ask you to do the same for π and w. Notice here, unlike the previous examples, there are multiple derivatives on the exogenous variables $(dw, d\pi, dT)$. Since we are interested in the effects of T, we are asking how c and l respond to a change in T, *holding everything else constant.* Holding everything else (w and π) constant means that w and π do not change. That is, we set $dw = 0$ and $d\pi = 0$ to indicate they do not change.

Setting $dw = 0$ and $d\pi = 0$, and rewriting our matrices, we have

$$\begin{bmatrix} -1 & -w \\ U_{lc} - wU_{cc} & U_{ll} - wU_{cl} \end{bmatrix} \begin{bmatrix} dc \\ dl \end{bmatrix} = \begin{bmatrix} 1 \\ 0 \end{bmatrix} dT.$$

Cramer's rule then gives us

$$\frac{dc}{dT} = \frac{\begin{vmatrix} 1 & -w \\ 0 & U_{ll} - wU_{cl} \end{vmatrix}}{\begin{vmatrix} -1 & -w \\ U_{lc} - wU_{cc} & U_{ll} - wU_{cl} \end{vmatrix}}$$

$$\frac{dc}{dT} = \frac{(U_{ll} - wU_{cl})}{-(U_{ll} - wU_{cl}) + w(U_{lc} - wU_{cc})}.$$

As before, the denominator is strictly positive because it is part of the second-order conditions necessary for a maximum, so to figure out how c reacts to T, we just need to sign the numerator, $U_{ll} - wU_{cl}$. Well, $U_{ll} < 0$, which means that the marginal utility of leisure is decreasing or, in other words, utility is increasing in leisure but at a diminishing rate. The sing of the cross–derivative, $U_{cl} > 0$, means that the marginal utility of consumption is increasing in leisure and both consumption and leisure are normal goods. Thus, $U_{ll} - wU_{cl} < 0$, which means that consumption falls with higher taxes. No surprise there at all. For dl/dT we have

$$\frac{dl}{dT} = \frac{\begin{vmatrix} -1 & 1 \\ U_{lc} - wU_{cc} & 0 \end{vmatrix}}{\begin{vmatrix} -1 & -w \\ U_{lc} - wU_{cc} & U_{ll} - wU_{cl} \end{vmatrix}}$$

$$\frac{dl}{dT} = \frac{-(U_{lc} - wU_{cc})}{-(U_{ll} - wU_{cl}) + w(U_{lc} - wU_{cc})}.$$

Again, we just need to figure out the sign of the numerator, $-(U_{lc} - wU_{cc})$. As before, $U_{cl} > 0$. w must be positive to make any economic sense. Finally, $U_{cc} < 0$ because we assume diminishing marginal utility of consumption. Putting all that together, we have that $-(U_{lc} - wU_{cc}) < 0$, which means $dl/dT < 0$. Leisure time also falls with taxes.

12.5   Exercises

12.5.1   Quick Check Answers

a) A is 3x2. $A^T = \begin{bmatrix} 1 & 3 & 3 \\ 4 & 1 & 0 \end{bmatrix}$, b) B is 3x4. $B^T = \begin{bmatrix} 5 & -2 & 0 \\ 4 & 1 & 0 \\ 1 & -3 & 0 \\ 6 & 9 & 7 \end{bmatrix}$

c) $A + B = \begin{bmatrix} 6 & 4 \\ 2 & -2 \end{bmatrix}$, d) $AB = \begin{bmatrix} 11 & 2 \\ 2 & 1 \end{bmatrix}$

e) $A^T A = \begin{bmatrix} 1 & 3 & 3 \\ 4 & 1 & 0 \end{bmatrix} \begin{bmatrix} 1 & 4 \\ 3 & 1 \\ 3 & 0 \end{bmatrix} = \begin{bmatrix} 19 & 7 \\ 7 & 17 \end{bmatrix}$

f) $\mathbf{B}^{-1} = \begin{bmatrix} 1/2 & -1/2 \\ -3/4 & 5/4 \end{bmatrix}$, $\mathbf{C}^{-1} = \begin{bmatrix} 1 & 0 \\ -1 & 1/4 \end{bmatrix}$.

g) $|A| = -7$, $|B| = 9$ 　　　h) $x_1 = 3$, $x_2 = 2$

i) $\dfrac{dK}{dr} = \dfrac{\frac{-2}{9} P K^{1/3} L^{-4/3}}{\frac{4}{81} P^2 K^{-4/3} L^{-2/3} - \left(\frac{-2}{9} P K^{-2/3} L^{-1/3} \right)^2}$

12.5.2   Practice Problems

The following matrices are for the five problems that follow:

$$\mathbf{A} = \begin{bmatrix} 2 & 3 & 5 \\ 1 & -2 & 0 \\ -2 & 3 & 4 \end{bmatrix} \qquad \mathbf{B} = \begin{bmatrix} 2 & 3 & -5 \\ 1 & -2 & 8 \\ -2 & 3 & 2 \end{bmatrix} \qquad \mathbf{C} = \begin{bmatrix} 1 & 2 & 3 \\ 0 & 1 & 2 \\ 0 & 0 & 1 \end{bmatrix}$$

1. Use the matrices to find $\mathbf{A}+\mathbf{B}$, $\mathbf{B}+\mathbf{C}$, and $\mathbf{A}+\mathbf{C}$.

2. Use the matrices to find $\mathbf{A}\text{-}\mathbf{B}$, $\mathbf{B}\text{-}\mathbf{C}$, and $\mathbf{A}\text{-}\mathbf{C}$.

3. Find the transpose of the matrices $\mathbf{A}$, $\mathbf{B}$, and $\mathbf{C}$.

4. Using the matrices, carry out the matrix multiplication for: $\mathbf{A} \times \mathbf{B}$, $\mathbf{A} \times \mathbf{C}$, $\mathbf{B} \times \mathbf{C}$, $\mathbf{B} \times \mathbf{A}$, $\mathbf{C} \times \mathbf{A}$, and $\mathbf{C} \times \mathbf{B}$.

5. Perform the following scalar multiplications using the matrices: $2 \times \mathbf{A}$, $10 \times \mathbf{B}$, $-3 \times \mathbf{C}$, and $\alpha \times \mathbf{A}$.

6. Perform the following matrix multiplication operations:

$$\begin{bmatrix} 2 & 3 \\ 1 & -2 \end{bmatrix} \begin{bmatrix} 3 & -5 \\ -2 & 8 \end{bmatrix} = ?$$

$$\begin{bmatrix} 1 & 2 \\ 0 & 1 \end{bmatrix} \begin{bmatrix} 4 & 57 \\ 85 & 1 \end{bmatrix} = ?$$

The following matrices are for the three problems that follow:

$$\mathbf{A} = \begin{bmatrix} 5 & 3 \\ 0 & -2 \end{bmatrix}, \mathbf{B} = \begin{bmatrix} -10 & 2 \\ 3 & 9 \end{bmatrix}, \mathbf{C} = \begin{bmatrix} 1 & 0 \\ 6 & 4 \end{bmatrix}, \mathbf{D} = \begin{bmatrix} 12 & 2 \\ -6 & -1 \end{bmatrix}.$$

7. Find the determinants of matrices $\mathbf{A}$, $\mathbf{B}$, $\mathbf{C}$, and $\mathbf{D}$.

8. For the matrices where the determinant is not zero, find the inverse.

9. Verify that the inverses you found are correct by multiplying the inverse by the orginal matrix to get back the 2x2 identity matrix.

10. Use Cramer's rule to solve for x and y in each of these systems:

$$a) \quad \begin{bmatrix} 5 & 3 \\ 0 & -2 \end{bmatrix} \begin{bmatrix} x \\ y \end{bmatrix} = \begin{bmatrix} 2 \\ 2 \end{bmatrix}$$

$$b) \quad \begin{bmatrix} -10 & 2 \\ 3 & 9 \end{bmatrix} \begin{bmatrix} x \\ y \end{bmatrix} = \begin{bmatrix} -12 \\ 30 \end{bmatrix}$$

$$c) \quad \begin{bmatrix} 1 & 0 \\ 6 & 4 \end{bmatrix} \begin{bmatrix} x \\ y \end{bmatrix} = \begin{bmatrix} 1/3 \\ 3 \end{bmatrix}$$

$$d) \quad \begin{bmatrix} 5 & -10 \\ -3 & 6 \end{bmatrix} \begin{bmatrix} x \\ y \end{bmatrix} = \begin{bmatrix} 2 \\ 2 \end{bmatrix}.$$

11. In problem #10, what problem did you have with (d)? Why?

12. Referring to the model in Section 12.5.4 with consumption and leisure, find $dC/d\pi$, $dl/d\pi$, dC/dw, and dl/dw.

13. Using the model, follow the Cramer's rule procedure and find dc_1/dy_1, dc_2/dy_1, dc_1/dy_2, dc_2/dy_2, $dc_1/d\beta$, $dc_1/d(1+r)$, $dc_2/d\beta$ and $dc_2/d(1+r)$. The following are the first-order conditions and the budget constraint from the two-period consumer model discussed in Chapter 7:

$$\frac{\partial \mathcal{L}}{\partial c_1} = \frac{1}{c_1} - \lambda = 0$$

$$\frac{\partial \mathcal{L}}{\partial c_2} = \frac{\beta}{c_2} - \lambda \left(\frac{1}{1+r} \right) = 0$$

$$y_1 + \frac{y_2}{1+r} - c_1 - \frac{c_2}{1+r} = 0.$$

(a) Combine the two first-order conditions to eliminate the Lagrangian multiplier, λ.

(b) Take the total derivatives of the condition you found in part a with respect to all the endogenous choice variables (c_1, c_2) and the exogenous variables $(y_1, y_2, \beta, 1 + r)$.

(c) Take the total derivatives of the budget constraint with respect to all the endogenous choice variables (c_1, c_2) and the exogenous variables $(y_1, y_2, \beta, 1 + r)$.

(d) Arrange both equations so that the derivatives of the endogenous variables (c_1, c_2) are on the left-hand side and the exogenous variables $(y_1, y_2, \beta, 1 + r)$ are on the right-hand side.

(e) Using your two equations from (c) and (d), put them into matrix form.

(f) Apply Cramer's rule. Remember to set the derivatives of the exogenous variables you are not immediately looking at to zero. For example, if you are trying to find dc_1/dy_1, set $dy_2 = 0$, $d\beta = 0$, and $d(1 + r) = 0$.

(g) Return to Chapter 10, problem 6. Do that problem, but use matrices and Cramer's rule to get the answers.

12.6   Math Appendix

Here, we list some key math definitions, theorems, and formulae in formal math for your reference.

Linearly dependent: A set of N vectors, x_1, x_2, ..., x_N, is *linearly dependent* if there exists N coefficients a_1, a_2,...,a_N not all equal to zero, such that $a_1 x_1 + a_2 x_2 + \cdots + a_N x_N$.

Linearly independent: A set of N vectors x_1, x_2, ..., x_N is *linearly independent* if there does not exist N coefficients, a_1, a_2,...,a_N, such that $a_1 x_1 + a_2 x_2 + \cdots + a_N x_N$ *except* for the case when $a_i = 0 \ \forall i \in [1, N]$.

Higher order determinants: The chapter text provided the formula for the determinant of a 2x2 matrix:

$$|\mathbf{A}| \quad = \quad \left| \begin{matrix} a_{11} & a_{12} \\ a_{21} & a_{22} \end{matrix} \right| = a_{11}a_{22} - a_{21}a_{12}.$$

For matrices of larger size, we need to introduce a few terms:
Minor matrix: Let $\mathbf{M}$ be a matrix with at least two rows and two columns. Let $\mathbf{M}_{ij}$ be the matrix formed from deleting row i and column j in matrix $\mathbf{M}$. For example, if

$$\mathbf{M} \quad = \quad \begin{bmatrix} a_{11} & a_{12} & a_{13} \\ a_{21} & a_{22} & a_{23} \\ a_{31} & a_{32} & a_{33} \end{bmatrix} \text{ then: } \mathbf{M_{12}} = \begin{bmatrix} a_{21} & a_{23} \\ a_{31} & a_{33} \end{bmatrix}.$$

Cofactor: Let $\mathbf{C_{ij}}$ be a cofactor of matrix matrix M defined as $C_{ij} = (-1)^{i+j}|\mathbf{M}_{ij}|$.

The **determinant** of any matrix is the sum of the elements of *any* row or column times each element's corresponding cofactor. The emphasis on "any" is there to help make clear that you will get the same answer for the determinant no matter which row or column you use. So, from our 3x3 matrix, that means computing the determinant by using row 1 or column 3, for example, will yield the same result:

$$\begin{aligned} |\mathbf{M}| \quad &= \quad a_{11}C_{11} + a12C_{12} + a13C_{13} & (6) \\ &= \quad a13C_{13} + a23C_{23} + a33C_{33}. & (7) \end{aligned}$$

Further note that any cofactor for a 3x3 matrix contains the determinant of 2x2 matrix for which you can use the formula provided. That formula is consistent with and comes from the general determinant defintion. For matrices greater than 3x3 in dimension, the cofactors will contain larger matrices, which need to be further broken down until all of them reach 2x2 in size to be evaluated.

Credits

Printed in the USA
CPSIA information can be obtained
at www.ICGtesting.com
LVHW082107091023
760616LV00019B/45